AF342466

Fatigue Life Analyses of Welded Structures

# Fatigue Life Analyses of Welded Structures

Tom Lassen

Naman Récho

First Published in Great Britain and the United States in 2006 by ISTE Ltd

Apart from any fair dealing for the purposes of research or private study, or criticism or review, as permitted under the Copyright, Designs and Patents Act 1988, this publication may only be reproduced, stored or transmitted, in any form or by any means, with the prior permission in writing of the publishers, or in the case of reprographic reproduction in accordance with the terms and licenses issued by the CLA. Enquiries concerning reproduction outside these terms should be sent to the publishers at the undermentioned address:

ISTE Ltd
6 Fitzroy Square
London W1T 5DX
UK

ISTE USA
4308 Patrice Road
Newport Beach, CA 92663
USA

www.iste.co.uk

© ISTE Ltd, 2006

The rights of Tom Lassen and Naman Récho to be identified as the authors of this work have been asserted by them in accordance with the Copyright, Designs and Patents Act 1988.

Library of Congress Cataloging-in-Publication Data

Lassen, Tom.
  Fatigue life analyses of welded structures/Tom Lassen, Naman Recho.
     p. cm.
  Includes bibliographical references and index.
  ISBN-13: 978-1-905209-54-5
  ISBN-10: 1-905209-54-1
  1. Welded joints--Fatigue. 2. Welded steel structures. I. Recho, Naman.
II. Title.
  TA492.W4L38 2006
  671.5'2042--dc22
                                                                    2006020963

British Library Cataloguing-in-Publication Data
A CIP record for this book is available from the British Library
ISBN 10: 1-905209-54-1
ISBN 13: 978-1-905209-54-5

Printed and bound in Great Britain by Antony Rowe Ltd, Chippenham, Wiltshire.

*"J'ai appris qu'une vie ne vaut rien, mais que rien ne vaut la vie"*

André Malraux

# Table of Contents

**Chapter 9. Physical Modeling of the Entire Fatigue Process** . . . . . . . . .    287



**Chapter 10. A Notch Stress Field Approach to the Prediction of
Fatigue Life** . . . . . . . . . . . . . . . . . . . . . . . . . . . . . . . . . . . . . . . . .    309



**Chapter 11. Multi-Axial Fatigue of Welded Joints**. . . . . . . . . . . . . . . .    319

# Abbreviations

| | |
|---|---|
| ACPD | Alternating Current Potential Drop |
| ALARP | As Low As Reasonably Practical |
| APL | Advanced Production and Loading |
| AW | As-Welded |
| BS | British Standards |
| BV | Bureau Veritas |
| CA | Constant Amplitude |
| CDF | Cumulative Distribution Function |
| COV | Coefficient Of Variation = V = SM/SD |
| CP | Cathodic Protection |
| CT | Compact Tension |
| CTOD | Crack Tip Opening Displacement |
| DC | Duty cycle, a specified number of load cycles |
| DFI | Design Fabrication Installation |
| DNV | Det Norske Veritas |
| DoE | Department of Energy |
| DTS | Damage Tolerance Supplement |
| ECA | Engineering Critical Assessment |
| EFMM | Empirical Fracture Mechanics Model |
| Eurocode | European steel standard |
| FC | Free Corrosion |
| FCAW | Flux-Cored Arc welding |
| FDF | Fatigue Design Factor |
| FEA | Finite Element Analysis |
| FEM | Finite Element Method |
| FLAWS-CG | Spreadsheet for crack growth calculations |
| FLAWS-CI | Spreadsheet for Calculating the crack initiation phase |
| FLAWS-S-N | Spreadsheet Rule-based S-N estimates |
| FMM | Fracture Mechanics Model |

| | |
|---|---|
| FORM | First Order Reliability Method |
| FPSO | Floating Production Storage and Offloading |
| FRF | Failure Rate Function |
| HAZ | Heat-Affected Zone |
| HB | Hardness Brinell |
| HS | Hot-Spot |
| HSE | Heath and Safety Executive |
| IPB | In Pane Bending |
| LCC | Life Cycle Cost |
| LEFM | Linear Elastic Fracture Mechanics |
| MCT | Markov Chain Technique |
| MPI | Magnetic Particle Inspection |
| MTF | Mechanical Transfer Function (between applied load and stress) |
| NDI | Non-Destructive Inspection |
| NORSOK | Standards developed by the Norwegian petroleum industry |
| N-SIF | Notch Stress Intensity Factor |
| OPB | Out of Plane Bending |
| PDF | Probability Density Function (or frequency function) |
| PFL | Predicted Fatigue Life |
| POD | Probability of Detection |
| PTM | Probability Transition Matrix |
| RFC | Rain Flow Counting |
| RFLM | Random Fatigue Limit Model |
| ROV | Remote Operated Vehicle |
| SAF | Stress Average Factor |
| SAL | Single Anchor Loading |
| SAW | Submerged-Arc Welding |
| SCF | Stress Concentration Function |
| SD | Standard Deviation, $(SD)^2 = Var(\ ) = s$ |
| SIF | Stress Intensity Factor |
| SIFR | Stress Intensity Factor Range |
| SM | Sample Mean, $SM = E(\ ) = \mu$ |
| SMAW | Shielded Metal Arc Welding |
| SNCF | Strain Notch Concentration Factor |
| SR | Stress relived |
| STP | Submerged Turret Production |
| TIG | Tungsten Inert Gas |
| TPM | Two-Phase Model |
| TSL | Target Service Life |
| VA | Variable Amplitude |
| WFM | Weight Function Method |

PART I

# Common Practice

# Chapter 1

# Introduction

## 1.1. The importance of welded joints and their fatigue behavior

Welding is today the most common joining method for metallic structures. Its industrial application is extremely important and many of the large structures designed and erected in the last decades would not have been possible without modern welding technology. Typical examples are steel bridges, ship structures, and large offshore structures for oil exploitation.

The strength analysis of welded structures does not deviate much from that for other types of structures. Various failure mechanisms have to be avoided through appropriate design, choice of material, and structural dimensions. Design criteria such as yielding, buckling, creep, corrosion, and fatigue must be carefully checked for specific loading conditions and environments. It is, however, a fact that welded joints are particularly vulnerable to fatigue damage when subjected to repetitive loading. Fatigue cracks may initiate and grow in the vicinity of the welds during service life even if the dynamic stresses are modest and well below the yield limit. The problem becomes very pronounced if the structure is optimized by the choice of high strength steel. The very reason for this choice is to allow for higher stresses and reduced dimensions, taking benefits of the high strength material with respect to the yield criterion. However, the fatigue strength of a welded joint is not primarily governed by the strength of the base material of the joining members; the governing parameters are mainly the global and local geometry of the joint. Hence, the yield stress is increased, but the fatigue strength does not improve significantly. This makes the fatigue criterion a major issue. The fatigue strength will alone give the requirements for the final dimensions of the structural members such as plates and stiffeners. To overlook this fact may result in fatigue facture and serious consequences.

## 1.2. Objectives and scope of the book

This book is confined to steel structures made by fusion welding and most of the examples are taken from the offshore industry. The book is divided into three parts which cover the following subjects:

– Part 1: common practice:

- the basic understanding of the fatigue behavior of welded joints based on theoretical considerations and experimental results (Chapters 2, 3 and 4),

- the S-N approach with reference to current rules and regulations (Chapter 5),

- the fracture mechanics approach with numerical computations (Chapter 6).

– Part 2: uncertainties in crack growth and life predictions:

- reliability modeling and risk assessments,

- the random nature of the fatigue damage process and stochastic modeling (Chapter 7).

– Part 3: recent advances in description of the fatigue behavior:

- recent advances in understanding the fatigue process and estimating the fatigue life (Chapters 8, 9, 10, 11 and 12).

The objective of this book is to disseminate, to practicing engineers, the knowledge possessed by the two authors. The main goal is to teach engineers how to cope with frequently occurring problems related to the fatigue design of welded structures. Hence, the scope of the book is primarily about practical problems in structural design and in-service inspection. For this purpose, industrial cases are included along with spreadsheets for carrying out both S-N calculations and fracture mechanics calculations. Although available models of fatigue behavior may not be perfect, they are very useful tools in engineering assessment if properly understood and used. In most practical situations, the shortcomings of the available fatigue models are less important than the problems related to the uncertainty in the parameters included in the models. Furthermore, fatigue design is experimental, empirical, and theoretical – and in that order. Without testing, fatigue analysis often remains an academic speculation. Hence, our agenda in Part 1 of the book is to put forward rather simple models that fit the experimental facts.

In addition to this strategy, it is important to disseminate knowledge on how to deal with uncertainty in a logical and unified manner. Fatigue life data exhibit considerable scatter even under controlled laboratory conditions and the standard deviation is equally as important as the mean value. Furthermore, typical in-service variable loading may be stochastic in nature and stress calculations may be uncertain. These considerations call for some insight into applied statistics and probability calculations. This is emphasized in Part 2 of the book.

Although the book emphasizes the practical aspects of fatigue life calculations and the assessment of crack growth and criticality, some advances in methods and models are included in Part 3 of the book. Chapter 8 focuses on the statistical background of the S-N curves, whereas Chapter 9 is dedicated to the fatigue process. Chapter 10 suggests a life model where the weld notch stress is replaced by the weld notch stress field as the key parameter for fatigue life. Chapter 11 outlines some recent advances in methods of stress calculation for cracked joints. Finally, Chapter 12 describes and models the effect of an overload. All these chapters present methods and models that deviate from the common practice in current rules and regulations. The proposed models are more in accordance with the realistic fatigue damage behavior of welded joints. The practical impact of the model on fatigue design and inspection planning is important.

The ultimate objective is to achieve optimized structures with respect to fatigue design, dimensions and inspection efforts without compromising reliability and safety.

## 1.3. The content of the various chapters

Chapter 2 provides basic understanding of the fatigue damage process with reference to some failure cases, and gives an overview of parameters influencing the process. Chapter 3 gives some insight into laboratory fatigue testing, which is the basis for rule-based S-N curves. This chapter also includes a brief overview of common statistical methods to cope with the scatter in fatigue life results. Chapter 4 treats the definition and description of the fatigue load spectrum. Accuracy in applied loading description is crucial for the credibility of fatigue life results. Both the time-series approach and the energy-spectrum approach are presented. After having read Chapter 4, the reader is prepared for an elaborate fatigue life calculation scheme based on the S-N model according to rules and regulations. This is presented in Chapter 5. The basis is the original S-N design rules from the Department of Energy, further developed in the Eurocode 3 design rules. To account for corrosive environments, the Norwegian NORSOK and DNV guidance developed for offshore structures in the North Sea is commented upon. Rules for ship structures are also reviewed. Chapter 5 also gives a qualitative assessment of what is a good detail-design of welded joints and how to obtain improvements in fatigue resistance though post-weld treatment. This chapter may in fact be read directly after Chapters 1 and 2, but we have chosen to present it at the end of the sequence as final practical guidance and to sum up.

The above reading recommendation is given for the practicing engineer mainly involved in detailed design and in choosing dimensions for welded joints. The goal is to achieve sufficiently long predicted fatigue lives compared to the target service life. This is the daily task for many steel structural engineers. If, however, decisions must be made regarding the acceptability of existing flaws or crack-like defects in the joints, then Chapter 6 should be included in the engineer's reading. Chapter 6 is an outline of applied fracture mechanics. In this chapter important questions, such as how fast a crack will grow during service loading and what is the critical crack size that leads to unstable rupture during extreme load, are treated. These are crucial questions to answer at a post-fabrication stage when cracks have been detected and the alternatives are repair or no repair. Furthermore, crack growth behavior is crucial information when planning scheduled in-service inspections.

Chapter 7 gives an outline of common and frequently used reliability methods. First we present some elementary life models, such as the Weibull model and the Lognormal model, for the fatigue life. The latter is the one most often used in rules and regulations. Then we proceed with the Monte Carlo simulation as our main method for reliability calculation. The reason for this choice is twofold: it is a good method from a pedagogical point of view; and it is a powerful method from a practical point of view. Markov chains are also treated in some detail.

Chapter 8 gives a critical reevaluation of the validity of the conventional S-N curves that are used in rules and regulations. Based on experimental fatigue life data, a new stochastic model is suggested. In this model both the fatigue life and the fatigue limit are treated as simultaneous random variables. The model results in a non-linear S-N curve for a log-log scale. These types of curves are in better accord with the experimental results than the conventional S-N curves.

Chapter 9 presents a two-phase model for the damage process in welded joints. The objective of this chapter is to emphasize the importance of the crack initiation phase in welded joints. It is the authors' opinion that this phase is significant for high quality welded components subjected to in-service loading. A model not accounting for this phase may lead to wrong decisions regarding both dimensions and scheduled inspections.

Chapter 10 presents an S-N model based on the weld notch stress field. The model is based on the fact that it is the stresses at the weld toe notch that are the agents for fatigue damage accumulation in the joint. However, the stress situation should not be characterized by the stress concentration factor alone as sometimes recommended in rules and regulations, but rather the entire local stress field in vicinity of the toe weld notch.

Chapter 11 outlines a method for stress calculation in cracked joints and gives a brief presentation of the multi-axial fatigue problem. This is an important issue that still has many unsolved enigmas and the chapter touches on some of the main topics.

Chapter 12 is devoted to the effect of an overload on the crack growth. A new, efficient method is developed for predicting the retardation of the growth rate.

## 1.4. Other literature in the field

First of all it should be emphasized that good textbooks exist in many of the fields related to the issues in this book. A good textbook on general fatigue is:

– J. Schive, *Fatigue of Structures and Materials*, Kluwer Academic Publishing, 2001

Regarding the fatigue of welded joints, some standard books are:

– T.R. Gurney, *Fatigue of Welded Structures*, Cambridge University Press, 1979

– A. Næss (ed), *Fatigue Handbook*, Tapir, 1985

– S.D. Maddox, *Fatigue Strength of Welded Structures*, Abington Publishing, 1991

The approaches of these books are mainly based on the S-N method, and the chapters dealing with applied fracture mechanics are rather short. Compared with these books, the present text gives the latest updates found in rules and regulations and a more thorough presentation of the fracture mechanics approach. Some computational models based on applied fracture mechanics are also included. Furthermore, the present text emphasizes stochastic modeling of both the S-N and the fracture mechanics approach. Finally, some recent advances that lead to a more precise description of the fatigue behavior of welded joints are included.

For further reading on the fracture mechanics approach, the reader may also consult:

– D. Broek, *The Practical Use of Fracture Mechanics*, Kluwer, 1989

– N. Recho, *Rupture par fissuration des structures*, Hermès, 1995

For a better understanding of the stochastic analysis and the reliability approach, the reader could start with books covering all the fundamental issues:

– J.R. Benjamin and CA Cornell, *Probability, Statistics and Decision for Civil Engineers*, McGraw Hill, 1979

– E.E. Lewis, *Introduction to Reliability Engineering*, John Wiley & Sons, 1994

The following rules and regulations are used for illustration purposes:

– Land-based structures

  - Eurocode 3, Steel Structures, 1993 (Fatigue in air)

– Offshore structures

  - NORSOK standard: Design of Steel Structures, Document N-004, Appendix C 1999, and the DNV Fatigue guidance CP-R for offshore structures 2005 (Fatigue in sea-water environment)

– Ship structures

   - DNV Rules for Ship Structures 2003

   - BV rules for ship structures 1998

– Special guide for damage tolerance assessments

   - BSI: "Guide on Methods for Assessing the Acceptability of Flaws in Metallic Structures" BS7910 (2005)

The authors are fully aware of the existence of post-processing tools dealing with fatigue life predictions and crack growth in finite element analysis (FEA) software. We will give examples about how this can be carried out in Chapter 11. It is, however, essential that the user of these programs has the required knowledge to be able to assess the validity of the results from these post-processing modules.

## 1.5. Why should the practicing engineer apply reliability methods?

As already stated, the fatigue behavior of welded joints is random by nature. Very few load-bearing details exhibit such large scatter in fatigue life as welded joints. This is true even in controlled laboratory conditions. As a consequence, it becomes an important issue to take scatter into consideration, both for the fatigue process and for the final life. Furthermore, the in-services stresses may often be characterized as stochastic processes.

There has been a trend the last two decades to treat the strength problem of a structure by applying statistics and probability calculations. As a consequence, the probability of failure is used as a criterion, instead of the more traditional safety factors, when checking various design criteria. The methods and tools for performing this type of analysis have become available and quite easy to use. The probability of no failure during a given time period is considered to be the reliability of the structure. The methods used for determining the probability are often called reliability methods. Furthermore, if the associated probability of failure is weighed against the consequence of failure we arrive at the risk concept. Achieving high reliability and low risk levels will maintain operational capability and secure life and assets. The more sources of uncertainty there are in a structural problem, the more appropriate will be the application of a reliability approach. For processes where damage is accumulating with time, the probability of failure will increase with service time, depending on decisions made for the design concept, configurations, dimensions, material properties, and in-service inspection strategy. For fatigue of welded joints the following sources of uncertainty are dominant:

   – the service stress history;

   – the global and local geometry of the joint;

– the material parameters; and

– the performance of the in-service inspection.

A reliability approach pinpoints the sources of uncertainty and treats them in a rational way based on probability models and statistics. An alterative is to hide the uncertainty by fixed parameters often based on "worst case assumptions". To optimize structures with respect to fatigue strength one should avoid using worst case assumptions. This may result in costly over-dimensioned structures. An optimization of design and member dimensions based on reliability calculations will give lightweight structures. Last but not least, inspection planning should be developed based on risk criteria to avoid unnecessarily costly inspection during service life. This leads to the concept of risk-based inspection.

## 1.6. How to work with this book

This textbook regards the S-N approach and the fracture mechanics approach as equally important tools. Furthermore, both approaches are enhanced by introducing more advanced statistical models for the S-N approach and adding an initiation phase to the fracture mechanics approach. The problem of uncertainty is recognized and dealt with in a rational and consistent manner using stochastic methods.

The first six chapters of the book can be regarded as a diffusion of basic understanding and practical skill, whereas Chapter 7 deals with uncertainty modeling. Chapters 8 to 11 present recent advances in the knowledge of the fatigue behavior of welded joints. In these chapters, new methods and models are proposed, which are based on experimental facts, and their practical consequences are discussed. These chapters are intended for readers who wish a deeper understanding of fatigue behavior and have ambitions towards research work within the field.

It is our belief that knowledge is gained through both reading the text and studying the cases and examples given. Calculations tools in an Excel spreadsheet format are included to give hands-on experience. These spreadsheets are so constructed that they may be used for real industrial cases. Another important use of these spreadsheets is in sensitivity analysis: which parameters have the strongest bearing on the crack growth rate and fatigue life. Hence, the application of various calculation methods through the use of the spreadsheets is essential to the learning process. Three spreadsheets are provided in a weblink which readers can use to apply their knowledge. These can be accessed at the following address: http://www.iste.co.uk/static/flaws.html.

## 1.7. About the authors

The two authors have a vast experience related to fatigue problems, within research projects, teaching, and industrial consultancy work.

Professor Tom Lassen was educated at the Norwegian University of Science and Technology, Trondheim in 1973. He graduated with a Masters in naval architecture with specialization in structural analysis and strength of materials. He submitted his PhD thesis entitled "Experimental investigation and stochastic modeling of the fatigue behavior of welded steel joints" at the Aalborg University, Denmark in 1997. He has published numerous articles in international journals on the fatigue of welded joints. He has been with Det Norske Veritas in Oslo and with Elf Aquitaine Research Centre in Pau, France. He has, for a long time, taught strength analysis and fracture mechanics at Agder University College in Grimstad. He also teaches aircraft maintenance for the Norwegian Royal Air Force. He has worked extensively with moored floaters for offshore oil exploitation in collaboration with Advanced Production and Loading in Arendal, Norway. For these types of installations, the fatigue behavior of welded joints is a major concern. He has recently been a visiting professor at University Blaise Pascal in Clermont, France. It was during this period the work with the present textbook was initiated in collaboration with Naman Recho. He is at present with the University of Stavanger in Norway.

Professor Naman Recho was educated at the French University of Pierre and Marie Curie in Paris. He graduated from this university as Doctor Ingenieur in 1980 and *Docteur d'Etat Es-Sciences Physiques* in 1987. He has been a university Professor since 1988. From 1978 to 1988 he was with CTICM (*Centre Technique Industriel de la Construction Métallique*) in Paris. Between 1988 and 1993, he was a Professor at the University of Haute Alsace in Mulhouse (France). Between 1993 and 2005, he was a Professor at the University Blaise Pascal in Clermont Ferrand, France. He has written two textbooks about fracture mechanics. He has worked extensively with conceptual and applied aspects of fracture mechanics. He has worked, in particular, with welded offshore structures, and reliability analysis of cracked structures. In the past ten years, he directed more than 15 theses dealing with fracture mechanics and fatigue design of welded joints. He participated in several research programs with industries such as CETIM-Senlis, Bureau Veritas in La Défense, Giat Industries in Borges, Michelin in Clermont Ferrand, CEA-Saclay, Trelleborg Industrie. In 1986, Naman Recho founded the program MMS (*Mécanique des Matériaux et Structures*) at EPF – Ecoles d'Ingénieurs in Sceaux, near Paris. He is still managing this program. Since 1985 he has also been teaching at CHEC (*Centre des Hautes Etudes de la Construction* – Paris). He is a Guest Professor at HUT (Hefei University of Technology in China). He is at present with CNRS (National Centre of Scientific Research) in Paris.

Chapter 2

# Basic Characterization of the Fatigue Behavior of Welded Joints

## 2.1. Introduction and objectives

This chapter gives an overview of fatigue behavior of welded joints. The basic mechanisms of the fatigue damage process are discussed and the usual method and models for fatigue life verification are outlined. The basic parameters that are important for the fatigue process in welded joints are pinpointed. The objective of the chapter is to give the reader basic insight into the fatigue process in welded joints, with emphasis on the aspects that makes these joints vulnerable to fatigue. The reader will be acquainted with the most common methods and models for describing the fatigue problem. These methods will be elaborated further in Chapters 5 and 6 where the S-N model and the facture mechanics approach are described in more detail.

## 2.2. Fatigue failures

Fatigue is defined as damage accumulation due to oscillating stresses and strains in the material. Therefore, fatigue cracks do occur in welded details that are subjected to repetitive loading. In significant structural items they may lead to failures with severe consequences. The Health and Safety Executive, UK, has listed the main causes of structural damage for installation in the North Sea (1974-1992, Ref [1]) as:

- fatigue                25%
- vessel impact          24%

– dropped objects        9%

– corrosion              6%

As can be seen, fatigue is the main cause of damage, followed by groups that can be designated as accidental damage. Corrosion damage is ranked as ranked as the fourth most frequent cause of damage. Figure 2.1 shows a fatigue failure of a propeller shaft in a shuttle tanker. The fracture occurred in the intermediate part of the shaft. The crack started from the surface of the shaft due to a weld arch strike. The fatigue surface is characterized by its smooth appearance with almost no plastic strain. At several stages during crack propagation, marks which are due to low stress variations are left as traces on the fatigue surface. These so-called beach marks correspond to changes in the fatigue loading; the crack front will make a mark during the time of slow growth due to smaller stress cycles. These marks are analogous to the dark winter rings found in the cross section of a tree. As can be seen, the beach marks have a typical semi-elliptical shape indicating the position of the crack front at various stages during the crack propagation. When the fatigue crack has reached the size of about three-quarters of the shaft diameter (D = 360 mm), the final fracture has occurred due to lack of the remaining ligament of the shaft cross section. It is a ductile fracture governed by the maximum occurring shear stress. The 45-degree share planes are easy to recognize for the final fracture. In the example in Figure 2.1, the fatigue failure leads to a severe leakage in the stern bearing tube and the blackout of the main engine. Nobody in the crew was injured.

In other cases, fatigue cracks from welded details have lead to severe consequences and loss of life. Figure 2.2 shows the semi-submersible *Alexander Kielland* that capsized in the North Sea in March 1980. Although the disaster was a consequence of several unfortunate circumstances, the root cause was a fatigue crack in an attachment weld at the surface of a brace member. The brace member is denoted D6 in Figure 2.2 and the crack has started from the fillet weld between a hydrophone support and the brace. After the crack had propagated through the wall thickness of the brace, it continued to grow along the circumference of the tube splitting up the main cross section of the brace. Before the final fracture the crack was over 1 meter long. The accident had 123 fatalities and initiated a large research effort on fatigue in Norway in order to obtain a better understanding of the fatigue damage problem of welded joints. Based on the increased knowledge of the fatigue behavior, improvements were made regarding both detailed fatigue design and inspection planning for offshore structures. A final failure case is shown in Figure 2.3. A large fatigue crack in the deck area in the mid-ships section of a tanker can be seen. These types of cracks may cause a rupture of the cross section of the hull beam. This may endanger the life of crew members and cause environmental damage to the sea and the shore.

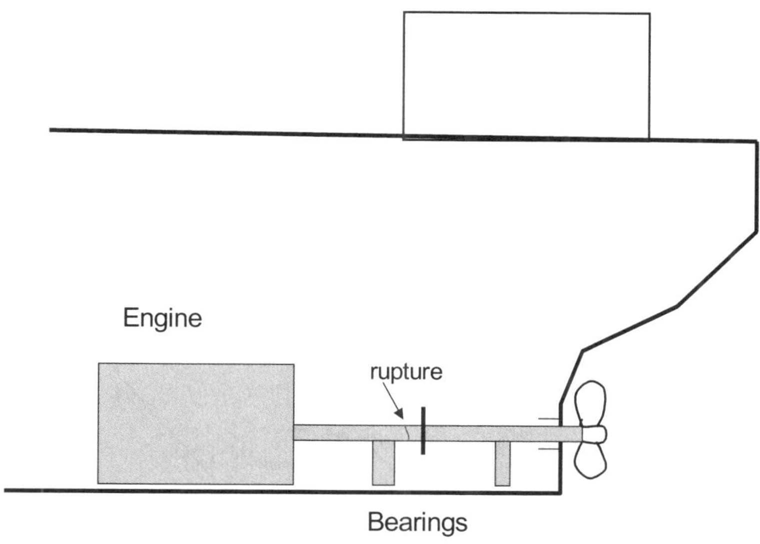

**Figure 2.1.** *Fatigue failure in a propeller shaft in a shuttle tanker. Crack has initiated from a weld arch strike at the surface*

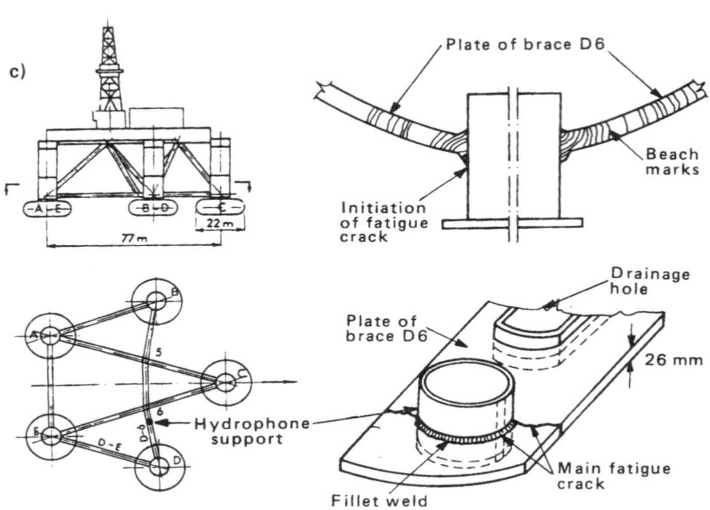


**Figure 2.2.** *The Alexander Kielland fatigue failure, Ref [2]*

**Figure 2.3.** *Large fatigue crack at the mid ships section of a tanker*

## 2.3. Basic mechanisms of metal fatigue

Before addressing the peculiarities of the fatigue behavior of welded joints, we will give a short overview of metal fatigue in general. Fatigue is defined as a damage process in the metal due to fluctuating stresses and strains. Although the stresses and strains may be well below the static resistance level of the metal, the damage is accumulating cycle by cycle and after a certain number of load fluctuations a failure will occur. For structures in service, the damage accumulation may take several years to reach a critical level with resulting failure. The fatigue process is usually divided into three phases:

– phase 1: crack initiation;

– phase 2: crack growth;

– phase 3: final fracture.

The crack initiation usually takes place on the surface of the metal in the vicinity of a notch. The mechanism is explained by a slip band mechanism at a microscopic level driven by the maximum shear stress. When the load is imposed, some grains will be subjected to plastic deformation involving the sliding of some of the crystallographic planes. The mechanism is limited to a few grains where these crystallographic planes have an unfavorable orientation with respect to the local maximum shear stresses. When the load is reversed, the planes will not slide back to their initial position due to the cyclic stain hardening effect. Hence, in the reversed part of the load cycle, it is the neighboring planes that will suffer yielding by sliding in the opposite direction. The final result is microscopic extrusions and intrusion on the metal surface. The intrusions act as a micro-crack for further crack extension during the subsequent loading cycles. The mechanism is schematically shown in Figure 2.4

After crack initiation has occurred within a few grains, subsequent microscopic growth will extend the crack to pass several grain boundaries. When the crack front reaches over several grains, the crack will continue to grow in a direction perpendicular to the largest tensile principal stress. This transition from microscopic to large-scale crack growth is also indicated in Figure 2.4. It is important to realize that whereas the initiation phase is related to the surface condition of the metal and governed by the cyclic shear stresses, the crack growth depends on the material as a bulk property and the crack is driven by the cyclic principal stresses. A more thorough presentation of the subject is found in Ref [3].

In the growth phase, the crack growth process is explained by a crack opening and front blunting mechanism followed by a subsequent crack closing and front sharpening mechanism during each load cycle. After one complete cycle, the crack front has advanced a small increment which may be traced by microscopy on the fatigue surface. This advancement corresponding to one load cycle is the distance between two so called striations. These striations are shown for a high-strength steel in Figure 2.5. The advancement depends on the range of the stress intensity factor (SIF), which will be explained below and elaborated in Chapter 6.

The final fracture in phase 3 will take place when the crack becomes so large that the remaining ligament of the cross section is too small to transfer the peak of the load cycle, or when the local stresses and strains at the crack front inflict a local brittle fracture. In the former case, it is the net section average stresses that are the driving force for the fracture. In the latter case, it is a local failure that is driven by the maximum SIF. This factor uniquely characterizes the magnitude of the stress field at the crack front under linear elastic conditions. Again, details will be given in Chapter 6.

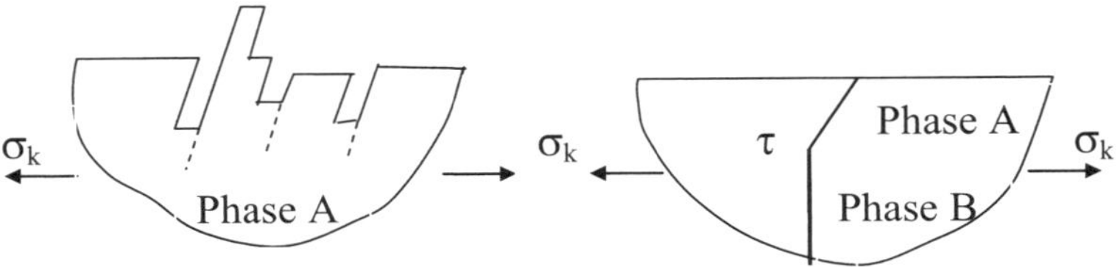


**Figure 2.4.** *Sketch of various phases of the fatigue process.*
*Left: Phase A with shear stress driven intrusion/extrusions.*
*Right: Phase A with transition to Phase B driven by normal stresses*

**Figure 2.5.** *Microscopy of crack growth in high strength steel*
*(SEM Magnification x500)*

## 2.4. Parameters that are important to the fatigue damage process

The fatigue process is quite complex and is influenced both by the nature of the external loading, the geometry of the structural item and its material characteristics. The following conditions and parameters are important to the damage process:
- external cyclic loading:
    - loading mode with reference to the actual structural item,
    - time history of the external forces,
- geometry of the item:
    - global geometry of the item,
    - local geometry at potential crack locus,
- material characteristics;
- residual stresses;
- production quality in general;
- surface finish in particular;
- environmental condition during service.

We will discuss the various points below. First of all, we must bear in mind that the fatigue damage starts as a local phenomenon. The cracks emanate from a small detail on the item, often in the vicinity of a notch. Therefore, before discussing the topics listed above it is important to be aware of the location of potential crack initiation spots in a structural item.

### 2.4.1. *External loading and stresses in an item*

The external forces may create normal, bending or torsion effects on a structural item with associated stress situations near a potential crack locus. These loading and response situations are often referred to as loading modes and stress modes. The latter concept is defined by the stress direction relative to the crack planes. The normal and bending loading mode will give rise to normal stresses that will act as the main agents for the crack initiation and growth. In this case, the crack planes will be moved directly apart by the normal stresses. This situation is denoted stress mode I. Therefore, these two loading modes inflict the same stress mode and damage mechanism when they create normal stresses in the same direction. If there are several stress axis involved, it is often assumed that one of them will be dominant to avoid a more complex multi-axial fatigue analysis. Shear stresses due to torsion may involve a fatigue mechanism that is different from mode I. The crack planes will be sheared to extension. However, this is a rarer mode than the stresses normal to the crack plane. Figure 2.6 shows a link from a mooring chain. When each link in a chain is subjected to pure tension, as shown, the side bars of the links are subjected to tension and bending loading modes. The highest tensile stress is found on the inner side of the bars. One of the side bars has a flash weld at the mid-section and this is a potential crack locus as indicated. This crack will be subjected to pure traction due to tension and bending, i.e. stress mode I.

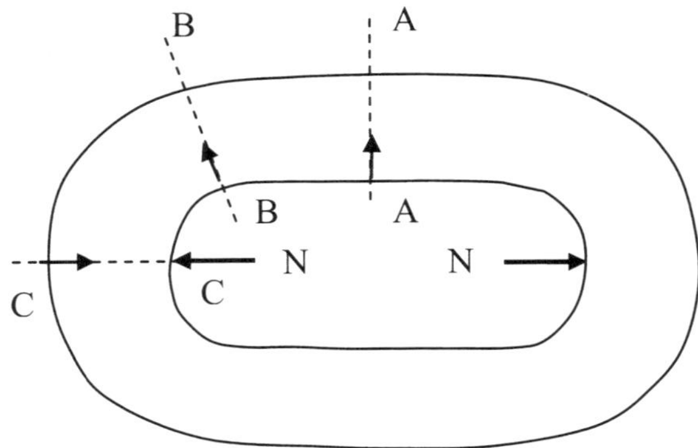


**Figure 2.6.** *Chain link with potential crack locus at the flash weld*
*on the mid-section of one of the side bars (section A-A)*

When it comes to force (or stress) variation over time, it is mainly the force variation and the number of cycles that matter. The frequency of the repetitive loading does not usually influence the damage process. If there are several external forces acting on the item, the problem of in- or out-of-phase between the applied loads arises. The external forces can have different frequencies and be more or less correlated in time. An example is shown, in Figure 2.7, of tension in the mooring line system for an offshore floating installation. The link described in Figure 2.6 will be subjected to such a load history. There is one low-frequency component that is given by the stiffness of the mooring system and the mass of the floater. In addition there is one load component induced by the ocean waves. When several mooring lines spread in different directions are connected to the floating structure, they will inflict different loading modes at various structural items. It goes without saying that all these matters are difficult to model and analyze at the design stage. This is due to the complexity in both space and time, and is also due to the simple fact that the use of the structure may change as service life elapses. We will come back to these considerations in some industrial cases that will be presented in Chapter 5 and 6.

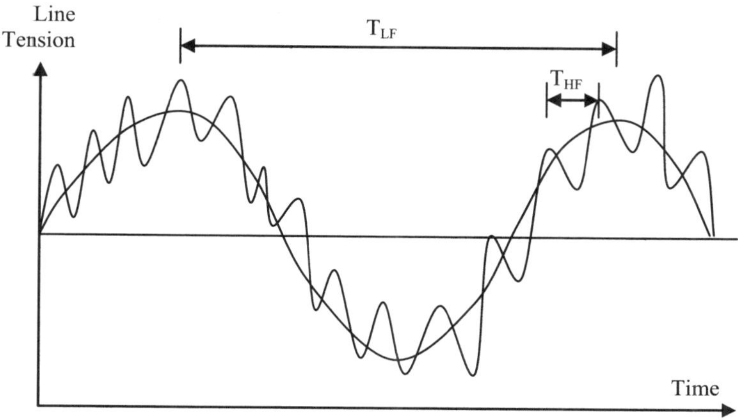


**Figure 2.7.** *Fatigue loading in one mooring line
for an offshore floating installation*

### 2.4.2. *Geometry, stress and strain concentrations*

The responses of structural items to the external forces are usually characterized by the stresses and strains associated with the forces. The general stress response is important, and even more so is the local stress rise caused by geometrical discontinuities. The geometry of the item is important and is one of the issues that must be addressed at the design stage to achieve improved fatigue durability. The main issue is to reduce the stress concentration factor $K_t$ in areas susceptible to fatigue. This will limit the stress and strain response inflicted by the external loads at potential crack spots. The stress concentration is defined as a local rise of stresses due to a geometrical change or discontinuity. A more or less abrupt change in geometry is often referred to as a notch. Typical examples are threads, cope holes, and welds. The stresses at the notch may be magnified locally by a factor of from 3 to 8 compared with the average nominal stresses in the item. This magnification factor is defined as the stress concentration factor $K_t$. This stress rise will of course be very harmful with respect to fatigue damage. The stresses can be reduced by increasing the general dimension of the item or improving the local geometry of the notch, typically the notch radius. Whereas the first approach will increase the weight of the item, the second improvement can often be achieved without any additional weight or costs. Furthermore, increasing the dimension of the item will also have an unfavorable effect on the fatigue resistance that is referred to as the size effect. This means that the benefits achieved by reducing the stresses will partly be lost due to poorer fatigue quality related to the increased thickness. All these matters favor a design where the local notch geometry is optimized.

Typical stress concentrations due to notches are shown in Figure 2.8. At the left of Figure 2.8 is shown a plate with two edge notches symmetrically on each side. As can be seen, the stresses will increase at the notched section due to the reduced cross section; but far more important is the stress rise caused by the local disturbance of the notch itself. This local effect very much depends on the notch radius ρ. Cracks may appear in the root of both of the notches. The same phenomenon is shown for a butt joint, at the right in Figure 2.8. In this case we have no reduction of the cross section, but overfill of the weld metal will act as a stress riser. The stress will rise at the transition between the base plate and weld metal. This area is often referred to as the weld toe. This is the potential crack locus. The local geometry can be characterized by its flank angle θ and radius ρ. The stress concentration factor is defined as:

$$K_t = \frac{\sigma_l}{\sigma_n}.$$

(2.1)

The reader should be aware of the fact that the nominal stress can be defined in the section away from the notch or the nominal stress in the notch section. This may vary in various handbooks, Ref [4].

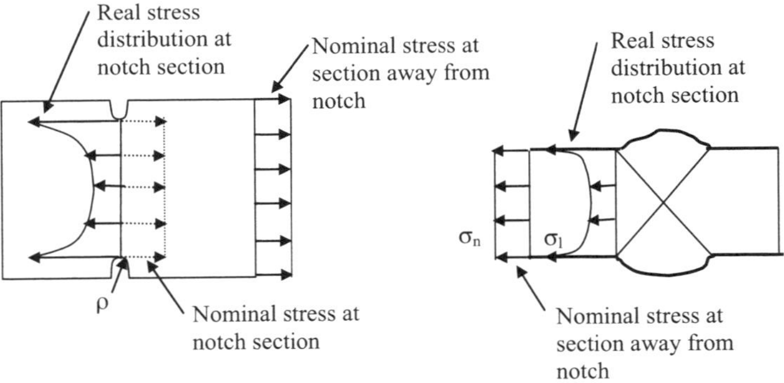


**Figure 2.8.** *Stress concentrations at notches. Left: plate with edge notches. Right: butt joint*

### 2.4.3. *Material parameters*

The common material parameters, such as yield strength, tensile strength, and module of elasticity, do have an impact on the fatigue strength of the metal. However, to characterize the fatigue resistance more explicitly, special tests with smooth specimens are often carried out to determine the fatigue life as a function of the applied stress range. In addition, a crack growth test with pre-cracked standard specimens can be carried out to reveal crack behavior. The first approach is to present the fatigue life in number of cycles as a function of the applied constant amplitude stress range as shown in Figure 2.9. The key parameter for the fatigue life is the stress range $\Delta\sigma$, but the results will also depend on the applied mean stress level. As can be seen from Figure 2.9, increased mean stress will result in shorter fatigue lives for a given $\Delta\sigma$. A linear approximation of the upper part of the curve for a log-log scale gives the following equation:

$$\log N = \log A - m \log \Delta\sigma \qquad (2.2)$$

or:

$$N = \frac{A}{\Delta\sigma^m} \qquad (2.3)$$

where $\log A$ is the curve intercept with the vertical axis, whereas $-1/m$ is the slope of the curve. When the stress range is sufficiently low, no fatigue fracture will occur. This stress range $\Delta\sigma_D$ is referred to as the fatigue-limit under constant amplitude loading. It can be regarded as a material parameter characterizing the fatigue resistance of the metal, but it will in fact depend both on the size and geometry of the specimens, as well as on the surface conditions. This is also the case for the parameters $A$ and $m$. We will learn more about the experimental background and application of the S-N curve in Chapters 3 and 5.

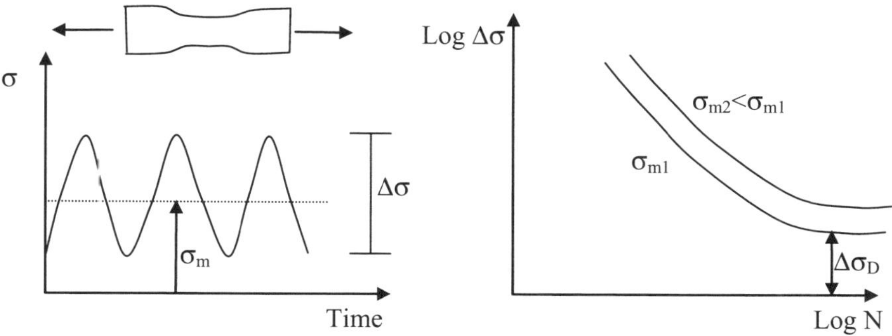


**Figure 2.9.** *S-N curve based on constant amplitude fatigue tests*

The crack growth test is based on tests with standard cracked specimens where the crack growth rate da/dN (m/cycle) is measured as a function of applied stress intensity factor range (SIFR) as shown in Figure 2.10. The specimen is a compact tension (CT) specimen with standard dimensions for a given thickness. The SIFR is a parameter that alone governs the local cyclic stress variation at the crack front:

$$\Delta K = \Delta\sigma_n \sqrt{\pi a} F \qquad (2.4)$$

where $\Delta\sigma_n$ is the applied nominal stress range and a is the crack size. $F$ is a geometry function accounting for the crack and component geometry and the loading mode.

As for the S-N tests with smooth specimens, there is a limit at which the crack will stop growing. This limit, $\Delta K_0$, is referred to as the fatigue threshold value. For $\Delta K$ values greater than this value, one will enter the region II in the growth-rate diagram where stable crack growth is taking place. If the peak value of the SIFR gets too high, one will enter into region III with the onset of unstable crack growth; see Figure 2.10. Again, the results will depend on the mean stress level. This level can be characterized by the R ratio defined as the minimum applied force $F$ as fraction of the maximum applied force. The growth rate will increases for a given $\Delta K$ if the R ratio increases. This is in agreement with the mean stress effect that was shown in Figure 2.9 for the S-N approach. In the region II, the growth results can be approximated by a straight line for a log-log scale:

$$\log\frac{da}{dN} = \log C + m\log\Delta K \qquad (2.5)$$

or:

$$\frac{da}{dN} = C\Delta K^m \qquad (2.6)$$

where $C$ and $m$ are material parameters that characterize the crack growth. If the entire fatigue life consists of crack propagation only, the exponent m coincides with the parameter m defining the S-N curve. We will study crack behavior in more detail in Chapter 6.

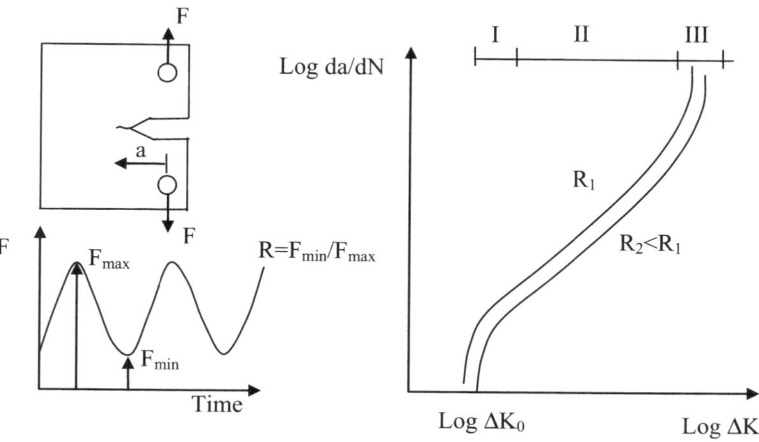


**Figure 2.10.** *Crack growth rate versus stress intensity factor range*

In addition to the S-N tests shown in Figure 2.9, and the crack growth test in Figure 2.10, it is possible to carry out tests on small specimens to study the time taken to crack initiation in the material. The tests are carried out on specimens that are so small that the fracture will occur after a crack of some tenths of a millimeter has initiated. For this type of test the strain range during a load cycle is often used as the key parameter for the time to crack initiation. This is a better key to the time to crack initiation than the stress range if the load cycle involves plasticity. A typical stress-strain cycle and the resulting strain-life curve are shown in Figure 2.11. For the left part of the life curve there is a significantly cyclic plastic strain involved and the number of cycles to fracture of the specimen is relatively short, typically 1,000 cycles or less. This is referred to as low cycle fatigue. In the right part of the life curve the stress-strain cycle is completely linear elastic and one will have typical fatigue life of 10,000 cycles or more. This is referred to as high cycle fatigue. As we have shown earlier, the results depend on the mean stress. The strain-life curve to the right is given by the Manson-Coffin equation with Morrow's mean stress correction:

$$\frac{\Delta \varepsilon_T}{2} = \frac{\left(\sigma'_f - \sigma_m\right)}{E}\left(2N\right)^b + \varepsilon'_f\left(2N\right)^c \tag{2.7}$$

Here $\Delta\varepsilon_T$ is the total local strain range and $\sigma_m$ is the local mean stress at the weld toe. The parameters $b$ and $c$ are the fatigue strength and ductility exponents, and $\sigma'_f$ and $\varepsilon'_f$ are the fatigue strength and ductility coefficients respectively. The local stress and strain behavior is given by the Ramberg-Osgood stabilized cyclic strain curve:

$$\Delta\varepsilon = \frac{\Delta\sigma}{E} + 2\left(\frac{\Delta\sigma}{2\,K'}\right)^{\frac{1}{n'}},$$

(2.8)

where $K'$ and $n'$ are the cyclic strength coefficient and strain hardening exponent respectively. This equation governs the hysteresis curve to the left on Figure 2.11. We will elaborate and apply these equations for the time crack initiation in welded joints when we present a two-phase model for the entire fatigue process in Chapter 9. In that chapter we will discuss how to derive the involved parameters for the conditions at the weld toe.

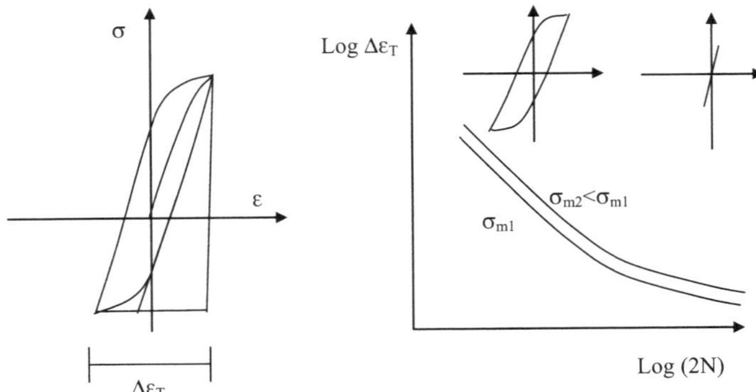


**Figure 2.11.** *Local strain cycle used as key to time to crack initiation based on tests with small-scale specimens*

### 2.4.4. *Residual stresses*

Residual stresses are defined as static inherent stresses present in the structural item before the external forces are applied. The residual stresses are often created by the fabrication procedure. They are self-equilibrated; if there are zones that have tensile stresses there must be other zones that have compressive stresses. The areas subjected to tensile stresses will be more vulnerable to fatigue. The simple explanation of this is the mean stress effect, which has been discussed above. The presence of large tensile residual stresses will increase the mean stress. Even compressive stresses caused by the external forces may, when added on to pre-existing static tensile stresses, effectively act as a tensile stress cycle in the material. The influence of residual stresses is shown in Figure 2.12 for a partly compressive external stress cycle.

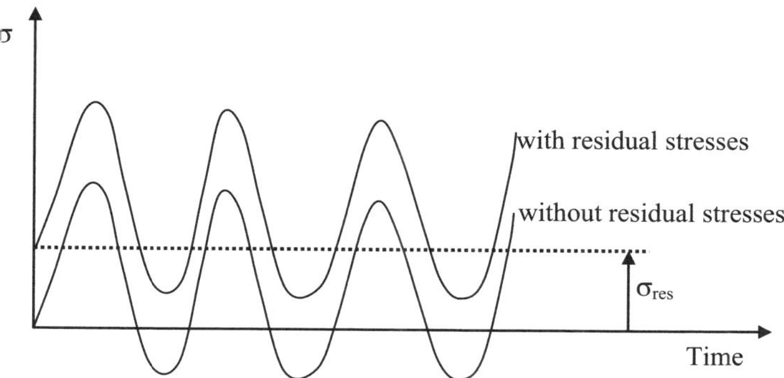



**Figure 2.12.** *The effect of residual stresses in the stress cycles*

### 2.4.5. *Fabrication quality and surface finish*

Fatigue predictions must not be entirely based on ideal design and dimensions taken from an engineering drawing. It is the quality of the actual as-built structural item that, in the end, governs the fatigue strength. Important topics are:

– actual dimension and alignment;

– radius of severe notches;

– surface finish;

– presence of sharp flaws.

Dimension control must be carried out to check that the dimensions are within the given tolerances. If misalignment occurs it may introduce secondary bending for an axial loading mode. Sharp flaws may act as starters for fatigue crack growth and in the worst cases the crack initiation phase is lost. We have already presented such a case for the propeller shaft failure in Figure 2.1. To ensure quality regarding these matters, dimension control and non-destructive testing should be carried out. A smooth surface will increase time to crack initiation.

### 2.4.6. *Influence of the environment*

Most fatigue material properties are given for normal air environment. Other types of environment may change the fatigue behavior and related parameters rather dramatically. One important example is the fatigue behavior of steel in seawater, where the steel is without any protection against corrosion. In this case there will be a synergistic effect between the corrosive deterioration mechanism and the fatigue damage process. Also, the load frequency becomes important as the corrosion process is a time-dependent electrochemical process. Therefore, fatigue laboratory tests cannot be accelerated in the same manner as for air environment. The fatigue life may decrease by a factor of 3 to 5 compared with dry air environment. If cathodic protection is provided, the fatigue life will lay between the air and free corrosion results. These matters are extremely important to ship and offshore structures.

## 2.5. Important topics for welded joints

### 2.5.1. *General overview*

The general issues described above are, of course, also important for welded joints. However, a welded joint has some peculiar features that make some of the subjects and parameters play a much more important role than others. Let us start with an S-N curve for a welded joint and compare it with other specimens. Figure 2.13 shows the S-N curves based on tests with one smooth plate, one plate with a bore hole, and one plate with a fillet welded transverse attachment (gusset). This type of fillet welded joint is often referred to as non-load-carrying due to the fact that the load is carried straight through the base plate. It is also noted that the applied force is transverse to the welding direction.

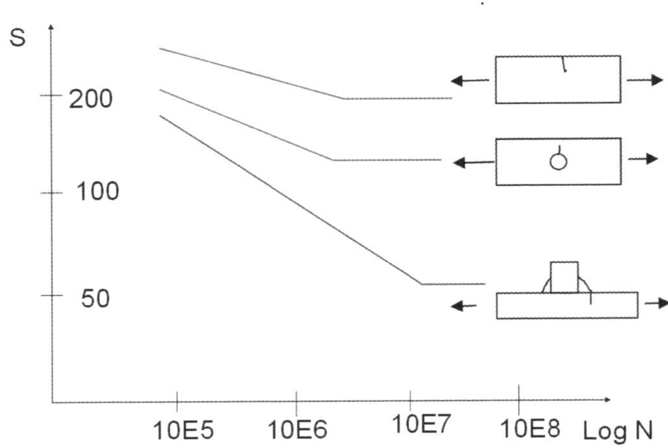


**Figure 2.13.** *Fatigue life curves for various details, reproduced after Ref [5]*

In the plate, the fatigue cracks can appear anywhere in the longitudinal edges of the plate, whereas they will appear at the inner edges of the bore hole transverse to the applied loading in the specimen with the hole. The latter phenomenon is due to the stress concentration at the edge of the hole. In the fillet welded joint, the cracks will appear at the weld toe and grow through the base plate in a direction perpendicular to the applied principal stresses. The cracks are indicated on the specimens in Figure 2.13. Various stages in the crack development for the welded detail are shown in Figure 2.14. The cracks may emanate from several spots along the weld seam. Small semi-elliptical-shaped cracks are formed, grow, and coalesce to become larger cracks. In the final stage, the remaining ligament of the plate section is too small to carry the peak load. As can be seen from Figure 2.13, the fatigue life for the fillet welded joint is significantly shorter than both the smooth plate and the plate with a bore hole. The difference between the S-N curve for the specimen with the hole and the S-N curve for the welded joint is rather surprising considering the fact that the bore hole creates a stress concentration factor close to 3 at the edge of the hole. The relatively short fatigue life of welded detail is explained, in the main, by three factors:

  − severe notch effect due to the attachment and the weld filler metal;

  − presence of non-metallic intrusions or micro-flaws along the fusion line;

  − presence of large tensile residual stresses.

We have discussed these topics in the foregoing section. The notch effect has already been shown by the stress concentration occurring at the weld toe for a butt joint in Figure 2.8. In the present case, the notch effect is created by a fillet weld and this effect is often more severe due to the fact that the fillet-weld has a steeper toe flank angle. In addition we have the stress concentration due to the gusset itself. Hence, the detail has in fact two stress raisers at the weld toe: the gusset and the weld metal. The latter is the most important one and can be characterized by the local toe geometry as shown in Figure 2.15. A favorable geometry is characterized by a low flank angle $\theta$ and a large toe radius $\rho$. One complication when defining the notch effect is that these two parameters may vary considerably along the weld seam. It is likely that the crack spots shown in Figure 2.14 coincide with unfavorable local toe geometry with high notch effect.

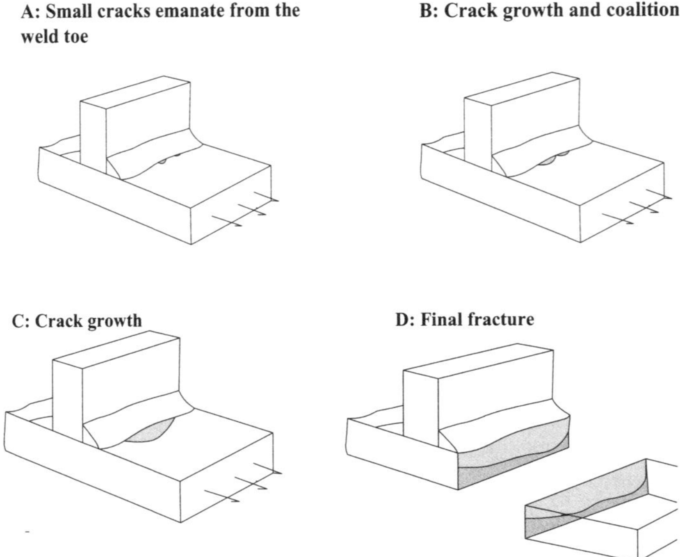


**Figure 2.14.** *Various stages of crack growth in the fillet welded joint*

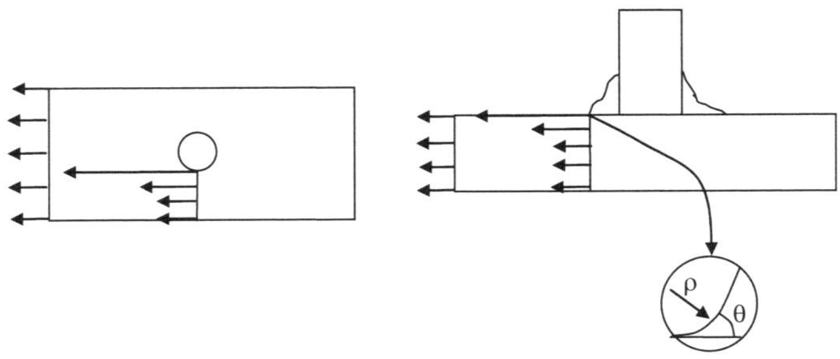


**Figure 2.15.** *Notch effect at the edge of a bore hole and at the weld toe*

If we now claim that the stress concentration due to the notch effect is at a comparable level for the bore hole and the welded attachment as indicated on Figure 2.15, then the explanation for the shorter fatigue life of the welded detail must be found from the differences in surface condition (micro-flaws) and residual stresses for the two specimens. The plate with the bore hole does not have any intrusions or micro-flaws near the edge of the hole where the stress concentration is greatest. For the welded detail, these types of imperfections are created during the welding process. It is an unfortunate coincidence that the weld toe area is subjected both to the highest stress concentration and to the presence of these imperfections. As a result, the weld toe becomes that most likely crack site and the fatigue life is significantly reduced compared to a machined detail. If we add the influence of the tensile residual stresses for the welded detail, the reason for the shorter fatigue life found in welded joints is fully explained. These stresses are introduced during the cooling process after welding has been carried out, and over the length of the weld seam they will appear as both compression and tensile stresses. Typical stress distributions are shown in Figures 2.16 and 2.17 for residual stresses that are normal to the welding direction and parallel to the welding direction respectively. The maximum stresses in the distributions can be close to the yield stress of the material. In the tension area the effective mean stress will increase and the associated fatigue life decrease as has been explained in the foregoing section. As a result, the fatigue life of welded joints has limited sensitivity to the applied mean stress caused by the external loading. When these stresses are added to the residual stresses, the stress cycle will have a maximum value near the yield stress regardless of the mean stress level associated with the applied loading (see Figure 2.12). This is true for joints in the as-welded condition. If the joints are stress relived by post-weld heat treatment, we will notice a dependency on the applied mean stress level as was shown in Figure 2.9. For such cases, a welded joint will not be so vulnerable if the stress cycles inflicted by external loading are partly compressive.

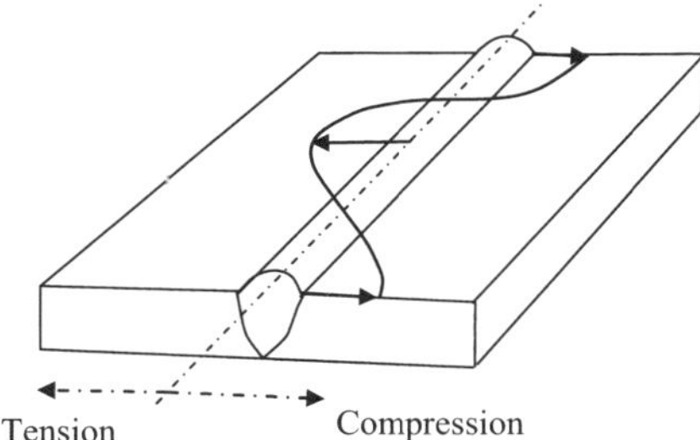


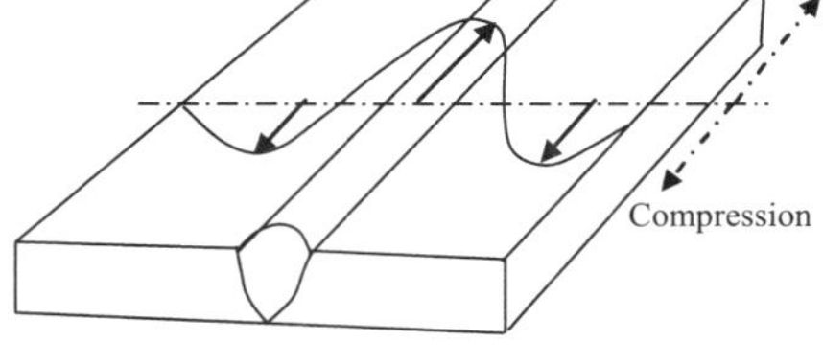


**Figure 2.16.** *Welding residual stresses transverse to the welding direction*

**Figure 2.17.** *Welding residual stresses parallel to the welding direction*

All the issues discussed above are now generally accepted and well understood, although there is still some dispute regarding the presence of intrusions or flaws. The traditional school of thought has been to assume the presence of a crack-like defect, with a depth of several tenths of a millimeter, right from the start. This assumption also comprises the effect of an undercut. Hence, the entire fatigue process in a welded joint has been regarded as large-scale crack growth. This point of view is often laid down in rules and regulations. However, this belief does not fit the experimental facts. Several test series, with welded details where the crack depth is measured during the course of the test, have shown that there is an initiation phase present in the damage process. This is especially true at low stress ranges. This means that initiation may be important under typical in-service loading conditions. Instead of speaking of crack-like defects, one may speak of micro-flaws or rough surface conditions at the weld toe. One may say that the welding process leaves fingerprints at the weld toe that reduces the incubation time before crack growth starts. The incubation time – although reduced – still exists. This is the rationale for choosing a two-phase model as a more correct approach to the fatigue behavior of high-quality welded joints. The crack initiation is modeled according to Figure 2.11 and the crack growth according to Figure 2.10. The obtained results should be consistent with the total fatigue life as given in Figure 2.13. One argument for this approach is that the quality of welded details has improved over the last decades due to better quality assurance and control. However, when discussing these topics one must bear in mind what we have said about the fabrication quality. There may be an appreciable difference between the size of initial flaws present in a single component welded in shop environment and a details found in a huge steel structures erected *in situ*.

The issues discussed in this section will be elaborated throughout the book.


## 2.6. Various types of joints

### 2.6.1. *Plated joints*

We have so far shown a full penetration butt joint (Figure 2.8) and a fillet-welded joint with axial loading transverse to the welding direction (Figure 2.14). Several other possible configurations and loading modes exist. However, even large complex structures may often be divided into a relatively low number of elementary joints. To verify the fatigue life and control the fatigue damage process during service of such structures, it is necessary to study in detail the fatigue behavior of these simple joints. When characterizing a welded detail the following topics are important:
   – geometry of the detail (welded plate stiffeners, attachments etc.);
   – loading direction (perpendicular or parallel to the weld seam);
   – joint configuration (e.g. butt joint, fillet welded joint);
   – welding procedure (full penetration, flat position);
   – loading mode (axial or bending);
   – production quality (tolerances, surface finish);
   – extent of non-destructive inspection (visual or detailed);
   – post-weld treatment (stress relieving, grinding);
   – potential crack locus (weld toe, weld root).

Rules and regulations recommend categories (classes) as a result of the characterization given above. Each category is assumed to have a given fatigue quality for a given locus for potential fatigue cracking. The fatigue strength is usually given as the number of cycles to fracture as a function of the nominal stress range under constant amplitude loading (see Figure 2.9). It is noticed that the steel quality does not enter our list of important topics. We have explained why: the fatigue quality is mainly governed by geometrical aspects and the local surface condition at the weld toe, together with the welding residual stresses.

Two examples of joint categories are shown in Figure 2.18. To the left is shown the full penetration butt joint (welded from both sides) with axial loading perpendicular to the welding direction. This is the most common high-strength joint found. To the right is shown a plate with a gusset attached by a fillet weld. The plate is subjected to bending loading mode, again perpendicular to the weld seam as for the butt joint. There are several important differences between these two details. The most obvious is that the butt joint has to transfer the loading through the weld itself, whereas for the fillet weld, the load path is through the main plate. The fillet weld is regarded as non-load-carrying.

The butt joint gives fewer geometrical changes in the direction of the load path. Hence, the stress concentration factor will be smaller than for the fillet weld which gives a larger geometrical disturbance. This is the main reason for the higher fatigue quality of the butt joint. For both these joints, a potential fatigue crack will emanate from the intersection between the weld and the plate. This weld toe region is the critical point for most welded joints, as we already have pointed out.

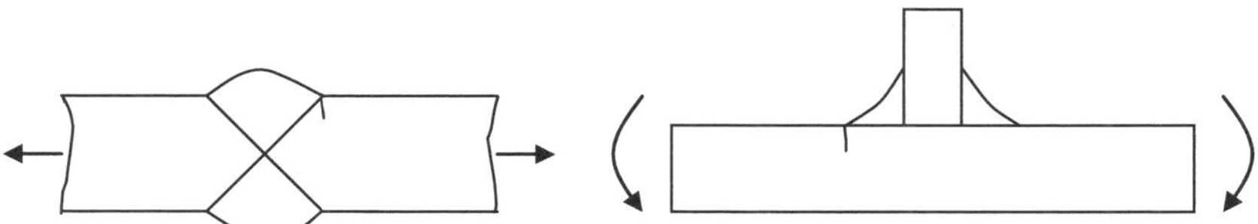

**Figure 2.18.** *Types of welded joints: left – double-sided butt weld between plates, right – fillet welded stiffener*

The fillet welded stiffener is shown in more detail in Figure 2.19 for the purpose of studying the geometrical parameters. It is convenient to distinguish between global and local geometry parameters. These are defined below:

– global parameters of the detail: the plate thickness T and the leg length L;

– local weld geometry: weld toe angle $\theta$ and weld toe radius $\rho$.

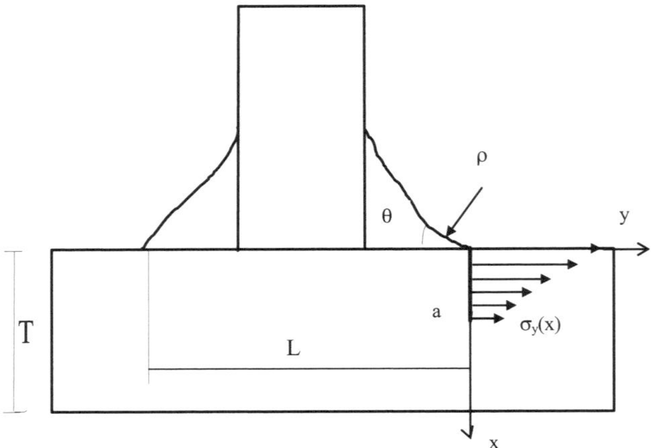

**Figure 2.19.** *Definition of geometrical parameters*
*of a fillet welded joint*

The geometrical parameters given above are by far the most important ones for the fatigue quality because they govern the stress concentration at the weld toe. If one adds the surface condition at the weld toe and the residual stresses, the fatigue quality is almost determined. All these considerations lead to the determination of a joint category with an associated S-N curve in rules and regulations. However, the categories are defined in less detail than we have discussed here. Local toe geometry and surface finish are not included in the definition of a category. Residual stresses are not controlled. As a result, the test population from which the S-N curve is obtained may exhibit a large scatter in fatigue life. If the population (defining a category) had been defined by local toe geometry, surface finish, and residual stresses then the fatigue life would exhibit less scatter and the design curve would be easier to define. In fact, more recent rules have defined a larger number of classes due to the issues we have discussed above. We will discuss this point further in Chapter 5. It is also likely that the material quality will play a role at low stress ranges. We will return to this issue in Chapter 9.

Figure 2.20 shows two other welded details with the weld direction transverse to the loading direction. On the left is shown a butt weld made from one side in a V-groove. The potential crack site will in this case be the weld root due to lack of penetration. The condition may improved by a backing bar as shown, but the fatigue quality is still much poorer than for the butt weld made from both sides in an X-groove. To the right is shown a cruciform joint with a fillet weld. Compared with the detail we studied in Figure 2.14, the load transfer now has to go through the weld metal. As a result this detail has two potential crack sites; one at the weld toe as discussed earlier and one at the weld root. The first crack will be driven by the stress in the plate section, whereas the crack from the weld root will be driven by the stresses in the weld throat. Hence, which one of them will be critical depends on the plate thickness T versus the leg length $\lambda$ of the weld. In this case the fatigue life of the joint must be checked for two possible crack locations by using two different categories and applied stresses. Again, we will learn more about this in Chapter 5. Both cases can also be treated by a fracture mechanics model, as we will show in Chapter 6.

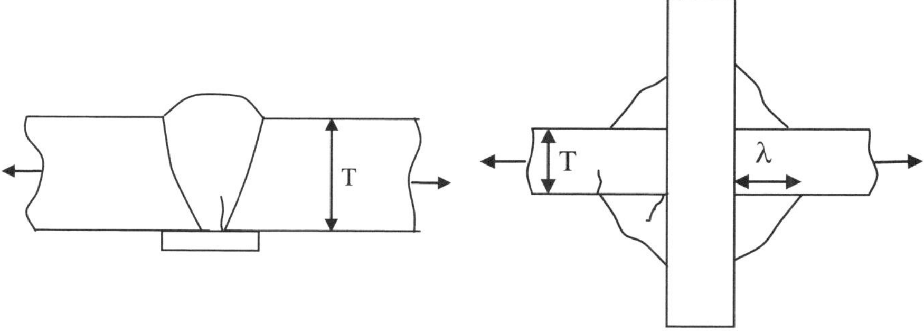


**Figure 2.20.** *Left: butt joint welded from one side with a backing bar in the root Right: load-carrying fillet welded joint*

All the welded details shown so far will usually be less vulnerable to fatigue damage if the loading direction is changed to be parallel to the weld seam (normal to the paper plane). The reason is that the stress concentrations caused by the weld toe will disappear as this notch no longer will be transverse to the direction of the stresses. For continuous longitudinal welds in a beam subjected to bending, the most likely crack site will be a ripple as shown to the left in Figure 2.21. It is assumed that the bending moment decreases towards the end of the beam so the terminations of the welds are in an unstressed area. For the attachments to the right in Figure 2.21, this is not possible and the critical spots will in this case be found in the start and the termination of the welds. This is due to the fact that the ends of the stiffeners give a severe stress concentration. The weld on the plate edge is usually more vulnerable than the weld in the middle of the plate.

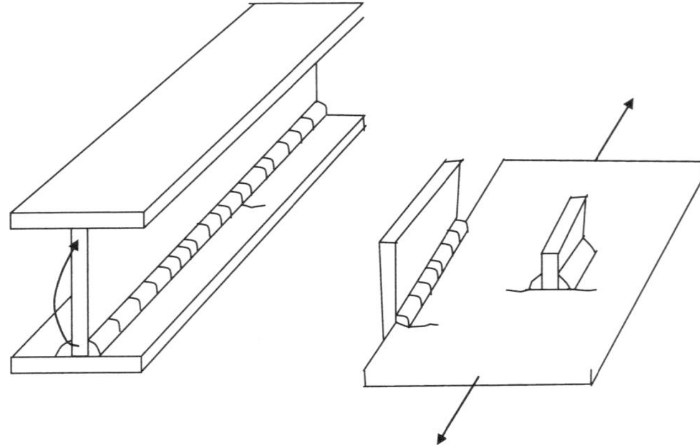

**Figure 2.21.** *Welds longitudinal to the direction of the applied loading*

We have emphasized above that fatigue of a welded joint is mainly a geometrical and production-quality problem. Welded detail made from high-strength steel does not behave appreciably better than details made from normal mild steel in an as-welded condition. This is one of the peculiarities of the fatigue behavior of welded joints. The other is that the applied mean stress hardly matters. It must however be born in mind that this is true under certain conditions, i.e. accelerated testing in the laboratory. At lower stress ranges, there is an initiation phase present in the joint, and the steel grade will play a role. Although this role is a minor one compared with the toe-notch effect and the surface condition at the fusion line, it should be modeled. Furthermore, the applied mean stress will obviously play a role if the welded joint is stress relieved. We will come back to these issues in Chapter 9.

### 2.6.2. Tubular joints

One group of joints that needs special attention is welded joints between hollow sections such as cylindrical or rectangular pipes. One simple example is the T-joint between two pipes as depictured in Figure 2.22. It consists of a chord member with diameter D and wall thickness $T_C$ and a brace member with diameter d and thickness $T_B$. The principal loading is the axial force and the in-plane bending moment acting on the branch. For these kinds of joints the stress concentration factor can be very high near the intersections mainly due to secondary plate bending in the pipe walls near the welds. Details are shown to the right in Figure 2.22 with the stress distribution through the chord thickness and the maximum surface stress at various distances from the potential crack site at the weld toe. The section shown is defined as the crown point of the joint. There is a similar phenomenon on the branch side.

The problem is that the stress concentration can be very different for different loading modes and thickness ratios, although these loading modes may give the same nominal stresses (elementary beam stresses) away from the critical intersection. Hence, for these types of joints, the nominal stresses are not an appropriate key to the fatigue life given by an S-N curve. It is much more logical to use the stress concentration at the weld toe, ignoring the part of the concentration created by the notch effect of the weld toe itself. The main argument for this choice is that whereas the geometrical stresses may differ significantly from one joint to another due to changes in loading mode and gross geometry, the local stress concentration due to the weld notch will remain the same. Hence, the weld notch effect is not explicitly taken account of, but will be inherent in the S-N curve. The so-called geometrical stress concentration or hot-spot stress has also gained popularity for plated joints and will be discussed in Chapter 5 as a strategy to reduce scatter when presenting fatigue results.

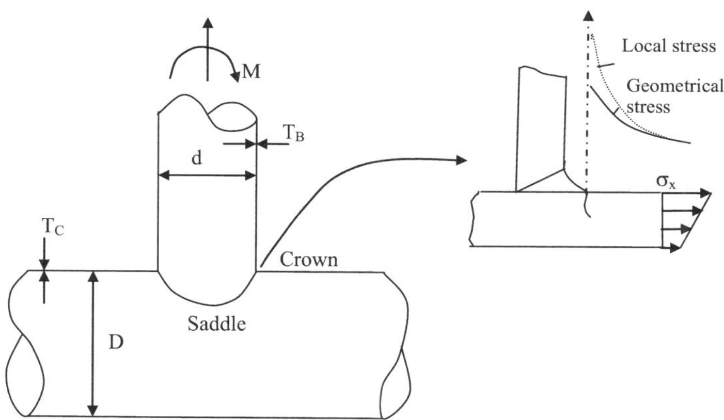


**Figure 2.22.** *Tubular T-joint with in-plane loading. Geometrical stress (fully-drawn line) and local stress (dotted line) at the weld toe on the chord crown point*

## 2.7. References


1    *Review of Repairs to Offshore Structures and Pipelines*, Publication 94/102 Marine Technology Directorate, UK, 1994

2    A. Almar-Næss (ed), *Fatigue Handbook*, Tapir, 1985

3    J. Schive, *Fatigue of Structures and Materials*, Kluwer Academic Publishing, 2001

4    A. Boresi and R.J. Schmidt, *Advanced Mechanics of Materials*, Wiley, 6th ed., 2003

5    S. Maddox, *Fatigue Strength of Welded Structure*, Abington Publishing, 1991

Chapter 3

# Experimental Methods and Data Analysis

## 3.1. Introduction and objectives

As pointed out in Chapter 2, fatigue of welded joints is a quite complex problem. All theories and models have to be verified and corroborated by experimental data. As for the S-N model, this model is nothing else but curve-fitting between applied stress range and experimental lifetime data. A fracture mechanics model has to be based on experimental data to determine appropriate values for the parameters involved. When these parameters have been determined, the model can predict crack growth and critical crack size for similar material and environments. The only theory involved is the development of the stress intensity factor concept. In fact, all models discussed in this book are semi-empirical models. The number of cycles to crack initiation, crack path history, and final fatigue life must be measured for typical steel materials, joint types, and loading conditions. Model parameters must be chosen to fit the experimental facts. In this chapter a brief overview of the common testing techniques is given. Finally, some often-employed statistical methods used to deal with the scatter in experimental results are also described.

The objective of this chapter is that the reader learns how to plan and carry out experimental work primarily related to the S-N model and the fracture mechanics model. The reader will also gain some knowledge on measurements techniques and how to analyze the results. Hence, life testing and crack growth testing will be shown in some details. More details are found in Ref [1].

## 3.2. Overview of various types of tests

In Chapter 2 we briefly touched upon the three most common types of fatigue testing:

– S-N testing using smooth specimens to characterize the base material;

– S-N testing using specimens to characterize a structural detail, e.g. a welded joint;

– crack growth testing using standard pre-cracked specimens;

– strain-life testing with small-scale smooth specimens.

One may also add that a large part of a structure can be tested, as for example a welded frame with large girders, but this is rarely the case. We will, however, present a test setup of a welded plate structure in Chapter 11. It should also be mentioned that S-N testing can also include crack growth measurement before final fracture. This will produce much more information than just the total fatigue life. Electrical-based methods have increased the possibility for crack growth monitoring during testing. However, crack growth testing with standard compact tension (CT) specimens is usually much more accurate than the growth measurement carried out on realistic details. The growth parameters can be determined much more precisely in CT specimens. The strain-life tests are carried out with small-scale specimens under constant strain amplitude. The strain amplitude is used as the key to explain the initiation life. The method is convenient to cope with cyclic plastic strain condition, i.e. low-cycle fatigue. Furthermore, the method is applicable to characterize the time to crack initiation under cyclic elastic strain condition. We will discuss the methods in some detail in what follows, with the emphasis being on the testing of welded joints

## 3.3. Stress-life testing (S-N testing) of welded joints

### 3.3.1. *Test specimens and test setup*

S-N testing is defined as testing using the stress range as the main explanation of the fatigue life. The life, given as the number of cycles to failure, is plotted as a function of the applied stress range. Details that have nearly the same geometry, welding quality, residual stresses and loading mode define one experimental population. This experimental population forms the basis for a detail-category (class) in the building codes. A straight-line relationship between logNT and log$\Delta\sigma$ is assumed in the finite life region, and the best-fit mean S-N curve is found by method of least squares, i.e. a linear regression analysis. As the involved scatter is one of the main characteristics for lifetime data of a welded joint, this subject will be addressed in some detail at the end of this chapter.

Hydraulic digitally-controlled testing machines are used to subject test specimens to repetitive loading. Although it is possible to simulate variable amplitude loading, most tests are carried out using constant amplitude loading. The testing is carried out under constant amplitude loading at various stress levels. The machine is usually in the load control mode, and it is recommended that the specimen stresses are monitored by strain gauges in addition to the information provided by the machine's control panel. Regarding the definition of stresses, there has already been a short discussion, at the end of Chapter 2, of the choice between nominal stress range and geometrical stress range. The fatigue test specimens should usually have a geometry and a loading mode that are representative for in-service condition. Smaller specimens are tested in standard servo-hydraulic testing machines, whereas full-scale testing of larger structural parts (e.g. tubular joints, large girders) are tested in specially built frames with adaptable hydraulic cylinders. A cruciform fillet-welded joint subjected to axial loading is shown in Figure 3.1. The specimen has a width of 60 mm and a thickness of 25 mm. The specimen is mounted directly between the two grips of an AMSLER hydraulic testing machine with a performance of 250 kN dynamic loading. It is the upper piston that inflicts the loading by vertical displacement. The main plate of the welded detail is in a vertical position and the specimen is subjected to an axial force perpendicular to the welding direction. The electrical cables seen on the photo pertain to unidirectional strain gauges in the loading direction located 10 mm from the weld toe. These types of test series form the basis for the F-class design S-N curve that will be discussed further in Chapter 5.

Figure 3.2 shows a tubular joint with an X configuration in the same type of testing machine. The diameter of the chord is 320 mm, whereas the diameter of the branches is 250 mm. The wall thicknesses of the tubes are 16 and 12 mm respectively. Due to the size of the joint, it is mounted in a frame on the table of the testing machine. The upper piston of the machine introduces a vertical force on the chord, with associated bending in the two branches through vertical movement of the chord. The lower photo also shows all the strain gauges that are attached to the tube surfaces to control the strain distribution in vicinity of the welded intersection between the cord and the branches. This type of joint geometry gives data for establishing the T-class S-N curve. As we have discussed, the geometrical stress range is used as the key parameter to determine the fatigue life for this case. We will, in what follows, briefly discuss the preparation and measurements carried out for the kind of tests shown in Figures 3.1 and 3.2.

**Figure 3.1.** *Cruciform fillet-welded joint in a hydraulic testing machine*

**Figure 3.2.** *Tubular joint with an X geometry subjected to in-plane bending of the braches. Upper part: setup in testing rig. Lower part: detail of branch-chord intersection at the crown point with strain gauges*

### 3.3.2. *Preparations and measurements*

Fatigue testing includes all the phases such as test planning, the preparation of the specimens and the statistical analyses of the results. It is important to keep track of all the parameters that have an influence on the test results. This is necessary to define a homogeny specimen population as a basis for the obtained design curve. If a population involves test specimens with large differences in these important parameters, the compiled test results will exhibit large scatter that can be difficult to explain and cope with. The practical results will be that high-quality joints will be penalized because they have been merged with joints of lower quality. For the populations defining categories in current rules and regulations, this is often the case and improvement in this area will lead to more accurate predictions due to reduced scatter. Thus, high-quality joints will get the fatigue life predictions they deserve. This is an important topic for advanced mechanical industries.

The following background information should be gathered:

– Before the test:

   - steel type, chemical composition, mechanical characteristics,

   - welding procedure, method, electrodes, number and sequence of passes, heat input,

   - manufacturing sequence,

   - global specimen geometry and local weld toe geometry,

   - axial or angular distortion,

   - non-destructive testing (NDT),

   - estimate of residual stresses in the specimens,

   - rolling direction of plates with regard to applied loading,

   - microscopy of the heat-affected zone (HAZ).

Most of this information is straightforward to obtain. The chemical composition and mechanical characteristic for a C-Mn steel is given in Table 3.1. The data are typical for medium-strength steels with nominal yield stress close to 350 MPa. This is one of the most widespread steel types used for welding. Both the fillet joint in Figure 3.1 and the tubular joint in Figure 3.2 are made of this steel. The welding has been carried out by shielded metal arc welding (SMAW). Regarding the given information for mechanical parameters, hardness measurements should be carried out after the welding has been completed. This gives important information regarding the base metal, weld deposit, and the HAZ. Figure 3.3 and Table 3.2 show typical results from hardness measurements carried out on the material in Table 3.1 after welding had been carried out.

| C | Si | Mn | P | S | Cu | Ni | Cr | Mo | Nb |
|---|----|----|----|----|----|----|----|----|----|
| 0.08 | 0.15 | 1.40 | 0.006 | 0.002 | 0.01 | 0.02 | 0.02 | 0.01 | 0.008 |

| Yield strength [MPa] | Tensile strength [MPa] | Elongation % |
|---|---|---|
| 416 | 501 | 25 |

**Table 3.1.** *Chemical composition in % and mechanical characteristics for a C-Mn steel*

| Spot | HB |
|---|---|
| Base 1 | 142 |
| Base 2 | 147 |
| Weld 1 | 175 |
| Weld 2 | 178 |
| Haz 1 | 213 |
| Haz 2 | 181 |
| Haz 3 | 173 |
| Haz 4 | 165 |

**Table 3.2.** *Hardness measurement*

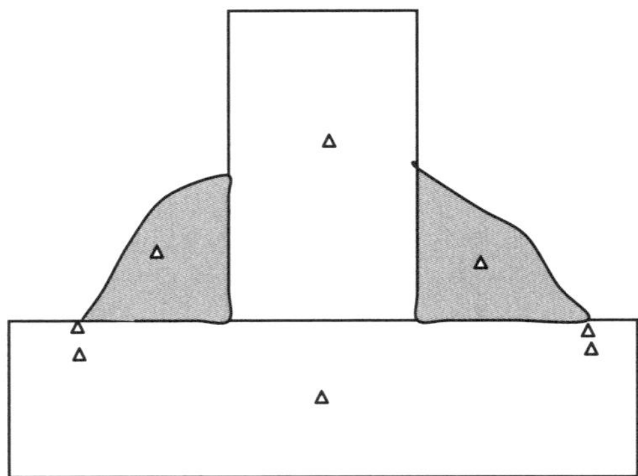

**Figure 3.3.** *Results for hardness measurements for cruciform plated joint (C-Mn steel in Table 3.1)*

The local toe geometry parameters are measured by applying replica material on the weld toe and inserting cuts of replica cross sections into a profile projector with a typical magnification of 10. The replica cuts must be made transverse to the weld seam direction. Table 3.3 shows the results from two different test series with specimens as shown in Figure 3.1. As can be seen, the toe profile of series 1 is much more favorable than for series 2. Series 1 has low toe angle (mean value 30 degrees) and larger radius (mean value 2.7 mm). Series 2 has a mean angle of 58 degrees and a toe radius of 0.75 mm only. The fatigue testing actually showed that the mean fatigue life of series 1 was twice as long as for series 2. If local toe geometry had not been measured, this difference in fatigue life would have been difficult to understand and explain.

|  | Weld toe angle (Degrees) | | Weld toe radius (mm) | |
|---|---|---|---|---|
|  | Series 1 | Series 2 | Series 1 | Series 2 |
| Number of recordings | 384 | 400 | 384 | 400 |
| Min value | 15 | 30 | 0.1 | 0.1 |
| Max values | 69 | 95 | 11.3 | 4.0 |
| Mean value | 30 | 58 | 2.7 | 0.75 |
| Std. Dev. | 9 | 9 | 2.2 | 0.44 |
| COV | 0.31 | 0.16 | 0.81 | 0.58 |

**Table 3.3.** *Statistical results from local toe geometry measurements of a fillet weld*

When the global and local geometry and possible distortion have been measured, a finite element analysis (FEA) should be carried out to determine the stress concentration factor at the weld toe which is the potential crack locus. A result of such an analysis is shown in Figure 3.4. As can be seen, the model is using a refined mesh of the weld profile and plates. This is the weld toe area of the test specimen in Figure 3.1 analyzed under plane strain conditions. As can be seen the stress concentration factor (SCF) is close to 3.0. In Chapter 11 we will present more advanced FEA modeling of full-scale structures including the presence of a crack.

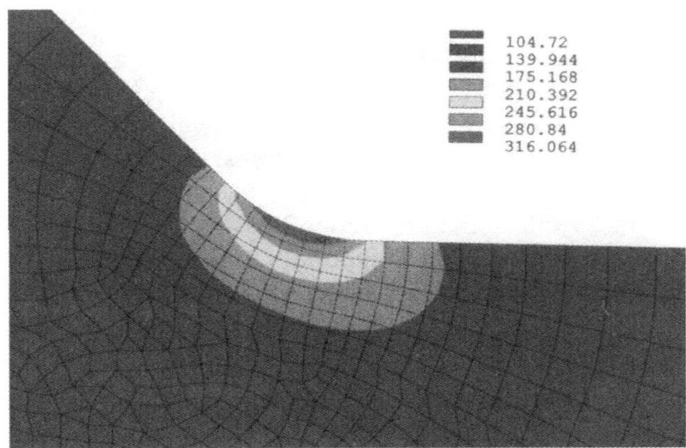


**Figure 3.4.** *Results from a local FEA model for a cruciform fillet-welded joint*

Non-destructive testing should be carried out at the same level as is typical for details entering into service conditions. Specimens with flaws that would have been rejected for in-service purposes should, of course, also be excluded from the test series. The most frequent method is magnetic particle inspection and no crack detection is the only acceptable result. Estimating the actual residual stresses is probably the most challenging part of the preparation work. In general, small specimens will have lesser residual stresses than larger specimens. If the specimens are properly stress relieved, the residual stresses may be set to zero. If not, they can be measured by boring a hole and measuring the change in stresses in vicinity of the hole by train gauges. This will reveal the residual stress gradient. The problem is that the residual stresses vary over the volume of the test specimen. Finally, metallographic study of the HAZ will reveal any abnormal phases or grain size.

During the test the following should be measured and registered:
- loading mode, load range and nominal stress range,
- the applied R-ratio for the applied stresses,
- load frequency,
- stresses by strain gauges monitoring,
- crack growth measurements by an electrical method,
- beach marking of the fatigue surface by temporarily decreasing the stress range,
- number of cycles to final failure.

The loading mode is usually uni-axial loading or bending. One should be aware of the fact that most S-N data are compiled without taking into account the difference between these two modes; it is only the maximum stress range that is assumed to matter. Most specimens are subjected to a positive load ratio to avoid problems with buckling. Typical values of R are in the range of 0.1 to 0.3. The loading frequency is assumed to play a minor role in air environment, and most tests are accelerated so that the test program will not consume too much time. A typical frequency of 6-10 Hz is often applied. However, special care should be taken for the environmental condition (e.g. corrosive seawater.) In such cases there is an interaction between the fatigue damage process related to the oscillating stresses and the time-dependant electro-chemical reaction at the crack front. Therefore, test frequency becomes important. The load frequency should usually not exceed 1 Hz.

Detailed knowledge of the strain distributions can be obtained by the use of electrical resistance strain gauges as was shown in Figure 3.2. These gauges operate on the principle that the resistance of a wire changes when its length changes due to stretching or compressing. Hence, when bonded to the surface of the specimen, the local strain at the surface of the material is captured. Some test specimens should be equipped with strain gauges to reveal the actual strain in vicinity of the weld toe during the test. Based on these measurements, the strain concentration can be determined and also the related stress concentration. The SCF obtained from FEA can be verified by these measurements in the hot-spot region near the assumed fatigue crack location. The stress should account for the gross stress concentration due to the global geometry of the joint, but not the local effect caused by the weld profile. The gauge distance from the weld toe should be large enough to avoid the local effect of the toe itself, typically 10-15 mm. The strain measurements are normally compared to stresses obtained by FEA analysis. FEA obtained stresses were shown in Figure 3.4.

Local strain distribution for a tubular joint is shown in Figure 3.5. The distribution is obtained with a rack of uni-axial gauges with 0.3 mm length and 2 mm spacing. As can be seen, the strains measured at the chord are higher than the measurements at the branches.

Furthermore, the gauges closest to the toe are affected by the weld notch as their reading deviates from a linear distribution. If a straight line is drawn through the other points, a hot-spot strain level close to $\varepsilon_{HS}$ = 450 $\mu$s is found at the weld toe on the chord side. If the strain is assumed to perpendicular to the weld seam, the strain concentration factor is defined as:

$$SNCF = \frac{\varepsilon_{HS}}{\varepsilon_N} \qquad\qquad (3.1)$$

where $\varepsilon_N$ is the nominal strain in the branch. If this hot-spot strain coincides with the direction of the principal stress at the hot-spot, the following transformation can be used between SCF and SNCF:

$$SCF = SNCF \frac{1 + v \cdot \varepsilon_2 / \varepsilon_{HS}}{1 - v^2} \tag{3.2}$$

where $\varepsilon_2$ is the measured strain parallel to the weld seam. Again, the SCF value based on measurements can be compared with the stress concentration found by FEA or by parametric formulae, Ref [3]:

$$SCF = 0.595 \cdot \gamma^{0.6} \cdot \tau^{0.8} \cdot (1.6\beta^{0.25} - 0.7\beta^2) \cdot \sin^{(1.5-1.6\beta)} \theta \tag{3.3}$$

where:

$$\beta = \frac{d}{D} \qquad \gamma = \frac{D}{2T} \qquad \tau = \frac{T_B}{T_C}. \tag{3.4}$$

The geometrical parameters above were defined in Chapter 2, Figure 2.22. The parameter $\theta$ is the brace inclination angle in radians, in our case $\pi/2$. The SCFs for both the brace and the chord side are to be multiplied by the nominal stresses in the brace. The stress concentration can be further verified by the use of acrylic model. This has led to a clearer definition of applied hot-spot stress.

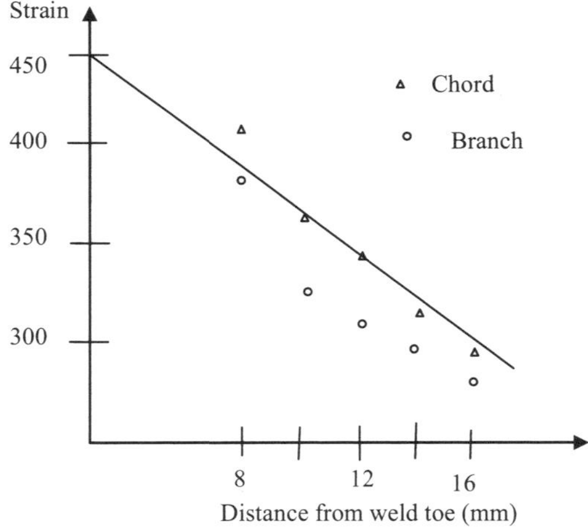


**Figure 3.5.** *Strain measurements at brace and chord at welded intersection (strain gauges length 0.3 mm spacing 2 mm, see bottom of Figure 3.2)*

### 3.3.3. *Test results*

Traditional fatigue life testing produced one result only: the number of cycles to failure. For small test specimens the failure is defined when a final fracture separates the specimen into two parts. For larger joints the failure state is a question of definition. It can be defined as a fracture separating the joint into two pieces or by a through-thickness crack. For the cruciform joint in Figure 3.1, these two time stages will coincide, whereas for the tubular joint in Figure 3.2, the joint will still have integrity when a through-thickness crack has appeared. Typical life data are shown in Figure 3.6 for small-scale testing with base material. The material is a high-strength weldable steel for mooring chains. For these small specimens, the failure occurred when the crack was less than 1 mm. As can bee seen, the 18 tests are carried out at three stress levels, six tests at each level. The tests were carried out in seawater without any protection against corrosion and the load frequency was 1 Hz. Some results in air are added for comparison. The mean and the design curves are drawn. We will return to how these curves are obtained at the end of the chapter. When this steel is used in a larger structural item, it must be born in mind that the life data in Figure 3.6 corresponds to early cracking of the item and to not final failure. Therefore, if the curves in Figure 3.6 are used directly as life data for a structural item, it will be overly pessimistic.

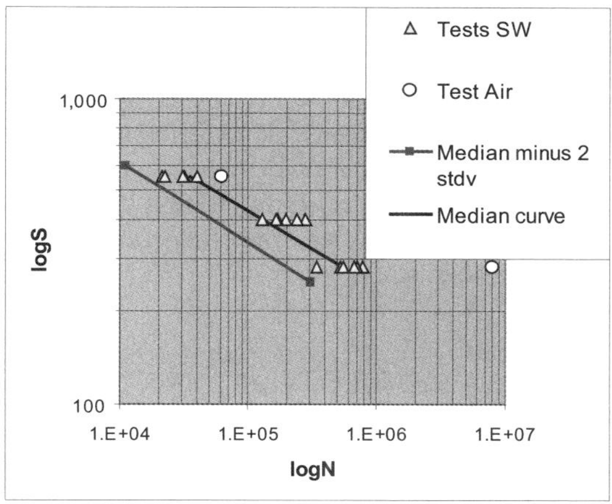


**Figure 3.6.** *Results from S-N testing with small specimens of high-strength steel in seawater*

Although the fatigue durability, given as the number of cycles to failure, is the most important result, one may say that great effort is made, but only limited information in produced. In the last decade, the use if of electrical methods has made it possible to obtain more information from the tests. Crack growth measurements are carried out at different time intervals during the test to reveal the entire crack growth history, and not only the number of cycles to failure. This is particularly interesting for full-scale tests where the crack growth plays and important part. The alternating current potential drop (ACPD) method has been used more and more frequently the recent years. A crack monitoring system based on this principle can be fully atomized; an example is shown in Figure 3.7 for a fillet welded joint. A constant AC current field is injected into the test specimen surface at a frequency of 6 kHz. The surface voltage distribution is measured by a crack micro-gauge that receives the signals from fixed pin probes spot welded to the test specimens. A micro-computer controls both the micro-gauge and the test machine loading unit through a multiplexer switching unit. The test data is displayed graphically on the computer and stored on a diskette for further analysis. The specimen has 10 logged stations equally spaced over the weld length. A detailed description of the system is given in Ref [4]. If the voltage measured over pins straddling the surface crack is Vcr and a nearby reference voltage is designated Vr, an estimate of the crack depth reads:

$$a_1 = \frac{e}{2}(\frac{V_{cr}}{V_r} - 1) \tag{3.5}$$

where e is the probe distance. The measurements can be calibrated against crack-front marking by blue ink penetrant injected at a chosen crack depth for each specimen. True crack depth can then be obtained from the fracture surface reading after the final failure. These readings also give valuable information about crack-shape development, i.e. the crack depth versus crack length. If several true crack depths are wanted on the same specimen, ink of different colors may be applied at different times. Alternatively, beach marking can be carried out. Beach marking was explained in Chapter 2 and is introduced during testing by reducing the stress range for a short while, typically 10,000 cycles. The stress range reduction is preferably done by increasing the minimum stress.

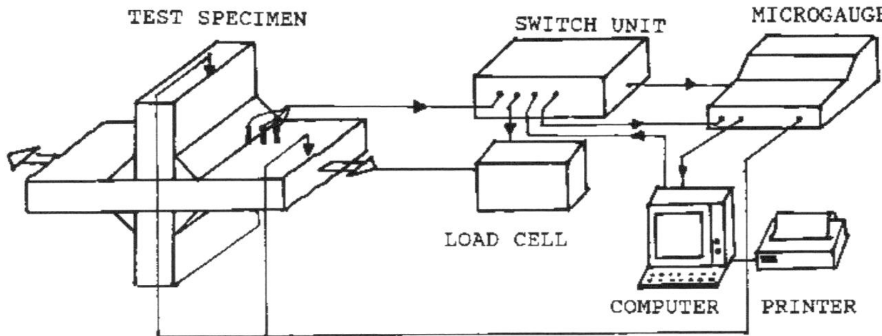


**Figure 3.7.** *Principal sketch for ACPD measurements on cruciform joints*

Typical a-N curves obtained from ACPD measurements are shown in Figure 3.8 for the cruciform joint in Figure 3.1. It is obvious that these curves bring much additional information, both on number of cycles to early cracking and crack growth rates. We will return to these results in Chapters 6, 7 and 9.

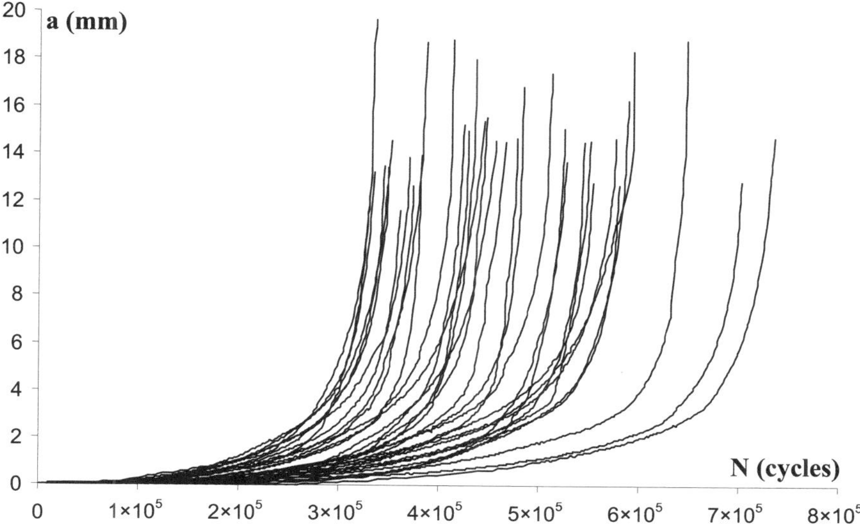



**Figure 3.8.** *Entire crack history in cruciform joints before final fracture is reached*

After a test has been carried out, fatigue and fracture surface microscopy should be carried out on some specimens. By applying a scanning electron microscope, it may be possible to trace initial flaws or intrusions that act as starters for crack initiation.

## 3.4. Testing to determine the parameters in the strain-life equation

We have established the equation for the strain-life in Chapter 2, equation (2.7). The tests are carried out with smooth, small-scale specimens instead of full-scale welded joints as discussed in the foregoing section. The test could, in fact, be carried out on the specimens that were used to obtain the results given in Figure 3.6 if the tests were carried out under strain control. These kinds of tests determine the four fatigue parameters given in equation (2.7) in Chapter 2. The parameters b and c are the fatigue strength and ductility exponents, and $\sigma'_f$ and $\varepsilon'_f$ are the fatigue strength and ductility coefficients respectively. To determine these parameters, the total local strain range is divided into one elastic and one plastic part. It can be shown that the number of reversals to failure can be written both in terms of the elastic strain range and the plastic strain range by the equations:

$$\frac{\Delta\varepsilon_e}{2} = \frac{\sigma'_f}{E}(2N_I)^b \tag{3.6}$$

$$\frac{\Delta\varepsilon_p}{2} = \varepsilon_f'(2N_I)^b . \tag{3.7}$$

We have in this case assumed zero mean stress. These two equations correspond to the two straight lines for a log-log scale in Figure 3.9. The sum of them corresponds to equation (2.7) in Chapter 2. In that chapter, the mean stress correction was included. Test results can be plotted on the two formats and linear regression can be carried out to determine b and $\sigma'_f$ from the elastic line and c and $\varepsilon'_f$ from the plastic strain line. The parameters are indicated on Figure 3.9.

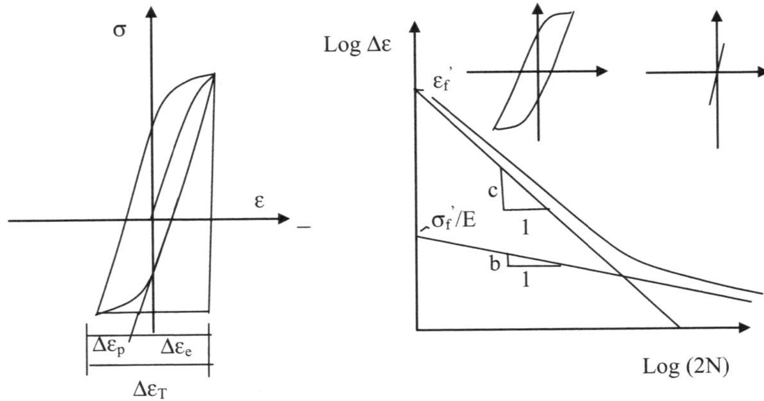


**Figure 3.9.** *Principal sketch for determining fatigue parameters in strain-life relationship*

As will be shown in Chapter 9, the test results can be used to describe the fatigue nucleation at the weld toe when the initiation phase becomes important. Finally, it should be mentioned that it is a common approximation to relate the parameters in equations (3.6) and (3.7) to the hardness measurements shown in Table 3.2. We will return to these issues in Chapter 9.


### 3.5. Crack growth tests – guidelines for test setup and specimen monitoring

Crack growth testing is based on the hypothesis that it is the stress intensity factor range (SIFR, $\Delta K$) that governs the crack rate. The SIFR uniquely determines the local severe stress field ahead of the sharp crack front under linear elastic conditions. We will come back to this concept in Chapter 6. Crack growth tests are usually carried out on standard test specimens for which the SIF (stress intensity factor) can be determined with great accuracy. One typical test specimen is the compact tension (CT) specimen shown in Figure 3.10.

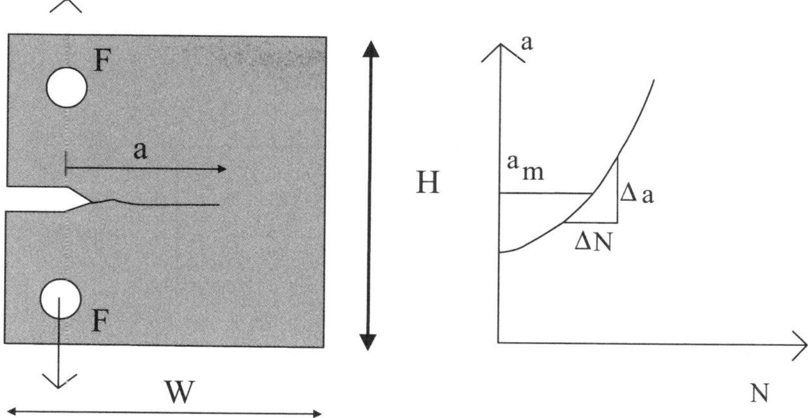



**Figure 3.10.** *Compact tension specimen for crack growth measurements*

The specimen is fabricated with specified dimensions (W and H) that are given once the thickness T of the specimen is chosen. As can be seen, the specimen has a sharp prefabricated notch from which the fatigue crack will grow. Definition of the loading is shown in Figure 3.11. The load F varies with constant amplitude between its maximum and minimum value. The corresponding SIF can be calculated and the SIFR is defined as the difference between them. The evolution of the SIFR with time is shown in Figure 3.12. As can be seen, there is a slight increase in $\Delta K$ for constant $\Delta F$. This increase is due to the increase in crack length.

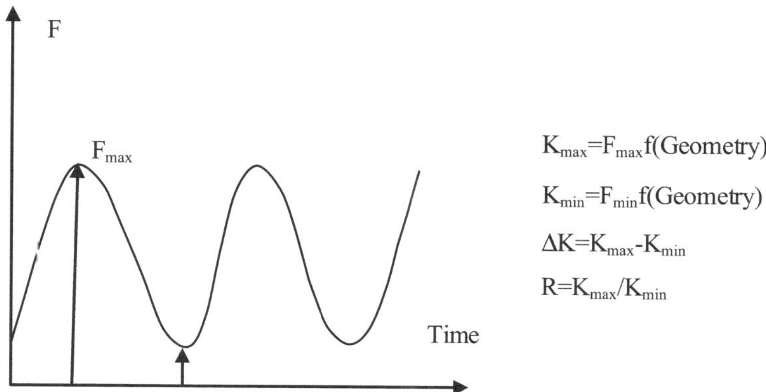



**Figure 3.11.** *Definition of loading and SIFR*

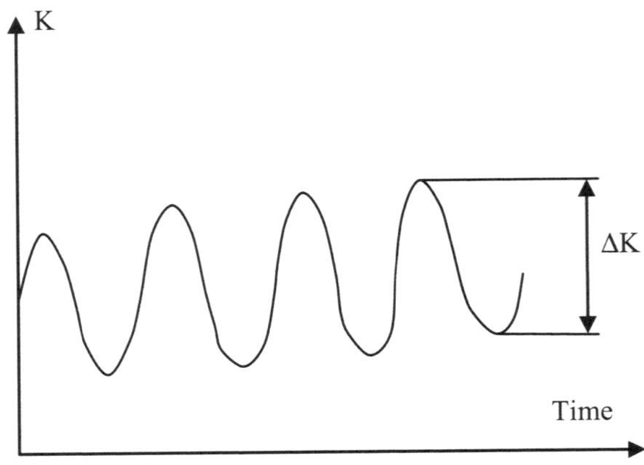



**Figure 3.12.** *SIFR as a function of time*

The crack depth is measured as a function of applied number of cycles, as shown on the left-hand side of Figure 3.13. The measurements can be made optically or by an ACPD technique, as we have just described, Ref [4]. At chosen stages, the crack growth rate $\Delta a/\Delta N$ in m/cycle is calculated and plotted against the actual range of the stress intensity factor $\Delta K$. This range has to be based on the mean crack size during the observation period $\Delta N$. The procedure is shown more detailed in Figure 3.13 where one point in the log $a/\Delta N$-log $\Delta K$ diagram is determined to the right on the figure. By following the same procedure, at several stages along the crack path, the crack growth rate parameters can be obtained as shown in Figure 3.14 for a log-log scale.

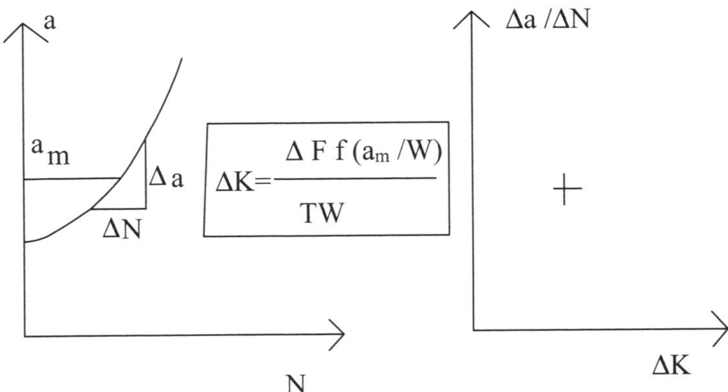



**Figure 3.13.** *Transforming measurements crack growth rates as functions of ΔK*

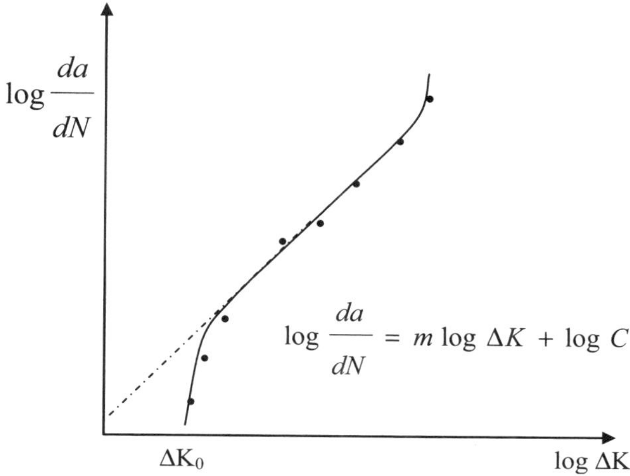



**Figure 3.14.** *Crack growth rate as function of ΔK for a log-log scale*

As can be seen from Figure 3.14, the results fall into three different regions depending on the magnitude of the $\Delta K$. This has already been discussed in Chapter 2. At low $\Delta K$ there is a threshold value $\Delta K_0$ below which a crack will stop to propagate. Furthermore, there is an almost linear relation between da/dN and $\Delta K$ for a log-log scale in an intermediate region. It is in this region that the results obey the Paris law. The slope m and parameter C defined by the intercept with the da/dN axis are obtained. Again, we see from the figure that the linear relation for a log-log scale is an approximation; deviations do occur. Hence, a linear regression analysis has to be carried out in this region in the same way as for the S-N data. The method will be addressed at the end of this chapter. At higher values of $\Delta K$, the maximum value of K is close to the critical value of what the material at the crack front can sustain. The consequence will be unstable fracture. As a result of our hypothesis that it is the SIFR that governs the growth rate, components with various geometries, but made of same material, may have different a-N curves, but when they are transformed into a da/dN-$\Delta K$ diagram, as in Figure 3.14, the curves will coincide. This is true provided the tested components have the same loading ratio R and the same environment. The great benefit is then that the curve in Figure 3.14 can be applied to predict the crack growth rate for various types of crack in any component if we are able to calculate the SIFR for the crack and component geometry in question. Hence, the da/dN-$\Delta K$ diagram and related equations are of great importance.

Figure 3.15 shows the test setup and the result for crack growth tests carried out in seawater. The CT specimens with 25 mm thickness are taken from high-strength steel in a mooring chain. Steel bars are forged and flash welded to obtain chain links from this type of steel. The results to the right in the figure pertain to a test in air, but other specimens were submerged in the small seawater basin as shown to the left in the figure. The tests were carried out with an applied SIFR in the range between 10 and 30 MPam$^{0.5}$. As can be seen, this gives growth rates from $10^{-5}$ to $10^{-4}$ mm/cycle. To reveal the threshold value, the testing has to be carried out at smaller SIFR. Typical values for $\Delta K_0$ are found between 2 and 8 MPam$^{0.5}$, dependent on the environment, Ref [5]. The threshold value can be defined when the rate da/dN is less than $10^{-7}$ mm/cycle.

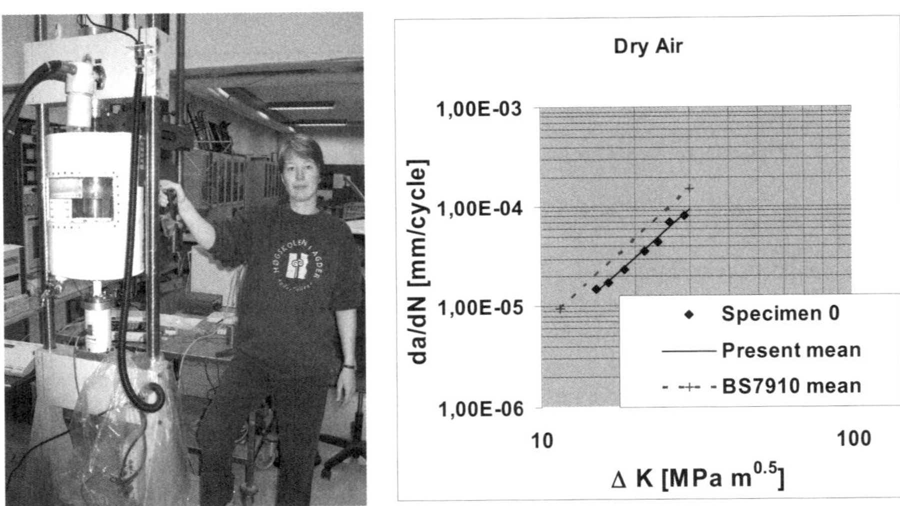


**Figure 3.15.** *Test setup and growth rate results with high-strength steel*

Figure 3.16 shows the final fracture of the CT specimen. This will occur when the either the net ligament ahead of the crack front is too small (global criterion) or if the crack front has too high local stresses (local criterion). In the latter case, the final fracture will be of the brittle type.

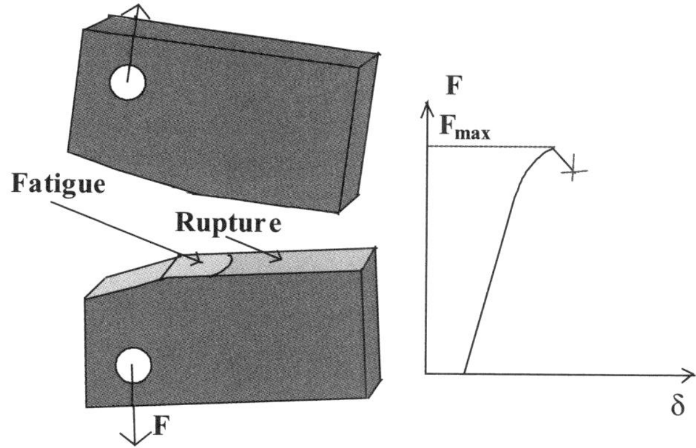


**Figure 3.16.** *Final fracture of a CT specimen after fatigue crack growth*

### Final comment

In this chapter we have presented common practice for crack growth and fatigue life testing. This type of testing forms the basis for the fatigue models and parameters found in rules and regulations. These results will be presented in Chapters 5 and 6. In Chapter 11, the reader will find examples of more advanced testing techniques, including multi-axial loading and full-scale tests of welded structures. Before ending the present chapter we will present the statistical method for data analysis.

## 3.6. Elementary statistical methods

As linear regression is important to almost all of the types of tests that we have described, we will briefly outline the basic equations for this method. We will treat stochastic analysis extensively in Part 2 of the book. It is, however, natural to include linear regression and the statistical uncertainty related to test results in the present chapter. Details for the method are found elsewhere, Ref [6].

### 3.6.1. *Linear regression analyses*

We have shown how the number of cycles N to failure is an approximately linear function of applied stress range $\Delta\sigma$ for a log-log scale. Likewise, the growth rate da/dN is almost a linear function of $\Delta K$ for a log-log scale with averaged values of $\Delta K$ in the actual intervals. However, both relationships are approximations and there are considerable deviations from the straight line. Let us treat the problem generally by letting a variable y be dependent on a variable x as shown in Figure 3.17.

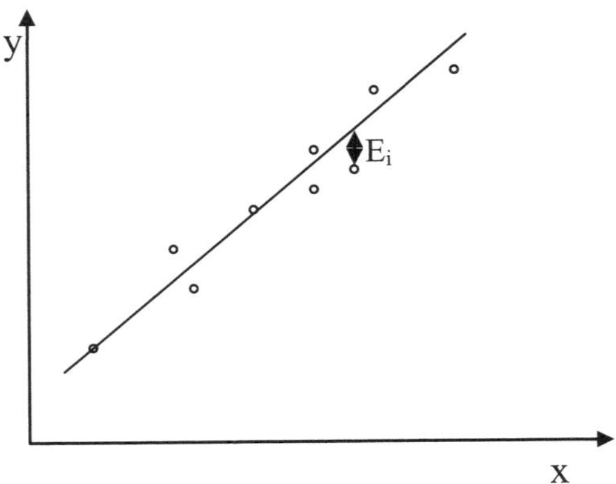


**Figure 3.17.** *The problem of an approximately linear relationship y = bx + a*

The mean value of y is given as a function of x by the straight line, but for each data point there is a deviation $E_i$. For a given x value, the y value is in fact a random variable designated by the bold letter y. The following equations apply:

$$E(\mathbf{y}|x) = A + Bx$$

$$y_i = A + Bx_i + E_i \tag{3.8}$$

where the parameter $A$ is the intercept where the regression line cuts the y-axis, whereas $B$ is the slope of the line. The $E_i$ gives the deviation from the mean regression line for data point no i. The parameters $A$ and $B$ can be estimated by the least squares method:

$$\hat{B} = \frac{s_{xy}}{s_x^2} \tag{3.9}$$

$$\hat{A} = \mu_y - \hat{b}\mu_x$$

The correlation factor between y and x is defined as:

$$\rho = \frac{s_{xy}}{s_x s_y}. \tag{3.10}$$

A high factor indicates good correlation, i.e. if $\rho = 1.0$ then all the data points will actually be found on a straight line. The following statistical parameters are needed:

$$\mu_x = \frac{1}{n}\sum_{i=1}^{n} x_i$$

$$\mu_y = \frac{1}{n}\sum_{i=1}^{n} y_i$$

$$s_x^2 = \frac{1}{n-1}\sum_{i=1}^{n}(x_i - \mu_x)^2 \tag{3.11}$$

$$s_y^2 = \frac{1}{n-1}\sum_{i=1}^{n}(y_i - \mu_y)^2$$

$$s_{xy}^2 = \frac{1}{n-1}\sum_{i=1}^{n}(x_i - \mu_x)(y_i - \mu_y)$$

where n is the total number of data points. The uncertainty in the prediction can be estimated as the square sum of the residuals:

$$Q^2 = \sum_{i=1}^{n}(E_i)^2 . \tag{3.12}$$

If the $E_i$ values are taken to be independent, then y is normal distributed for a given x. As a consequence, it can be shown that $Q^2/s^2$ is chi-square distributed by $n - 2$ degrees of freedom. The variable $s$ is defined as the standard deviation of y for a given x. Hence, the point estimate for $s$ reads:

$$\hat{s}^2 = \frac{Q^2}{n-2} . \tag{3.13}$$

The confidence level for $s$ can be determined from the chi-square distribution:

$$P(s^2 \leq \frac{(n-2)\hat{s}^2}{\chi_{1-\alpha,n-2}^2}) = 1-\alpha \tag{3.14}$$

where $\chi_{1-\alpha,n-2}^2$ is the tabulated value of the chi-square distribution which is exceeded with a probability of $1 - \alpha$. The standard deviation for y is used to define percentile curves for y given x. This is done by subtracting a certain number k of standard deviations from the right-hand side of equation (3.8). Hence, we use the equation:

$$y_k = A - ks - Bx \ .$$ 

(3.15)

There will be a known probability for obtaining a lower value than $y_k$, as shown in Table 3.4.

| K | Probability of smaller values |
|---|---|
| 0 | 0.50 |
| 1 | 0.16 |
| 2 | 0.023 |
| 3 | 0.0014 |

**Table 3.4.** *Probability of obtaining lower values than given percentile curves*

In our case of fatigue life data, $y$ is equal to log N whereas x is equal log $\Delta\sigma$. The stress range is the explaining variable, whereas the number of cycles is the result variable; see Figure 3.8. The S-N presentation is made with log N at the horizontal axis, but this does not change our consideration for the explaining and the result variables. The design rules usually apply k = 2, i.e. mean curve minus two standard deviations as a design curve. Hence, from Table 3.4, it can be seen that this corresponds to a probability of failure of 2.3%. The standard deviation has often been based on the point estimate for $s$. However, there has been a shift in the last decade to use 1.5 standard deviations and to require a $1 - \alpha = 75\%$ confidence level on the applied standard deviation. This will penalize curves obtained from few data points. This is shown in Table 3.5 where the 75% confidence estimates are listed as a function of the point estimate $\hat{s} = s_{50}$ for various numbers of test specimens in the test series.

| Number of specimens n | $\chi^2_{0.75,n-2}$ | $s_{75}$ (75% confidence level) | $1.5\ s_{75}$ |
|---|---|---|---|
| 12 | 6.73 | 1.22 $s_{50}$ | 1.83 $s_{50}$ |
| 24 | 12.74 | 1.13 $s_{50}$ | 1.69 $s_{50}$ |
| 36 | 28.14 | 1.10 $s_{50}$ | 1.65 $s_{50}$ |

**Table 3.5.** *Point estimates compared with 75% confidence estimates for standard deviation given as fraction of Q*

The columns to the right in Table 3.5 show the 1.5 standard deviations based on the 75% confidence level. These values are often used to construct design curves for fatigue life. As can be seen the difference between 2 $s_{50}$ and 1.5 $s_{75}$ is small when the number of specimens are small.

*Example*

A test series containing 24 specimens with a fillet joint (F-class detail, see Figure 3.1) gave the following results:

$$Q = 1.06, A = \mu_{logN} = 12.154, B = -m = 3.$$

Hence, according to equation (3.13), the point estimate gives $s_{50} = 0.198$.

According to Table 3.5 we will have $s_{75} = 0.224$.

According to equation (3.15) the design curve based on subtracting 2 $s_{50}$ from the median line will read:

$$logN = 12.15 - 0.390 - 3log\Delta\sigma = 11.754 - 3log\Delta\sigma$$

The design curve based on subtracting 1.5 $S_{75}$ from the median line will read:

$$logN = 12.15 - 0.446 - 3log\Delta\sigma = 11.815 - 3log\Delta\sigma$$

The predicted fatigue life will be 15% longer when using the curve obtained by subtracting 1.5 $s_{75}$ from the median line compared to the curve obtained by subtracting 2 $s_{50}$. If the number of specimens in the test series had been fewer, the difference between the two life estimates would have been even lower.

The linear regression method has its limitations if the data points include run-outs, i.e. if some tests are stopped before failure is reached. In this case, the maximum likelihood method is more appropriate, Ref [1]. We will return to this in Chapter 8. For the growth-rate test, the y variable is equal to log da/dN and the x variable is equal to log $\Delta K$; see Figure 3.15. The statistical analysis will be analogous to the one shown for S-N results. The same is true for crack initiation data pertaining to Figure 3.9.

Further details on these issues are found in Ref [7]. For S-N data in particular, we will propose a new type of statistical analysis Chapter 8.

## 3.7. References


1   H.P. Lieurade, Méthodologie d'essais de fatigue sur composants soudés. CETIM, Conference proceedings, November 1992 at Senlis

2   T. Lassen, "The effect of the welding process on the fatigue crack growth in welded joint" *Welding Journal* 69 (2): 75-s to 85-s, 1990

3   A. Almar Næss, *Fatigue Handbook*, Trondheim, Tapir 1985

4   American Society of Testing Materials: D1141 and E 647 specifications ASME

5   T. Lassen, J.L. Arana and L. Canada, "Crack growth in high strength steel subjected to fatigue loading in a corrosive environment", *The 24th International Conference of Offshore Mechanics and Artic Engineering (OMAE)*, Halkidiki, Greece, June 2005

6   W.H. Press *et al.*, *Numerical Recipes*, Cambridge, Cambridge University Press, 1987

7   C.R.A. Schneider and S.J. Maddox, "Best practice guide on statistical analysis of fatigue data", IIW-XIII-WGI-114-03, 2003.

Chapter 4

# Definition and Description of Fatigue Loading

## 4.1. Introduction and objectives

The determination of the loading and related stress histories that a structural detail will be subjected to during service life plays an important – if not the most important – role in achieving a reliable fatigue life estimate. It must be borne in mind that the number of cycles to failure is a function of the stress range level raised to a power of at least 3 for welded steel joints. This means that if the stresses are underestimated by, say, 20%, the fatigue life will be close to 100% overestimated.

There are three major topics regarding service life stress history:

– the loading scenarios have to be assumed at the design stage for say 20 years into the future;

– the loading is often random by nature, e.g. wave loading for structures at sea;

– the response calculation and stress analysis leading from the applied forces to the stress response in the vicinity of a weld is often based on approximations.

Regarding the first point, it is obvious that movable structures may possibly change operating area and operational conditions during the course of their service life. An example is large tankers for oil transport at sea. There is a large difference between the wave-induced loading in the Pacific Ocean and the North Atlantic during wintertime. The second point is usually treated by stochastic methods which we will include in this chapter. The last point is a part of load modeling and stress analysis. It is a too broad area to include in the present context. We will give a description of how to deal with the problem of defining the stress response due to external loading by applying a finite element analysis (FEA) in Chapter 5. In the present chapter we will show various methods of treating service stress (or strain) in a component as a function of time. Any load component can be treated in the same manner. We will not focus on how to transfer load to stress response.

The objective of the present chapter is that the reader understands the importance of defining the fatigue stress spectrum for a given time period. The reader will be familiar with cycle counting of time series and the stochastic energy spectrum approach used to describe fatigue loading.

## 4.2. Constant amplitude loading

Very few structures are actually subjected to constant amplitude loading during service life. The importance of constant amplitude loading is related to fatigue testing in the laboratory environment. Hence, the goal that laboratory tests should simulate realistic service condition as closely as possible is abandoned in this important area. The best approach would be to subject the structural detail in question to the actual variable amplitude load history that will occur during service. The problem is, however, that there are various possible load histories for the details found in different type of structures. It is usually too laborious and costly to simulate real load histories with variable amplitude. There are few examples of important details where a service variable load history has been standardized and actually used in testing. The inner wing section of transport airplanes is such an example. Efforts have also been made to establish a standard load spectrum for tubular joints in jacket structures in the North Sea; but these examples are exceptions. In most cases the engineer is left with fatigue test results that are based on constant amplitude loading as defined in Figure 4.1.

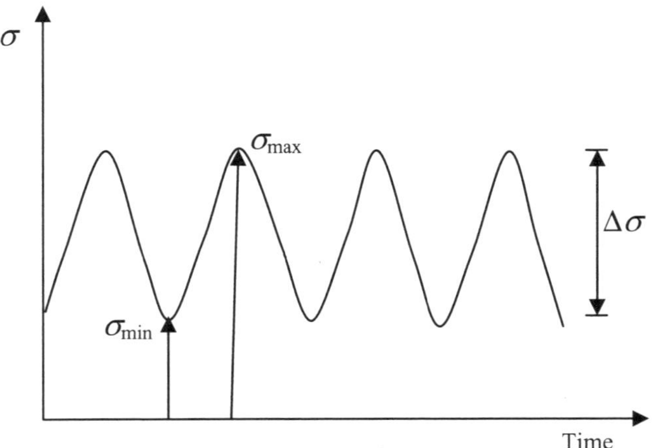


**Figure 4.1.** *Definition of constant amplitude loading*

The two most important parameters are the stress range $\Delta\sigma$ and the load ratio R:

$$\Delta\sigma = \sigma_{max} - \sigma_{min} \tag{4.1}$$

$$R = \frac{\sigma_{min}}{\sigma_{max}}$$

The choice of carrying out almost all fatigue testing at constant amplitude leaves the engineer with the problem of how the obtained fatigue resistance determined under such loading could be "translated" into the more realistic variable amplitude loading conditions. This has led to the Miner summation rule which is based on linear interaction assumption. Many engineers have spent much time investigating the truth of this translation and we will return to this problem in Chapter 5.

## 4.3. Variable amplitude loading

### 4.3.1. *Overview*

A typical in-service load history may appear as is given in Figure 4.2.

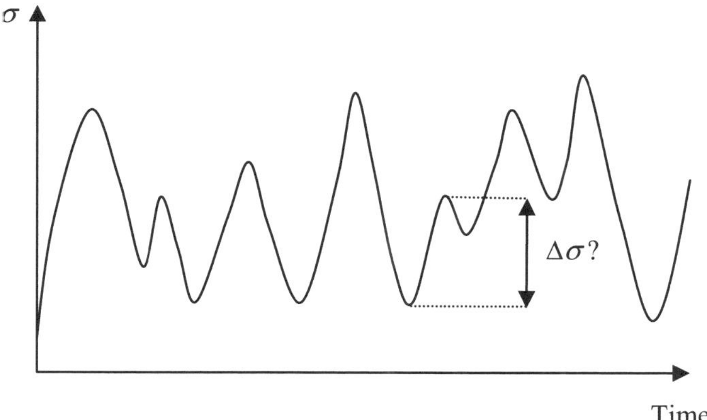


**Figure 4.2.** *Stress time series under variable amplitude loading*

As can be seen, the time series is quite irregular and random, although there are some dominating frequencies. We have two possibilities of how to treat this stress history:

– *Times series cycle counting*: the time series consist of several stress ranges which can be counted and presented in groups with the same stress range. The most common counting method is the so-called rain-flow counting technique. The results can be given in a histogram or a diagram of exceedance. This gives a clear definition of the load spectrum. The corresponding fatigue damage can then be estimated by Miner's summation rule.

– *The energy spectrum approach*: under stationary loading conditions, the time series can be split into harmonic components. The analysis is carried out using a Fourier transformation technique. The stress history can then be characterized entirely by the parameters defining the related energy spectrum.

Both approaches are equally important and we shall summarize the methods in the following sections.

### 4.3.2. *Rain-flow cycle counting of time series*

If one looks at the stress time series in Figure 4.2, it is obvious that defining a complete cycle and its stress range is not straight forward. One problem is that the stresses appear in half cycles between reversals. A reversal is defined as a point where the first derivative changes sign, i.e. a valley or a peak. As can be seen, a given half cycle may contain smaller half cycles. The counting can become rather involved and we need some guidance on how to undertake it. As a general rule, large stress cycles must not be fragmented into smaller ones as this will lead to an underestimation of the fatigue damage. Small stress cycles should be regarded as temporary interruptions of larger stress reversals. Let us take the stress history in Figure 4.3 as an example.

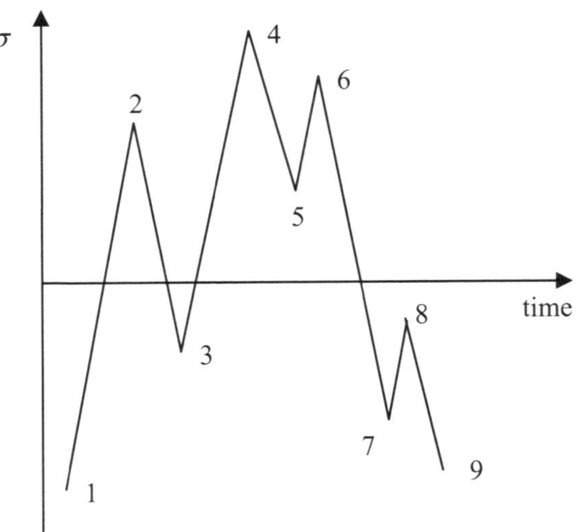


**Figure 4.3.** *Example of stress time series*

The stress cycles can be counted by the so-called rain-flow counting method. If the time axis is oriented vertically downwards, one could imagine that the stress history forms a number of "pagoda roofs" on which the rain is poring down. Cycles are then defined by the manner in which the rain is allowed to drip or fall down the roofs. A number of peculiar rules are imposed in order to carry out the counting. To computerize this type of cycle counting we will define a simpler rule based on the work by Dowling, Ref [1]. Let us analyze the two cases shown in Figure 4.4. The left-hand case will not define a stress cycle, but the right-hand case will. The right-hand side case will define a cycle due to the fact that, after having reached the valley 2, one passes the former peak 1 on the way up again. This occurs at the point denoted 2'. It does not occur in the left-hand case.

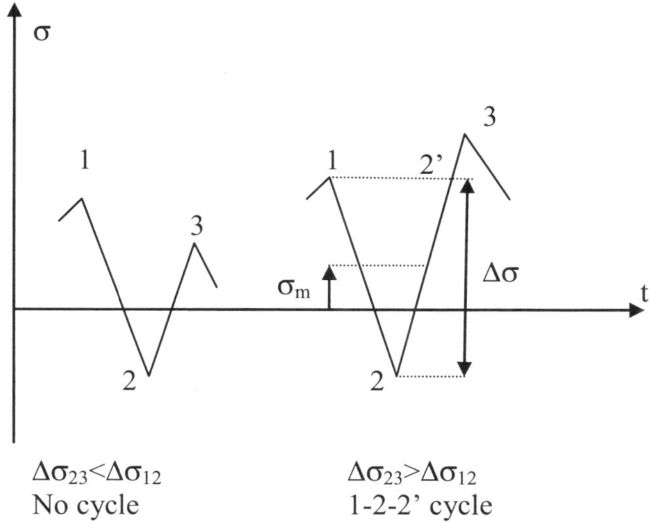


**Figure 4.4.** *Definition of stress cycles based on rain-flow method, Ref [1]*

We also see from the right-hand side of the figure that the defined cycle 1-2-2' (or briefly cycle 1-2) is characterized by its:

– stress range $\Delta\sigma$,

– mean stress $\sigma_m$ (or R-value).

Our simple counting rule is now as follows:

– when the stress history passes through a valley, a cycle will be counted when the stress level passes the former peak on its way up; and vice versa,

– when the stress passes through a peak, a cycle will be counted when the stress level passes the former valley on its way down.

The argument for this rule is that such external load cycles will correspond to a closed hysteresis loop for the local stress-strain found at the weld notch. Readers who wish to delve more deeply into this field are referred to Refs [1, 2]. The method is applicable for high cycle fatigue as treated by the S-N approach in rules and regulations. We obtain what is essential for life calculations: number of cycles at given stress range levels. In Figure 4.5 we have applied the technique for the complete stress history in Figure 4.3.

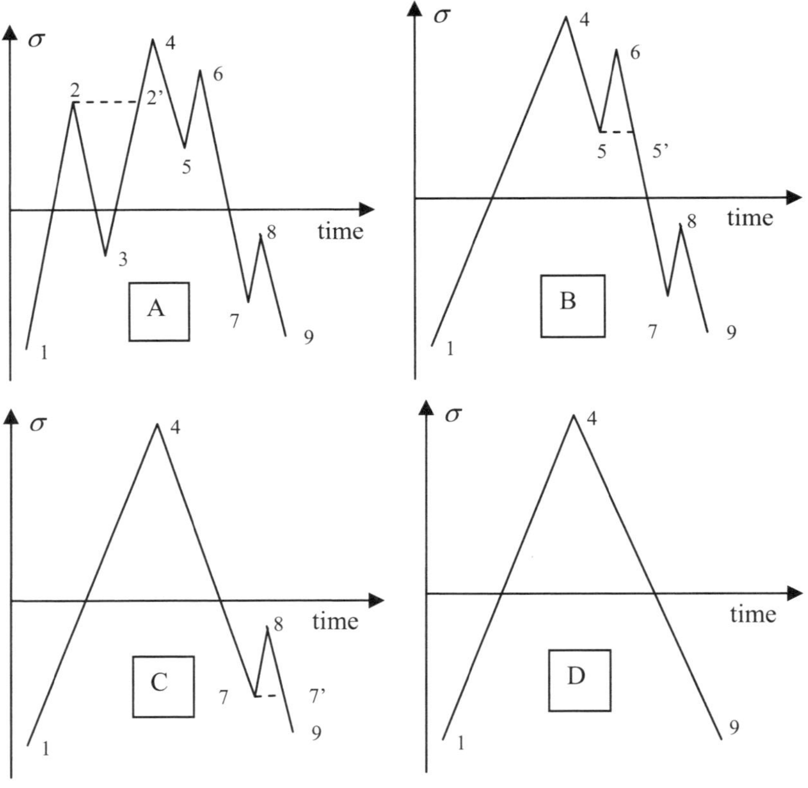

Counted cycles: 2-3-2', 5-6-5', 7-8-7' and 1-4-9

**Figure 4.5.** *Example of cycle counting for a stress time series*

As can be seen from Figure 4.5[A], we start by counting cycle 2-3-2' according to the given rules. When the stress rises from point 3 towards point 4, we notice that the former peak 2 is passed. Hence, cycle 2-3-2' is defined. To make the subsequent counting easier, we remove cycle 2-3-2' before proceeding. We then count cycle 5-6-5' as shown in Figure 4.5[B]. Then this cycle is removed and stored before we count cycle 7-8-7' in Figure 4.5[C]. Finally, the large cycle 1-4-9 is counted in Figure 4.5[D]. The procedure gets easier if one starts and finishes with the extreme values for the stresses, in our case point 1 and 9. If the time series does not start and stop with the extreme values it could be rearranged. The counted cycles are given at the bottom of Figure 4.5. As can be seen, the time series contains four complete cycles. The small cycles have been identified and registered without missing the large cycle 1-4-9. When time series are long, the cycle counting must be computerized and cycles are counted each time the ranges of two following reversals satisfy the inequality, Ref [3]:

$$\left|\sigma_i - \sigma_{i-1}\right| \ge \left|\sigma_{i-1} - \sigma_{i-2}\right| \tag{4.2}$$

where $i$ is the index identifying the reversal.

The stress range and mean stress for the identified cycle are given by:

$$\Delta\sigma_{i-1} = \sigma_{i-1} - \sigma_{i-2}$$

$$\sigma_{m,i-1} = \frac{\sigma_{i-1} + \sigma_{i-2}}{2} \tag{4.3}$$

When a cycle is identified and counted it is removed from the time series and stored, as we have shown graphically in Figures 4.5. The counting procedure subsequently starts at point 1 and is repeated until all the cycles have been registered. The reader can use the algorithm given by equations (4.2) and (4.3) to count the cycles in Figure 4.5. Figure 4.6 shows the stress-strain hysteresis loops pertaining to the counted cycles.

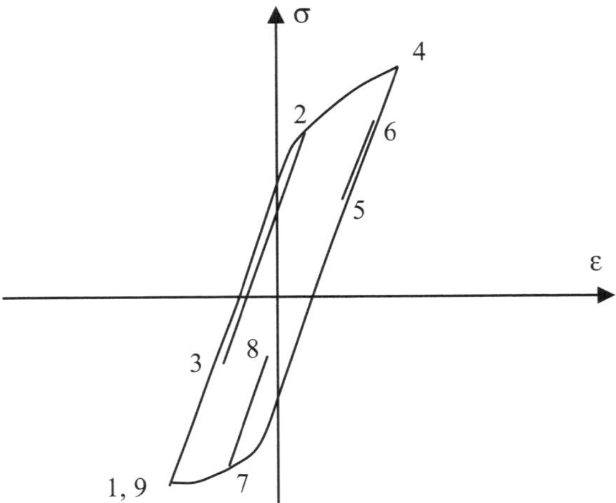

**Figure 4.6.** *Material stress-strain response at the weld for the given stress history*

For a longer time series, the results can be presented in a histogram. It may also be convenient to fit a probability density function (frequency function) to the histogram; see Figure 4.7. One frequency function that is widely used in such cases is the Weibull model:

$$f(\Delta\sigma) = \frac{h}{q}\left(\frac{\Delta\sigma}{q}\right)^{h-1} \cdot e^{-\left(\frac{\Delta\sigma}{q}\right)^{h}} \qquad , \Delta\sigma \geq 0$$

$$f(\Delta\sigma) = 0 \qquad\qquad , \Delta\sigma < 0 \qquad\qquad (4.4)$$

where $q$ is the scale parameter, whereas $h$ is the form parameter. In Figure 4.7 the frequency function to the left pertains to $h = 2.0$, whereas the function fitted to the right pertains to $h = 1.0$. The first type of presentation is often used for short-term fatigue loading description defined by a stress time series with some hours of duration. This can be the stress time series occurring during a special mission for a welded boogie frame or the stress time series occurring during a given sea state for an offshore structure. Short-term means that the statistics are representative for a time period when the operational conditions are stationary, e.g. a gale at sea. As can be seen, all the stress cycle can be relatively high and none are close to zero for such a severe short-term condition. The time period is usually three to six hours and the shape parameter h will be close to 2.0.

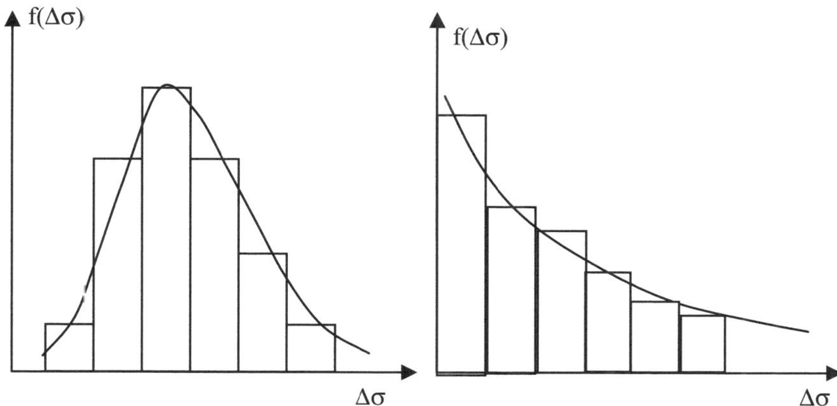

**Figure 4.7.** *Left: short-term stress distribution (h=2).*
*Right: long-term stress distribution (h=1)*

The right-hand diagram in Figure 4.7 illustrates a long-term stress distribution. The long-term distribution must be based on statistics where all possible missions and loading conditions are included. For weather-induced loading, all seasons must be included, i.e. at least one year's statistics. As can be seen for this case, this function will predict many low-stress ranges close to zero that pertain to less severe conditions.

One appealing feature of the Weibull load model is that if one uses the frequency function in the Miner summation rule in conjunction with the S-N method, a closed-form solution for the damage ratio accumulation can be found. This will be shown in Chapter 5. The scale parameter q can be related to the maximum stress range occurring during the time series. The maximum stress range is found by extreme value statistics for the n cycles in the time series. The relation is:

$$q = \frac{\Delta\sigma_{max}}{(\ln n)^{1/h}} \, . \tag{4.5}$$

### 4.3.3. *The energy spectrum approach*

The energy spectrum approach is faster and easier to use than the cumbersome, stress-cycle counting method. On the other hand, in some cases, one may loose some accuracy in the defined stress spectrum approach results compared to cycle counting results. The obtained stress cycles are not directly related to closed hysteresis loops for the stress-strain behavior in the material as was the case for the rain-flow counting technique. The mean stress pertaining to each stress cycle is not revealed. The spectrum approach can only be used if we have stationary loading conditions. The method is based on the assumption that any stationary time series can be split into an infinite number of harmonic components. Hence, the time series in Figure 4.2 can be written in the form of:

$$\sigma(t) = \int_0^\infty \sigma_\omega \sin(\omega t + \varphi_\omega) d\omega \qquad (4.6)$$

where $\sigma_\omega$ is the amplitude of the harmonic stress component that has a frequency $\omega$. The parameter $\varphi_\omega$ is a random phase distortion between the various harmonic stress components. The task is to find the amplitudes $\sigma_\omega$ as a function of $\omega$. This type of analysis can be carried out by a Fourier transformation technique. Although the theoretical background will not be discussed here, the reader is referred to Refs [4, 5] for details. The solution is shown in Figure 4.8. We have assumed a process with zero mean stress as it is only the stress amplitude or ranges that are our concern. As can be seen, the practical result is that we have left the time-axis presentation and replaced it by a frequency axis. The energy in the stress spectrum is given as a function of the frequency to the right in the figure. The energy in a narrow frequency band is proportional to $1/2\ \sigma_\omega^2$ of the corresponding harmonic stress component. Hence, the energy spectrum is defined by:

$$S(\omega)d\omega = \frac{1}{2}\sigma_\omega^2 d\omega \qquad (4.7)$$

A typical shape for $S(\omega)$ can be as shown in Figure 4.8.

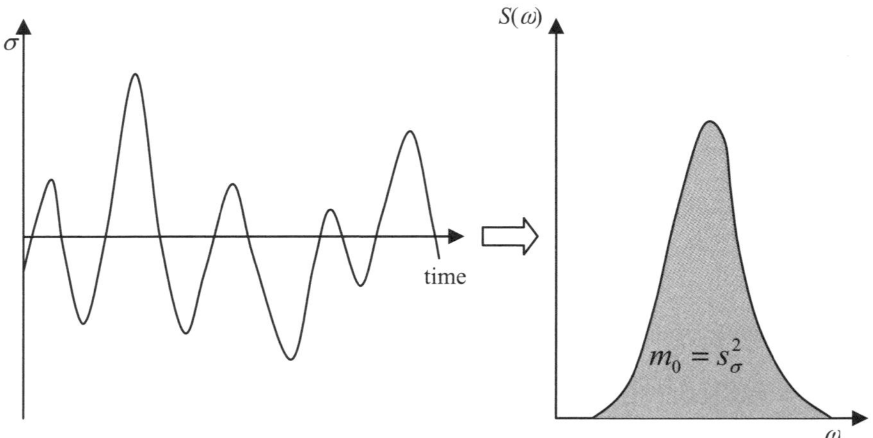


**Figure 4.8.** *Energy spectrum for a stress time series*

From the obtained spectrum we can make the definition of the moment function as follows:

$$m_n = \int_0^\infty \omega^n S(\omega)d\omega \qquad (4.8)$$

$m_n$ is a general moment function of the spectrum. In particular, $m_0$ corresponds to the area under the spectrum curve (see Figure 4.8) and equals the total energy in the load spectrum according to equation (4.7). From the second-order moment the average up-crossing time period can be determined as follows:

$$T_z \approx 2\pi \sqrt{\frac{m_0}{m_2}} . \qquad (4.9)$$

The stress level deviation from the zero mean value at an arbitrary time can be shown to follow a normal distribution as follows:

$$f(\sigma) = \frac{1}{2\pi s_\sigma} \exp(-\frac{1}{2}\left(\frac{\sigma}{s_\sigma}\right)^2 \qquad (4.10)$$

The stress history is said to be a Gaussian process with a zero mean and a standard deviation designated $s_\sigma$. This standard deviation of the process can be related to the energy of the spectrum in the following way:

$$s_\sigma^2 = m_0 . \qquad (4.11)$$

As we have seen before, we are not in particularly interested in the arbitrary stress level given in equation (4.10), but in the stress amplitudes or ranges. It can be shown that if the energy spectrum is narrow, i.e. a narrow band of frequencies is dominating, the stress ranges are Rayleigh distributed:

$$f(\Delta\sigma) = \frac{\Delta\sigma}{4m_0} \exp\left(-\left(\frac{\Delta\sigma}{2\sqrt{2m_0}}\right)^2\right) \qquad (4.12)$$

The Rayleigh distribution is a special case of the Weibull distribution that we studied earlier in equation (4.4). The Weibull distribution may be written in the following form:

$$f(\Delta\sigma) = \frac{h}{q}\left(\frac{\Delta\sigma}{q}\right)^{h-1} \exp\left(-\left(\frac{\Delta\sigma}{q}\right)^{h}\right)$$

(4.13)

This equation is identical to equation (4.12) when:

$$h = 2 \quad \text{and} \quad q = 2\sqrt{2m_0}\,.$$

(4.14)

Hence we can obtain the Weibull distribution for the short-term stress ranges pertaining to a stress time series either by cycle counting as was shown in section 4.3.2 or by the energy spectrum approach as outlined above.

The mean value of the 1/3 largest stress cycles is denoted as the significant stress range level and can be written as follows:

$$\Delta\sigma_{sign} = 4\sqrt{m_0}\,.$$

(4.15)

For marine structures with wave-induced loading, a linear relation between wave height and stresses in the structure is often assumed. Hence, the definition in equation (4.15) will correspond to the significant wave height which, in addition to the mean up-crossing period $T_z$, is the most-used parameter to characterize a stationary sea state. The corresponding probability of exceeding a given level of stress range is found by integration of equation (4.12) as follows:

$$Q(\Delta\sigma) = \exp\left(-\left(\frac{\Delta\sigma}{2\sqrt{2m_0}}\right)^{2}\right)$$

(4.16)

If we separate the one-third of the stress cycles that contains the largest ones, the mean value of this group is denoted by setting $Q(\Delta\sigma) = 1/n$, where n is the total number of cycles in the series. We then obtain:

$$\Delta\sigma_{max} = 2\sqrt{2m_0 \ln n}\cdot$$

(4.17)

Equation (4.16) can be presented by an exceedance diagram as we shall see in Chapter 5.

## 4.4. References


1    N.E. Dowling, *Estimating Fatigue Life. ASM Handbook Fatigue and Fracture*, ASM International, 1996, pp 250, 262

2    J. Banrantine, J. Comer and J. Handrock, *Fundamental of Metal Fatigue Analysis*, College of Engineering, University of Illinois, 1987

3    G. Glinka, *Fatigue Life Predictions of Notched Components: A Crack Initiation*, Waterloo, Canada, University of Waterloo, 1990

4    A. Almar Næss, *Fatigue Handbook*, Trondheim, Tapir, 1985

5    D.E. Newland, *Random Vibration and Spectral Analysis*, Longman, 1984

# Chapter 5

# The S-N Approach

## 5.1. Introduction and objectives

When designing a structure against the fatigue-limit state, the aim is to ensure, to an acceptable level of probability, that the integrity of the structure is satisfactory throughout its planned service life. The structure should be unlikely to fail from fatigue or to require extensive repair of damage caused by fatigue (Ref [1]). To carry out this type of *safe life* analysis, the designer must verify the fatigue life for each critical welded detail in the structure. Fortunately, even large, complex structures may often be divided into a relatively low number of elementary joints. The fatigue strength for these elementary joints must be known from experience or laboratory testing. Based on laboratory tests, the most common type of analysis for fatigue life prediction is based on the S-N method. For each critical joint it is verified that the predicted fatigue life (PFL) is superior to the target service life (TSL) of the structure, and there should be with a required safety margin. The stress spectrum must be estimated for the planned service life and the S-N curves characterizing the fatigue strength of the various critical details of the structure must be at hand.

As outlined in Chapter 2, the S-N curves are based on a simple relationship between applied stress range $\Delta S$ and the number of cycles N to failure. For constant amplitude loading, the curves are assumed to be linear for a log-log scale until the curves reach a fatigue-limit. For a stress range below this limit it is assumed that the fatigue life is infinite. Each curve is established from laboratory tests with a given type of welded detail with a given geometry, fabrication quality, environmental, and loading condition. The loading is usually uni-axial with constant amplitude. In the finite life area above the fatigue limit, the curve is determined by linear regression as described in Chapter 3. When the detail is subjected to variable amplitude loading in service, some type of damage accumulation formula has to be applied.

In Chapter 3 we discussed the test and the statistical analyses carried out to establish an S-N curve for a given detail. The main results from this analysis are the design S-N curves presented in rules and regulations. It is however an advantage that the designers who use these curves know the experimental background for the curves and how the numerical analyses are carried out to define the design S-N curve. Most rules and regulations only specify the design curve, whereas the mean value and the standard deviation pertaining to each curve are hidden for the user. We think that this is a pity because this information would give the design engineer more insight into the fatigue problem he or she is dealing with. Furthermore, rules and regulations allow for testing to establish S-N curves for special details not found in the regulations. In such cases the statistical analysis must be carried out by the engineer before obtaining the design S-N curve.

Useful information on the presented subjects is found in Refs [2, 3].


## 5.2. Method, assumptions and important factors

### 5.2.1. Statistics for the S-N approach, median and percentile curves

The problem of scatter in observed fatigue life has already been treated in Chapter 3 where details on how to carry out a linear regression analysis were outlined. Based on the obtained parameters, the total fatigue life at a given stress range level is assumed to be a stochastic variable $N_T$ which is lognormal distributed (i.e. $\log N_T$ is normal distributed). The probability that $N_T$ is less than a life $N_T$ is given by:

$$P(\mathbf{N_T} < N_T) = \Phi\left(\frac{\log N_T - \mu_{\log N}}{s_{\log N}}\right) \tag{5.1}$$

where $\mu_{\log N}$ and $s_{\log N}$ is the mean and standard deviation for $\log N_T$ at the given stress range level. The standard deviation $s_{\log N}$ is assumed to be constant when the stress range changes. The confidence level for $s_{\log N}$ was discussed in Chapter 3. The mean value $\mu_{\log N}$ is strongly dependent on the applied stress level:

$$\mu_{\log N} = \log A_0 - m \log \Delta\sigma . \tag{5.2}$$

The probability of failure at different $N_T$ can be found from tables for the standard normal distribution. The curve pertaining to a probability of failure of 0.5 ($\Phi(0)$) is designated the median curve and reads as follows:

$$N_{T0} = \frac{A_0}{\Delta\sigma^m} \tag{5.3}$$

where $N_{T0}$ ($\log N_{T0} = \mu_{\log N}$) is the median fatigue life defined by a 50% probability of survival.

This corresponds to the median S-N curve in Figure 5.1. The design curve is often defined by the logarithmic mean curve minus two standard deviations. According to equation (5.1) this gives $\Phi(-2)$. This corresponds to a probability of failure of 2.3%. In building codes this design curve is normally given directly in a logarithmic format, without explicitly applying equation (5.1), as:

$$\log N_{T2} = \log A_2 - m \, \log \Delta\sigma . \tag{5.4}$$

When $N_{T2}$ is used by the designer, it corresponds to a probability of failure of 2.3%, as discussed above. If the consequences of fatigue failure are unacceptable, this probability of failure may be reduced by subtracting more than two standard deviations for obtaining the constants in equation (5.4) before predicting the fatigue life. Alternatively, equation (5.4) can be applied without changes, but an additional safety factor is demanded on the predicted fatigue life $N_{T2}$ compared to TSL. This approach is the most common one in rules and regulations. As $N_{T2}$ is widely used in fatigue design it will be designated simply as N in what follows.

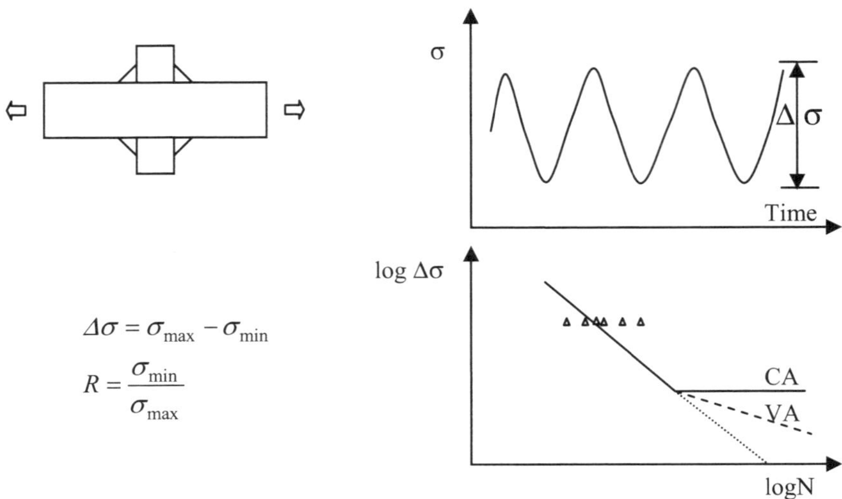


**Figure 5.1.** *Definition of loading and median S-N curve for fillet welded joints*

### 5.2.2. Discussion of S-N curves-important factors

#### 5.2.2.1. The threshold phenomenon

The bilinear S-N curve has a horizontal lower-line segment corresponding to the so-called fatigue limit under constant amplitude (CA) loading; see Figure 5.1. It is assumed that if the applied constant stress range is below this limit, no damage will ever occur; the detail will be able to withstand an infinite number of loading cycles. There are very few data that support this assumption. The present authors are somewhat skeptical of the existence of such a fatigue limit; if the test runs long enough, any detail will eventually fail. We tend to believe that the long lives observed at these low stress ranges are due to a long crack initiation period. It is not really a threshold phenomenon. We will treat this issue in more detail in Chapters 8 and 9.

Meanwhile, we accept the fatigue limit as prescribed in rules and regulations. However, as can bee seen from Figure 5.1, the assumption of the existence of a fatigue-limit has to be modified when the detail is subjected to variable amplitude (VA) loading. We will explain this in more detail later in this chapter.

### 5.2.2.2. Mean stress and loading ratio

The S-N curves for welded joints have just one key parameter to the fatigue life: the stress range. For a machined component, the S-N curve is peculiar for a given stress ratio R in the way that a low stress ratio is favorable with respect to fatigue life. If some of the stress variation is partly in the compressive side (R < 0), this will not contribute to the fatigue damage to the same extent as variation on the tensile side. It is only tensile stress variation that will tear the crack open and propagate the crack, as we shall discuss in the next chapter. The reason for not taking the R ratio into account for welded joints is because large tensile residual stresses are usually present in the joint when they are in as-welded (AW) condition; see Chapter 2, Figures 2.16 and 2.17. This is the most common case for joints in welded structures. Hence, an external load that causes partly compressive stress variation will inflict an entirely cyclic tensile stress response in such a joint when superimposed on large, static, residual tensile stresses. As can be seen from Figure 5.2, the maximum stresses may approach the yield stress $\sigma_y$ of the material in such cases.

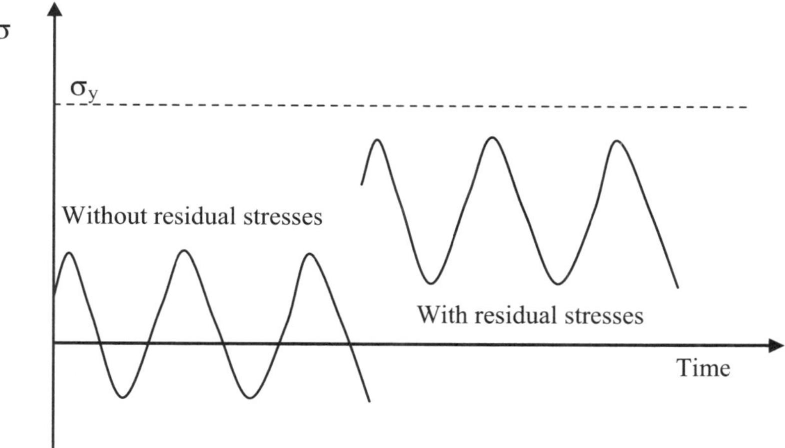


**Figure 5.2.** *Effect of residual stresses on the stress range*

### 5.2.2.3. Stress relieving

If the joints are stress relieved, it may be appropriate to take advantage of the fact that compressive stress variation is less damaging than tensile stress variation as explained above. Most rules and regulations allow for this, as we shall see below. The usual recommendation is to multiply the compressive part of the stress range by a reduction factor of f = 0.6.

### 5.2.2.4. *The thickness effect*

There is a reduction in fatigue life for the same applied nominal stress range if the thickness of the plate is increased. This effect can be caused by various factors:

– a notch effect, i.e. an increase in the stress concentration factor at the weld toe;

– a scale effect, i.e. the local stress field at the crack tip of, say, a 1-mm-deep crack will be more severe in a thick plate than in a thin plate;

– a statistical volumetric effect, i.e. larger material volume will have greater probability of containing flaws;

– a metallurgical effect, i.e. the steel microstructure of welded thicker plates, may be of poorer quality than that of thinner plates.

The various factors are ranged in the order of their importance, in the opinion of the authors. To understand the first effect one must bear in mind that fatigue starts as a local phenomenon, usually near the notch of the weld toe. The stress concentration at this notch is strongly dependent on the weld toe geometry, i.e. the toe angle $\theta$ and the toe radius $\rho$. It can be shown that the SCF (stress concentration factor) is particularly sensitive to the ratio between the toe radius $\rho$ and the plate thickness $T$; see Chapter 2, Figure 2.19. When $\rho/T$ decreases, the SCF increases and, consequently, fatigue life is reduced. Both the initiation and propagation parts of the life are reduced. The problem is due to the fact that when the plate thickness is increased, the variable toe radius is not controlled. The toe radius is related to the geometry of the weld pass, close to fusion line. This radius is likely to have the same value regardless of the chosen plate thickness. Hence, the $\rho/T$ ratio will inevitably decrease when $T$ increases.

This phenomenon is also observed for machined component. The designer of a shaft must take account for this effect. If the diameter D of the shaft is increased it is good fatigue design practice to increase any notch radius (e.g. at shaft shoulders) correspondingly to avoid a decrease in $\rho/D$ and the corresponding increase in the SCF. As explained above, the strategy is not possible for welded joints unless post-weld treatment is carried out. The weld notch can then be machined so the toe radius follows the thickness increase.

The scale effect points at the fact that the stress field in the front of a crack will be more severe in thicker plates than thinner plates for a given crack depth a, say a = 1 mm. This is clearly shown by the fact that the stress intensity factor (SIF) for a crack is a function of $a/T$ and not the absolute value of a. The SIF determines the stress field at the crack front. This will be shown in Chapter 6 where the fracture mechanics approach is explained.

It is the authors' opinion that the two last factors (statistical volumetric and metallurgical effects) are of less importance than the effect of the notch effect and the scale effect.

The stress increase due to the thickness effect, often found in rules and regulations, reads as follows (Ref [1]):

$$\text{SCF} = \left( \frac{T}{t_{ref}} \right)^{k} \tag{5.5}$$

where $T$ is the actual thickness of the joint in question, whereas $t_{ref}$ is the reference thickness. The reference thickness is representative for the joints to which the S-N data belong. Usually the reference thickness is close to 25 mm and the parameter $k$ may vary from 0.25 to 0.33 according to various rules and regulations.

Before closing this section, it should be pointed out that there is also a size effect related to increased length of the weld seam, i.e. the width of a test specimen. The explanation of this effect is that the probability of having an extremely unfavorable weld toe geometry increases as the weld length increases. The flank angle $\theta$ and the toe radius $\rho$ of the weld are, in fact, random variables along the weld seam and the effect of the length can be demonstrated by extreme value statistics. This will be shown in Chapter 9. It is also likely that residual stresses are higher in larger joints.

### 5.2.2.5. Misalignment

Due to fabrication imperfections, both out-of-plane distortion and angular distortion may appear in welded structures; see Figure 5.3.

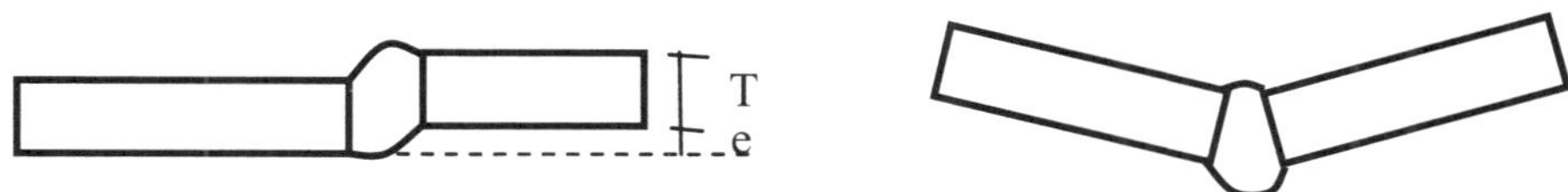


**Figure 5.3.** *Out-of-plane distortion (left) and angular distortion (right)*

These distortions give rise to secondary stresses. If, for example, the welded detail that has an out-of-plane-distortion (see Figure 5.3 on the left) is subjected to axial loading, the eccentricity e will cause secondary bending and a stress concentration given by the following approximation, Refs [1, 4]:

$$\text{SCF} = 1 + \frac{3e}{T}.$$ 
(5.6)

### 5.2.2.6. *Post-weld improvement techniques*

High quality welded joints are characterized by a favorable global geometry and a gradual transition at the weld toe. Furthermore, it is essential that the potential crack loci are proven to be crack free and that they have a smooth surface after fabrication. This can only partly be controlled by the choice of the welding procedure, whereas it can be controlled by various post-weld treatment methods. Various types of improvement techniques are (Ref [5]):

– burr grinding;

– TIG dressing;

– hammer peening.

The first two methods can be classified as weld profile improvement techniques. In the first case, this is obtained mechanically; in the second case it is obtained by re-melting the weld toe area. The hammer peening technique is the building in of compressive residual stresses in the weld toe area of the joint. The local toe burr grinding is illustrated in Figure 5.4. The grinding is carried out in such a manner that the weld profile is improved and sharp crack-like defects are removed. It was originally estimated that this type of grinding doubled the fatigue life compared with AW joints. The grinding should be up to 1 mm deep and the head of the grinding tool should have a diameter of at least 5 mm for a 20-mm-thick plate. If the plate thickness is increased, the tool head should be increased correspondingly for the reasons that were explained for the thickness effect. Non-destructive inspection should be carried out after grinding in cases where originally buried cracks have been brought to the surface. Additional grinding must then be carried out. A weld that has been ground will have less stress concentration and a smoother surface. Originally, the grinding procedure was the only post-weld treatment accepted to in the rules and regulations to give increased life. However, in the new guidance given by DNV, Ref [6], both hammer peening and TIG dressing are assumed to increase fatigue life. We will give the in a later section of this chapter. Specifications on how to carry out the improvement techniques are found in Ref [5].

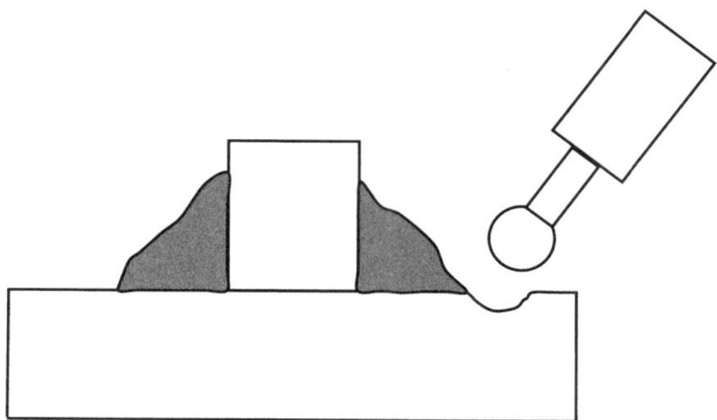

**Figure 5.4.** *Local toe burr grinding*

### 5.2.2.7. *Corrosive environment*

When welded joints are subjected to repetitive loading in a corrosive environment there is a synergy effect between the mechanical-fatigue damage process and the electro-chemical corrosion process. The corrosion may result in surface pits that shorten the crack initiation period. Furthermore, the corrosion process aggravates the condition within a crack near the crack front and may therefore significantly speed up the growth rate. Designers of welded structures for seawater have learned lessons from this synergy effect. Welded structures in seawater should always have some sort of corrosion protection. This is usually provided by cathodic protection and/or protective coating.

The influence of cathodic protection on the fatigue behavior of a welded structure depends, in a complex manner, on the interaction of mechanical, chemical, and electro-chemical parameters that affect both crack initiation and crack propagation. The best way to improve fatigue performance is to delay the crack initiation for as long as possible. Smooth-shaped welds, post-weld improvements, and maintenance of a moderate cathodic potential is the best way to fight the synergy effect. During crack growth, a moderate cathodic potential is the best measure to avoid growth acceleration. A cathodic potential between −850 and −1,050 mV with reference to a standard AgCl cell is preferable. For high-strength steel, cathodic protection may inflict damage mechanisms such as hydrogen embrittlement. Good articles on these issues are found in Refs [7, 8].

The principal differences in fatigue resistance between the air environment and cathodic protection (CP) and free corrosion (FC) are shown in Figure 5.5. As can be seen, it was previously believed that the cathodic protection gave the same fatigue life as for a joint in dry air. However, it was assumed that the fatigue limit was lost and the curves were drawn with the same slope all the way down towards zero stress range. More recent research has proven that it is the other way around; at high stress levels fatigue life under cathodic protection is close to 2.5 times shorter than in air, whereas for low stress ranges the cathodic protection is very efficient. At small stress ranges, fatigue life is very close to the life found in dry air and the assumption of a fatigue limit for CP is acceptable. The reason for this behavior is that the corrosion process may blunt the crack front at low stress ranges and inflict calcareous deposit in the wake of the crack front. The last effect may lead to crack closure. In FC, the curve gives significant shorter fatigue lives at all stress levels, as can be seen in Figure 5.5.

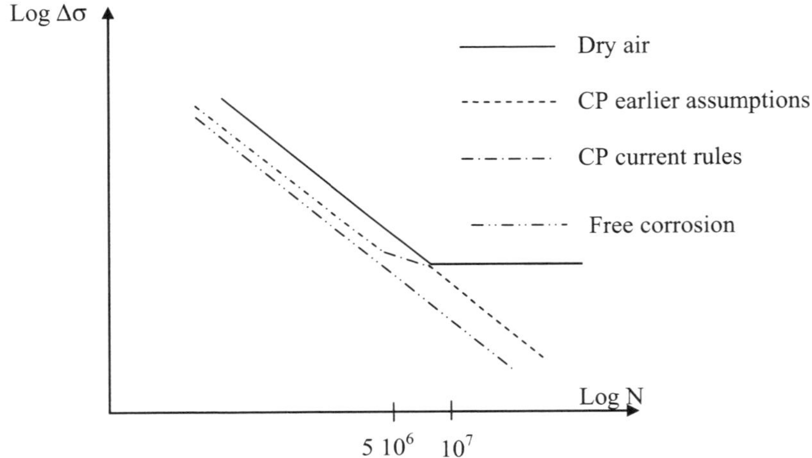


**Figure 5.5.** *Corrosion fatigue S-N curves in seawater, constant amplitude loading*

## 5.3. Mathematics for damage calculations

### 5.3.1. *Linear damage accumulation; load spectrum on a histogram format*

As already mentioned, S-N curves are based on CA test data, whereas a welded detail when appearing in a structure usually will be subjected to variable amplitude loading. Based on cycle counting, as outlined in Chapter 4, it is often possible to present the stress spectrum on a histogram format, i.e. in terms of stress blocks where each block is defined by its stress range $\Delta\sigma_i$ and corresponding number of cycles $n_i$; see Figure 5.6. The associated histogram is shown to the left on Figure 5.7 with five stress blocks. The spectrum can be valid for a given period of time or be representative for the entire service life. In the first case we are talking about short-term load spectrum, in the latter case a long-term load spectrum.

The problem then arises that we do not have the fatigue strength for the detail when subjected to the load spectrum in question; we have to use the S-N curve that is based on CA data only. In other words, the fatigue behavior under constant amplitude has to be transformed to that under variable amplitude loading to make life predictions. To cope with this problem it is usually assumed that each stress block contributes to the fatigue damage according to its damage ratio $n_i/N_i$. The nominator $n_i$ is the number of cycles actually occurring, whereas the denominator $N_i$ is the number of cycles to failure according to the S-N curve for the actual stress range. This ratio must obviously be less than 1.0 for each stress block. It is further assumed that the total damage caused by all stress blocks accumulates linearly.

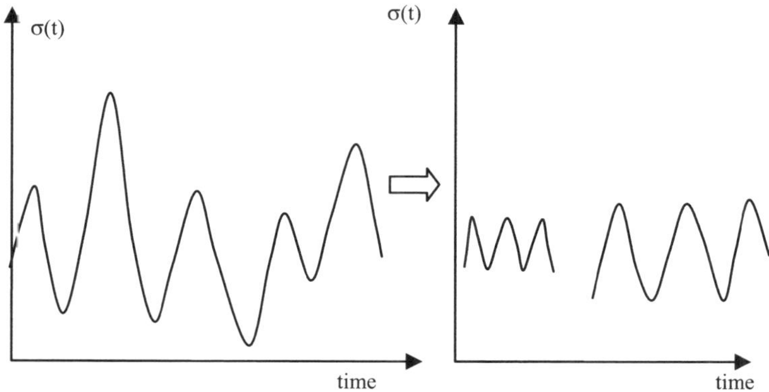


**Figure 5.6.** *Cycle counting of variable amplitude loading*

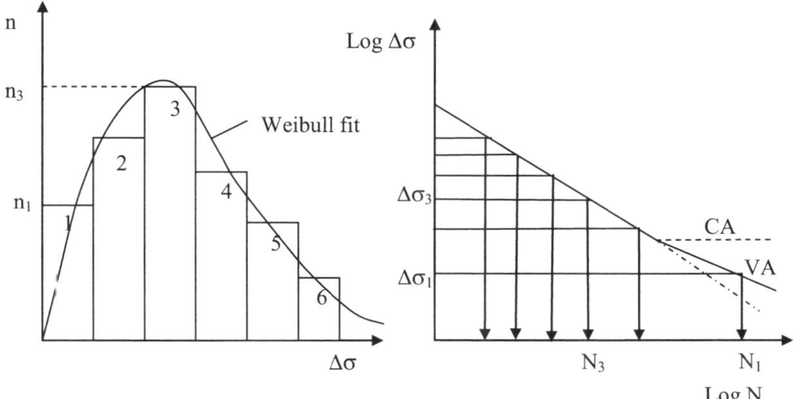


**Figure 5.7.** *Load spectrum given in six stress blocks and corresponding fatigue lives*

The damage accumulation is then given by the Miner summation rule:

$$D = \sum_{i=1}^{i=k} \left( \frac{n_i}{N_i} \right)$$
(5.7)

where $N_i$ is calculated from equation (5.4) at the actual stress range $\Delta\sigma_i$. The fatigue fracture criterion is $D = 1.0$. If the histogram pertains to a time span L (say one year long-term load spectrum), then the predicted fatigue life (PFL) will read:

$$L_P = \text{PFL} = \frac{L}{D}.$$
(5.8)

The fatigue design factor (FDF) is defined by:

$$\text{FDF} = \frac{\text{PFL}}{\text{TSL}}.$$
(5.9)

This is a safety margin that comes in addition to the safety margin inherent in the design S-N curve. For a given TSL (say 20 years), the required FDF will be a result of the consequences of fatigue failure and assessment of whether or not the detail in question is accessible for inspection and repair. Various rules have, as we shall see, different requirements as to this issue.

### 5.3.2. Discussion of the validity of the linear damage accumulation

All the damage calculations presented in this chapter are based on the assumption of linear damage accumulation as described above. The validity of this assumption has often been questioned. One of the consequences of this assumption is that the order of the stress blocks does not matter. As more variable amplitude-testing data have become available it has been shown that the chronological order of the stress blocks is important. The standard case is that a stress block with a low stress range may, in the beginning, be inferior to the CA fatigue-limit thus not contributing to fatigue damage. Stress block 1 in Figure 5.7 is such an example. However, if this stress block appears after several of the other more severe stress blocks, these blocks may have created a crack and the fatigue limit is no longer valid. The detail has become more vulnerable to fatigue damage and stress block 1 may obviously contribute to the fatigue damage at this stage. As a consequence, a CA S-N curve cannot be used in the fatigue-limit area for variable amplitude loading. A conservative approach is to neglect the fatigue limit all together and draw one line from the finite-life area down towards the zero stress range without changing the inverse slope m. However, based on the work of Haibach, Ref [9], a curve has been drawn to give reasonable results when applying Miner's rule for common stress spectra. The slope of the curve is set to 2m-1, where m is the slope pertaining to the fatigue data in the finite-life area. This approach has been adopted in rules and regulations. However, it is worth mentioning that the approach is a rather crude approximation, and the slope of the lower line will in fact be dependent on the cycle-counting method and on the type of load spectrum used. A good overview of the problem is given in Ref [10]. For various cases of VA loading, the actual damage ratio D at fracture was registered. It was found that the criterion D = 1.0 has only 10% confidence, i.e. it is over 90% probable that the specimen subjected to stress blocks of different stress range levels will fail before the Miner's sum reaches 1.0. The median value was in fact close to 0.3 and the 90% confidence level was as low as 0.1. Hence, there is a factor of 10 between the damage sum pertaining to the 10% confidence level and the sum pertaining to the 90% confidence level. This factor is used as a measure for the scatter band for the fracture criterion. The large value pinpoints the uncertainty in the Miner's criterion. The investigation involved all types of metallic materials; no results peculiar for welded joints were given. For steel material only, the Miner's sum was used both as a criterion for early cracking and as a criterion for final failure. In the first case, the number of cycles to early cracking is determined by the local stress strain approach, whereas the failure criterion is based on the conventional S-N curves. It was found that the nominal stress approach had a median value of D = 0.18 with a scatter band of 9.2, whereas the local stress strain approach had a median value of D = 0.48 and a scatter of 12.3. Hence, the local strain approach is closer to the theoretical median value of 1.0, but has a larger scatter. The conclusion is that the uncertainty in the Miner's summation criterion is much greater than hitherto presumed. In Chapter 9 we shall present an approach for welded joints that treats crack initiation and crack propagation separately. We shall further assume linear accumulation for each of these processes, but the accumulation of the total damage will be non-linear. This approach is based on more physical insight of the damage mechanism and we believe that the method will provide a more reliable fatigue fracture criterion.

Before closing this section, it should be mentioned that single overloads may in some cases create notch plasticity and compressive residual stresses after the overload. These stresses may appear at the weld notch or at the crack front if a crack is present. These residual stresses are favorable and will increase the fatigue resistance. This effect is not accounted for in the presented calculation scheme, but will be discussed in Chapter 12.

### 5.3.3. *Definition of the equivalent stress range*

The Miner's summation rule can also be applied by using the so-called equivalent stress concept. This is a fictitious, constant stress range $\Delta\sigma_e$ that, when applied with the total number of load cycles $n$, by definition causes the same Miner's sum as the actual stress spectrum. The following equation applies:

$$\sum_{i=1}^{i=k}\left[\frac{n_i}{\dfrac{A_2}{(\Delta\sigma_i)^m}}\right]=\frac{n}{\dfrac{A_2}{(\Delta\sigma_e)^m}} \tag{5.10}$$

and for a bilinear curve:

$$\Delta\sigma_e=\left[\frac{A_2\left(\dfrac{1}{A_{2l}}\sum_{i=1}^{i=j}n_i\Delta\sigma_i^{ml}+\dfrac{1}{A_{2u}}\sum_{i=j+1}^{i=k}n_i\Delta\sigma_i^{mu}\right)}{n}\right]^{1/m} \tag{5.11}$$

where j is the number of stress blocks with stress ranges below the knee point of the VA S-N curve, whereas j + 1 to k are the number of stress blocks with stress ranges above the knee point of the S-N curve. The VA S-N curve is the fully-drawn curve to the right in Figure 5.7. The parameters $A_2$ and m are kept at $A_{2u}$ and $m_u$ if the equivalent stress range is above knee point. Otherwise, they are set equal to $A_{2l}$ and $m_l$. The fatigue life can now be verified directly by using the equivalent stress range as the key to the S-N curve. Alternatively the equivalent stress range can be compared with the fatigue strength of the S-N curve for a given number of cycles.

### 5.3.4. *Load spectrum on the format of a Weibull distribution*

If the stress histogram with stress blocks as shown to the left in Figure 5.7 is fitted to a Weibull distribution, the Miner's sum will read:

$$D=\int_{\Delta\sigma=0}^{\infty}\frac{n\cdot f(\Delta\sigma)d\Delta\sigma}{A_2/\Delta\sigma^m} \tag{5.12}$$

where $f(\Delta\sigma)$ is the frequency function fitted to the histogram and $n$ is the total number of applied loading cycles. The Weibull model frequency function reads:

$$f(\Delta\sigma) = \frac{h}{q}\left(\frac{\Delta\sigma}{q}\right)^{h-1} \exp\left[-\Delta\sigma/q\right]^{h} \tag{5.13}$$

where $h$ is the shape parameter and $q$ is the scale parameter in the distribution. As we have seen in Chapter 4, the Weibull distribution can be obtained both by cycle counting and by the energy-spectrum approach. The integral in equation (5.12) can be solved by introducing the auxiliary variable $t = (\Delta\sigma/q)^{h}$. The integral can then be determined by using the well-known Gamma function:

$$D = \frac{n}{A_2}q^{m}\Gamma(1+m/h) \tag{5.14}$$

where $\Gamma(\bullet)$ denotes the Gamma function. This function can be found in standard tables and Excel spreadsheets. Equation (5.14) is valid for single slope S-N curves. In case of a bilinear S-N curve the damage ratio will read:

$$D = \frac{n\cdot q^{m_1}}{A_{2u}}\Gamma\left(1+\frac{m_u}{h};\left(\frac{S_0}{q}\right)^{h}\right) + \frac{q^{m_l}}{A_{2l}}\gamma\left(1+\frac{m_l}{h};\left(\frac{S_0}{q}\right)^{h}\right) \tag{5.15}$$

where:
$A_{2u},m_u$ = fatigue parameters for the upper S-N line segment
$A_{2l},m_l$ = fatigue parameters for the lower S-N line segment
$\Gamma(x,y)$ = complementary Gamma function
$\gamma(x,y)$ = incomplete Gamma function.

The scale parameter $q$ can be related to the most likely maximum stress range $\Delta\sigma_0$ occurring during a given number of cycles $n_0$:

$$q = \frac{\Delta\sigma_0}{(\ln n_0)^{1/h}}. \tag{5.16}$$

The most likely maximum stress range has by definition a probability of exceedance equal to $1/n_0$. The number of cycles $n_0$ need not be equal to $n$ in equation (5.12), but the number must be large enough to characterize the loading process so that the scale parameter $q$ becomes constant. If this expression is inserted into equation (5.14) or (5.15) and $D$ is set equal to a target value, the permissible limit for the maximum stress range can be found. This leads to the simplified approach where the fatigue-limit state can be checked by an extreme load cycle provided that the shape parameter h for the long-term load spectrum is known. With this simplified approach the fatigue-limit state criterion can be checked as easily as the yield criterions or buckling criterion. This is very convenient at an early design stage; limitation on admissible stress range in the extreme load cycle can be given without going into detailed fatigue analyses that includes the less severe loading cycles. These conditions are in fact accounted for by the choice of the shape parameter $h$. A high value for the shape parameter $h$ (say 1.1), indicates that the extreme stress cycle is accompanied by many other severe stress cycles. A low value of $h$ (say 0.7) indicates that the extreme load cycle is accompanied by quite a number of low stress range cycles. As an example, fixed steel structures in the North Sea subjected to wave-induced loading have $h$ in the range between 0.8 and 1.2 with a typical value close to 1.0. If the admissible level of the largest stress range is too low compared to results obtained from stress analyses, redesign has to be carried out. One possibility is to increase the global dimensions so that the occurring maximum stress range decreases. To avoid the weight increase, a better solution may be to improve the design of the critical welded details to obtain a better S-N curve. In this case the weight and the extreme nominal stresses may remain the same, but the admissible stress range is raised due to the fact that $A_2$ in equation (5.14) is increased.

So far we have assumed that one single Weibull distribution is valid for a long-term load spectrum. In other cases, several distributions must be used to characterize several short-term periods during service life, as discussed in Chapter 4. In this case the total damage accumulation is found as the sum of the damage accumulated in each of the steady state short-term periods:

$$D = \frac{n}{1} \sum_{i=1}^{N_{load}} p_i \left[ \frac{q_i^{m_u}}{A_{2u}} \, \Gamma(1 + \frac{m_u}{h_i}, (\frac{S_0}{q_i})^{h_i}) + \frac{q_i^{m_l}}{A_{2l}} \gamma(1 + \frac{m_l}{h_i}; (\frac{S_0}{q_i})^{h_i})\right] \quad (5.17)$$

where:

$N_{load}$   = the total number of short-term load spectra considered
$p_i$     = fraction of design life in load short-term condition i
$h_i$     = Weibull stress range shape parameter for load condition no i
$q_i$     = Weibull scale parameter for the load condition no i
$n$     = number of load cycles during service life.

Again, the knowledge of the shape parameter $h_i$ is crucial for the calculation.

As we illustrated in Chapter 4, the short stress spectrum can also be determined by the energy-spectrum approach. The scale parameter $q$ for the Rayleigh distribution can in this case be related to the variance of the process:

$$q = 2s_\sigma h^{1/h} \tag{5.18}$$

Hence equation (5.14) that is valid for a one-slope S-N curve will read:

$$D = \frac{n_0}{A_2}(2s_\sigma)^m \, h^{m/h} \, \Gamma(1+m/h)$$

(5.19)

Or with $h = 2$, $s_\sigma^2 = m_0$ and $q = \left(2\sqrt{2m_0}\right)$ (see Chapter 4) we get:

$$D = \frac{n_0}{A_2}(2\sqrt{2m_0})^m \, \Gamma(1+m/2)$$

(5.20)

For a bi-linear S-N curve, equation (5.15) can be written in a similar form as equation (5.20).


## 5.4. S-N curves related to various stress definitions

We have discussed the statistics for the S-N curves and the mathematics for damage calculations without being very clear about the definition of the stresses. Originally, fatigue life estimates based on S-N curves were carried out using the nominal stress range as the key parameter to the fatigue life. The nominal stress was defined as the principal stress in the plate at some distance from the discontinuity of a welded attachment and the weld bead itself. The geometry of the attachment and the weld did not affect the applied nominal stress level. Furthermore, no distinction was made between membrane forces and pure plate bending. The nominal stresses for the two conditions are shown in Figure 5.8.

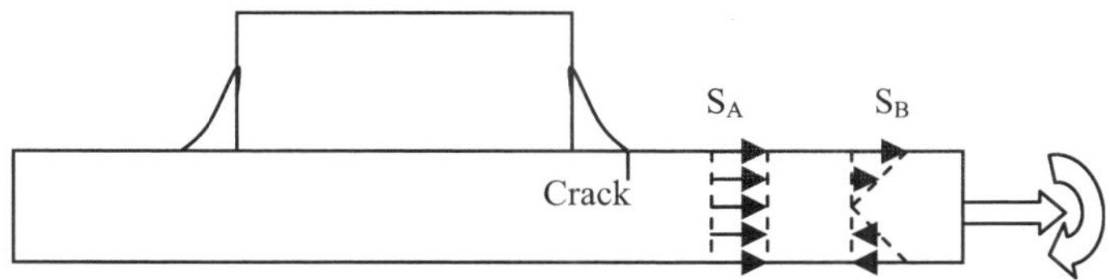


**Figure 5.8.** *Definition of nominal stresses for a welded detail under in plane tension and plate bending*

The approach may seem awkward, and one may argue that it is more logical to apply the stress range actually occurring at the potential crack locus in the notch area, i.e. at the weld toe where the crack is drawn in the Figure 5.8. It is the magnitude of this stress range that acts as an agent for the fatigue damage process; consequently it should be used as the key to fatigue life predictions. It is a temptation to compare the situation with the man who walked under a lamppost searching for a €50 note that he had lost. He searched underneath the lamppost not because he had lost the note there, but because the light made it easier to look for it. It is true that the nominal stress range can be easily determined by simple calculation for a test specimen. However, when a joint is appearing in a welded structure, stresses are usually calculated by finite element analysis (FEA) and it may in fact be a problem to define the actual magnitude of the nominal stress for the welded joint.

Based on these considerations, it has been a trend during the last decade to use the geometrical stress range $\Delta\sigma_g$ as the key parameter to fatigue life instead of the nominal stresses. This approach explicitly takes into account the stress magnification at the potential crack locus caused by the global geometry of the structural detail. The notch effect of the weld bead is, however, not included. If one takes one step further and includes the latter effect also, one can apply the local weld notch stress range $\Delta\sigma_w$ as a key to fatigue life. As already argued, this is the vehicle that drives the damage process. In this case, both the stress concentration effect due to the global geometry of the joint and the notch effect caused by the geometry of the weld are accounted for. This approach will make the designer more conscious about the parameters that local increase the stresses because these variables now are explicitly taken into account. They are no longer inherent and hidden in the S-N curve as is the case for the nominal stress approach. Due to the current interest and need to apply the geometrical and the weld notch stress approach, we shall use discuss and define these stresses in the following sections. We will also show how the stresses can be used for fatigue life predictions. We shall start by illustrating some important definitions related to the various stress concentrations.

### 5.4.1. *Nominal stress, geometrical stress and weld notch stresses*

Let us clarify the various levels of stress concentrations by using the example of a simple welded detail. We will consider a plate with a gusset that is welded to the plate, as shown in Figure 5.9. The plate is subjected to stresses in the same direction as the orientation of the gusset. The plate thickness is denoted $T$, and the length and height of the gusset are $L$ and $H$ respectively. The local geometry of the weld bead is given by the toe angle $\theta$ and the radius $\rho$. We know from experience that the potential crack locus for this detail is the weld toe at the end of the gusset. The potential crack line is dotted as A-A in Figure 5.9. We have chosen the pure membrane loading mode, but the considerations that follow are also true for the bending loading mode.

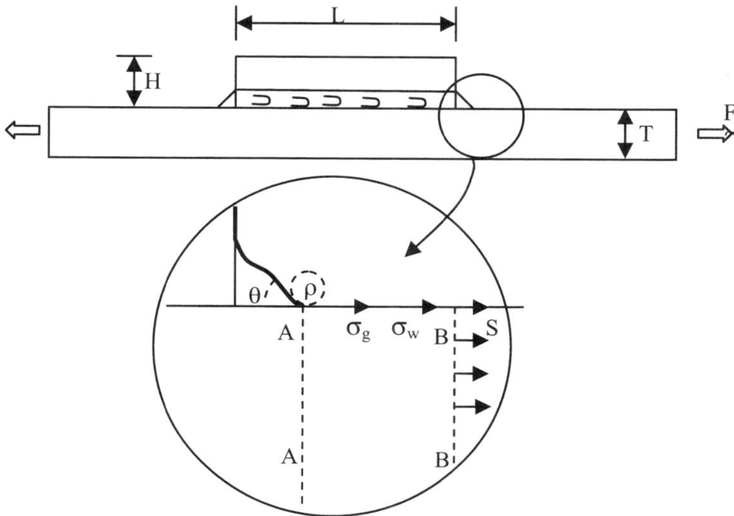


**Figure 5.9.** *Definition of various stress levels in a welded detail*

As can be seen from the above figure, there are three characteristic stresses for the detail:

– $S = \sigma_n$ – nominal stress the in main plate at some distance from the weld (section B-B);

– $\sigma_g$ – geometrical stress at the potential crack initiation locus (top of section A-A);

– $\sigma_w$ – weld notch stresses at the potential crack initiation locus (top of section A-A).

It is in fact only the nominal stress $S$ and the notch stress $\sigma_w$ that really appear in the material; the geometrical stress $\sigma_g$ is only a theoretical definition.

The relation between the nominal stress $S = \sigma_n$ and the two other stresses are given by the following equations:

$$\sigma_g = K_g S \tag{5.21}$$

and:

$$\sigma_w = K_w \sigma_g \tag{5.22}$$

and, consequently:

$$\sigma_k = K_g K_w S = K_t S \qquad\qquad (5.23)$$

$K_g$ is denoted the geometrical stress concentration whereas $K_w$ is denoted the notch (or weld) stress concentration. All of these equations are base on the assumption that the stresses are in the linear elastic region of the material behavior. This is not always the case, at least not for the weld notch stress $\sigma_w$. The first stress concentration given by $K_g$ is due to the effect of the gusset alone. The gusset is a geometrical disturbance that increases the stress from the nominal value $S$ to the geometrical stress $\sigma_g$ at the location of the weld toe. This magnification will also take place when the weld itself is absent. If this geometrical stress appeared in practice one could imagine that the gusset was glued to the plate. In this case one would be able to measure the geometrical stress with a strain gauge at the locus where the weld toe is to be located. The factor $K_g$ can more easily be revealed by a FEA of the detail. This type of analysis will, however, require special qualities for the model regarding element type and mesh, as we shall discuss later. $K_g$ will depend on the height $H$ of the gusset, and even more so on the gusset length $L$. Typical values of $K_g$ will be between 1.2 and 1.5. The longer the attachments length L is, the higher will the stress concentration be, see Figure 5.9. Due to the fact that $K_g$ is determined by the global geometry ($T$, $H$ and $L$), it is denoted as the geometrical stress concentration factor. In some literature it is also referred to as the hot-spot stress concentration factor.

If we now, as the next step, suppose that the weld bead is added onto the plate, then the stress at the weld toe will further increase due to the discontinuity of the weld bead itself. The notch stress concentration factor is denoted $K_w$. This concentration factor will depend on the toe angle $\theta$ and particularly the ratio $\rho/T$. Low values of $\theta$ and high values of $\rho/T$ are favorable. The stress can be approximately determined by measurement using small strain gauges located at the weld toe as was shown in Figure 3.2, Chapter 3, for a tubular joint. We have earlier also shown a local model of the weld toe area in Figure 3.4, Chapter 3, for a fillet welded joint. Such a refined analysis is usually not included in a FEA model for a global structure.

From a design point of view it is obviously important to keep both $K_g$ and $K_w$, as given in equations (5.21) and (5.22), as low as possible (with a target value of 1.0 for both). We will look into the details of this issue at a later stage.

Before finishing with this example, let us suppose that we carry out fatigue testing with the detail shown in Figure 5.9. Let us suppose that one test has given a number of cycles $N^*$ to failure. We must also have other test results to draw an S-N curve, but let us assume for simplicity that the data point we have chosen falls on the median line for the S-N curve. The result can alternatively be given as a function of $\Delta S$, $\Delta\sigma_g$, or $\Delta\sigma_w$ as we have discussed. This is illustrated in Figure 5.10.

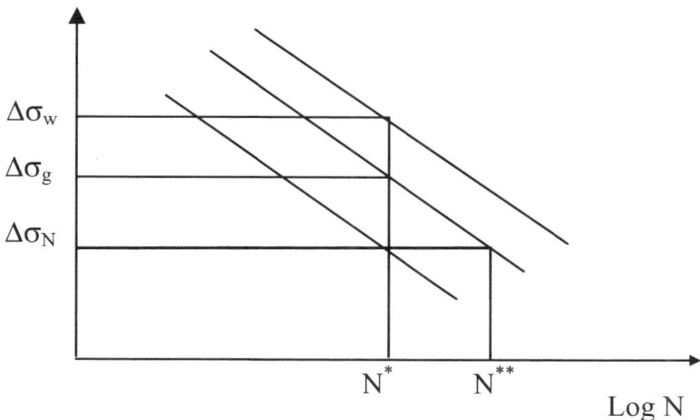


**Figure 5.10.** *Alternative presentation of test results based on nominal stress, geometrical stress, and notch stress*

We have presented the result by the three different keys on the vertical stress axis, as shown. Regardless of what approach is used, it is essential that the user of the S-N curve is aware of which stress concept has been applied and that they must use the same stress concept. If the same stress concept is not used, erroneous results may be the consequence. If the designer uses the nominal stress for a S-N curve that is based on the geometrical stress, this will give an overestimation of the fatigue life from N* to N**, as is shown by the dotted line in Figure 5.10. This consideration aside, one could argue that it is a matter of choice which key should be preferred as they are proportional. However, if we test similar details with somewhat different global geometry (given by *T, H* and *L* in Figure 5.9), the nominal stress concept will not reveal the associated stress variation from specimen to specimen, whereas the geometrical stress concept will. As a consequence, for an S-N curve based on nominal stress, all the data points will be plotted at one given stress level, whereas for the geometrical stress approach the points will be plotted at slightly different stress levels. This is shown in Figure 5.11 for five data points numbered from 1 (shortest fatigue life) to 5 (longest fatigue life). As can be seen, the scatter in fatigue life when using the nominal stress plot is appreciable, whereas the scatter is reduced when using the geometrical stress. It is to be noted that the shortest fatigue life $N_1$ is associated with the highest geometrical stress, whereas the longest fatigue life $N_5$ is associated with a smaller geometrical stress, but not the smallest; see Figure 5.11.

The smallest geometrical stress occurs for test number 4, but this test does not result in the longest fatigue life. The explanation for this can be related to unfavorable weld toe geometry and/or toe surface condition for test number 4. The same reasons explain the difference in fatigue life between test 2 and 3 which have the same geometrical stresses. Hence, even with the geometrical stress approach, we still have some variables and associated scatter that we do not control. These variables are mainly related to the weld toe profile and surface conditions, if we assume that the residual stresses are relieved.

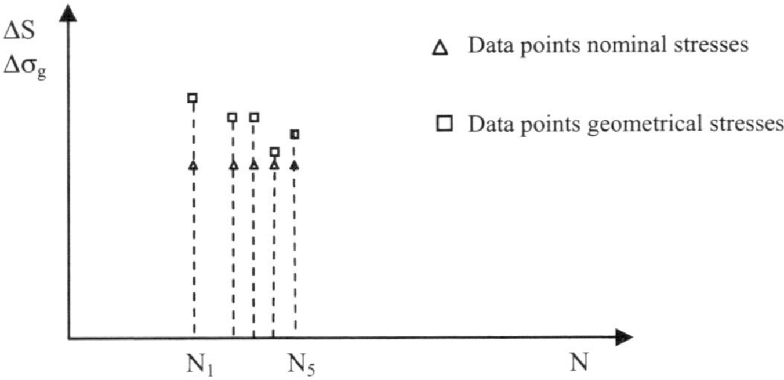


**Figure 5.11.** *Data points plotted against nominal and geometrical stresses*

In conclusion, the nominal stress approach will give larger scatter in fatigue life at the same given nominal stress level if the global geometry is not kept constant. The geometrical stress approach will in such cases give reduced scatter because longer lives will follow from reduced geometrical stresses whereas shorter lives will be related to increased geometrical stresses. Hence, the latter approach is more precise because it captures the actual geometrical stress concentration for each tested detail. Furthermore, the method can be used with great accuracy in conjunction with FEA of welded structures due to the fact that these stresses obtained by the analyses can be regarded as geometrical stresses. This requires that the FEA model fulfils some important criteria that we will discuss later.

### 5.4.2. *Geometrical stresses in tubular joints*

Some welded joints have such a variety of complicated geometrical configurations that the geometrical stress approach becomes a necessity. One of the most typical examples is welded tubular joints. These joints have so many possible configurations and loading modes that the tested fatigue life will exhibit an enormous scatter if it is plotted as a function of the nominal stress range. Hence, this type of joint serves as an excellent example to illustrate the advantage of the geometrical stress approach.

The problem is illustrated in Figure 5.12 where a T joint is shown to the left, and a Y joint is shown to the right. The former joint is subjected to axial forces on the branch member, whereas the latter is subjected to in-plane bending in both branches. Let us suppose that the external loading is such that the nominal stress ranges in the branches are the same for the two cases. One would then expect comparable fatigue lives in the two cases if one trusted the nominal stress range as the key parameter to life predictions. However, there is a severe stress increase in the weld toe region at the crown point and at the saddle point due to local plate bending effects for both types of joints. Hence, this is the potential crack loci in the two cases. Examples for stress concentration formulae were given in Chapter 3, equations (3.2) and (3.3). The cracks will emanate from the weld toe either on the branch or chord side, wherever the stress concentration is largest. From FEA it can be shown that the geometrical stresses close to the weld toes are significantly different for the two cases. The differences in the two cases are partly due to different global geometry (T joint-Y joint), as well as partly due to loading mode (axial in-plane-bending in-plane mode). The stress concentrations are dependent on the diameters and thicknesses in the chord and branches for a given loading mode; see equation (3.3), Chapter 3. It can be shown by FEA that the T joint to the left will have a higher geometrical stress concentration and thus in most cases will have a shorter fatigue life. Consequently, if the tested fatigue lives are plotted as a function of the geometrical stress range, the differences in geometrical stresses will become explicit and the S-N curve data will exhibit less scatter. However, the scatter will still be great due to the fact that two tubular joints with same geometrical stresses at the hot-spot can have very different stress distribution around the welded intersection between members. If the local bending stresses are relatively high all around the welded circumferential, this is unfavorable. This source of scatter is in addition to the explained sources of scatter due to weld toe variability and weld toe surface conditions with possibilities of micro-flaws. As a result, the S-N curve for tubular joints, designated the T curve, has the largest scatter among all the joint classes found in rules and regulations; see section 5.6.2.

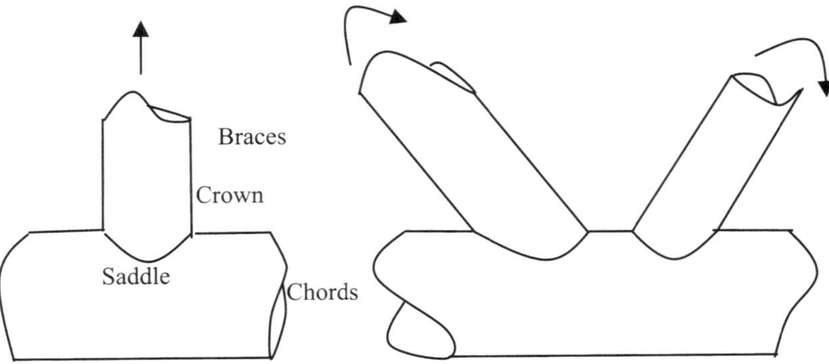


**Figure 5.12.** *Tubular joints with different geometry and loading modes: (a) T-joint with axial load in the branch, (b) Y-joint with in-plane-bending in branches*

### 5.4.3. *Fatigue life estimate based on the weld notch stress approach*

In light of what has been outlined above one may ask why one does not relate the fatigue life directly to the notch stress at the weld toe. This is the real appearing maximal principal stress range at the potential crack locus and thus the driving vehicle for the fatigue damage process. As we have discussed, there are several good reasons that support this approach. Apart from the rational and logical basis, the approach will force the designer to explicitly account for the stress concentrations $K_g$ and $K_w$ in the calculations. This will make the engineer conscious of good design detail, welding procedure, and fabrication quality. This urges the engineer to choose favorable global dimensions for the detail (e.g. $H$, $L$, and $T$ in Figure 5.9) in order that $K_g$ is as small as possible. The analysis must be based on FEA or available parametric formulae found in standard catalogues with $K_g$ for standard details. As we shall see in the section about rules and regulations, DNV has given such catalogs for details often found in ship structures.

We can account for the effect of the global geometry and, in addition, we may take into account the effect of the weld notch. Joints having an abrupt toe geometry will be penalized, whereas there will be a reward for smooth-toe transition geometry when fatigue life estimates are carried out. This will motivate good workmanship.

A theoretical objection to the approach is that it is based on the notch stress at one point only; it does not take the thickness of the plate into the stress gradient trough. A more practical drawback is that the toe geometry is highly variable even within a short length of the weld seam. Thus, the question that arises is what values to apply for the toe angle and radius. Furthermore, the notch stresses are not readily available from a FEA, as are the geometrical stresses. Consequently, the notch stress approach, although appealing, is burdened with some theoretical doubt and practical obstacles. As a result, it has gained less popularity than the geometrical stress approach. We shall give examples on both of the approaches in the following sections.

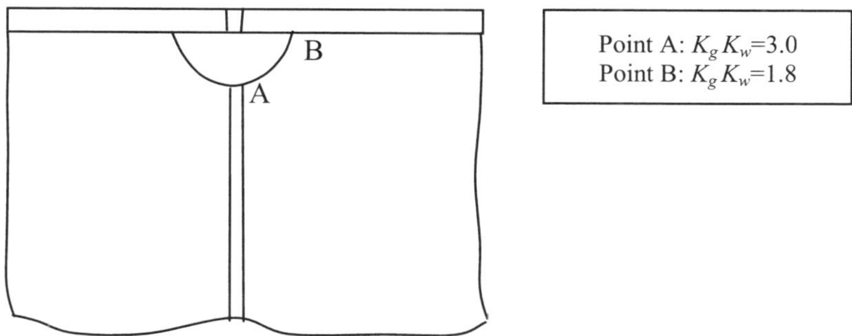


**Figure 5.13.** *Stress concentrations in a large welded girder joint with use of a scallop hole, Ref [18]*

Figure 5.13 shows the joint between two large girders with a scallop hole in the web plate to avoid the transverse weld in the flange plate and the longitudinal weld in the web plate crossing. Potential crack loci, point A and B, are given, as are $K_g$ and $K_w$. If the two weld strings had crossed, this would have given a three-axial residual stress situation. However, the disadvantage of the scallop hole is the stress concentration that follows with it and the related reduced fatigue resistance. In heavy-duty joints this consideration usually overrides the disadvantages of weld crossing. Consequently, scallop holes are avoided. If they are present, the stress concentration has to be accounted for. If drawings with stress concentrations for details as shown in Figure 5.13 are not available, one will have to carry out FEA in order to determine the stress concentration. If the model contains both the global geometry and the weld beads geometry, the stress results will reveal both $K_g$ and $K_w$. This will require a refined model with volumetric elements. For larger slender structures the model will usually be built up by thin shell elements and the welds will not be included. If the model is properly constructed the stress results will reveal $K_g$. To determine $K_w$ one can use the formulae given in Table 5.1.

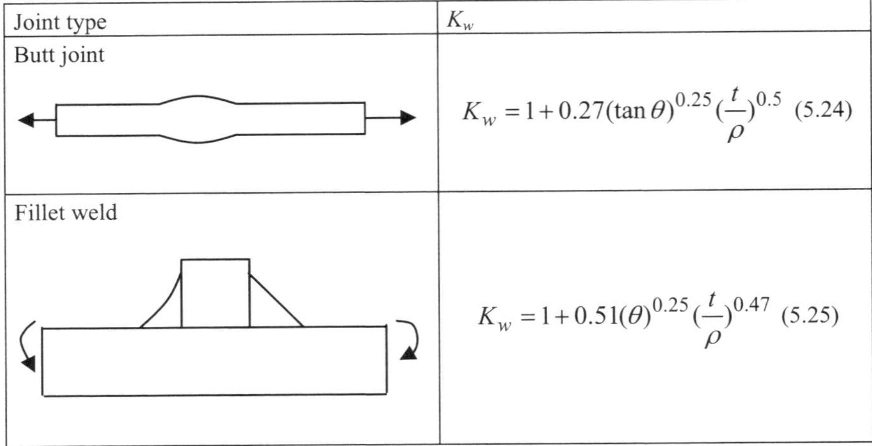

| Joint type | $K_w$ |
|---|---|
| Butt joint | $K_w = 1 + 0.27(\tan\theta)^{0.25}(\dfrac{t}{\rho})^{0.5}$   (5.24) |
| Fillet weld | $K_w = 1 + 0.51(\theta)^{0.25}(\dfrac{t}{\rho})^{0.47}$   (5.25) |

**Table 5.1.** *Notch stress concentration for butt and fillet welds, Refs [11, 12]*

The problem with using the formulae in Table 5.1 is that the weld toe geometry is highly variable along the weld seam. This is a serious objection to the method; the weld bead is so irregular that it escapes a specific profile characterization. Many suggestions have been made as to how to determine the toe profile, amongst them practical coin tests in combination with wire cables as shown in Figure 5.14. It should not be possible to drag a piano wire of 1 mm diameter underneath a coin with radius 10 mm.

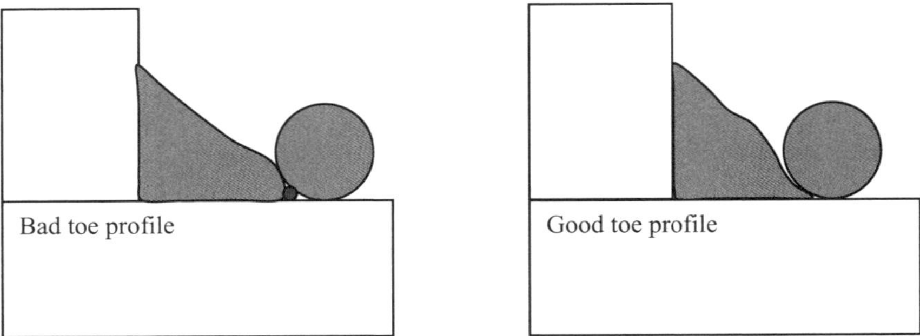


**Figure 5.14.** *Control of the toe profile by using a coin and a piano wire*

Due to the irregularity, the only rational solution is to treat the geometrical parameters as stochastic variables varying randomly from one type of joint to another and randomly along the weld seam for any given type of joint. For a given weld category the geometry parameters can be given by their mean values or expected extreme values. In Chapter 9 we shall advocate the latter approach based on the hypothesis that cracks will emanate at the spots with the most unfavorable weld geometry for a given geometrical stress. Once the extreme values have been found we can use equation (5.24) and (5.25) to determine the weld notch factor. Let us use the formulae for a plate thickness $T = 16$ mm. A good butt joint will typically have a mean radius of $\rho = 1$ mm and a mean angle $\theta = 30$ degrees. This gives:

$$K_w = 1 + 0.27(\tan 30)^{0.25}(16)^{0.5} = 1.92 . \tag{5.24}$$

For a fillet weld it is reasonable to assume the same radius, but an increased angle of 45. This gives the following figure:

$$K_w = 1 + 0.51(\frac{45}{180}\pi)^{0.25}(16)^{0.47} = 2.76 . \tag{5.25}$$

The angle has to be given in radians. As can be seen, a typically butt weld has a $K_w$ in the range from 1.5 to 2.0 depending on $T/\rho$ and $\theta$, whereas a fillet weld has typical values in the region 2.0-3.0. The difference is mainly due to differences in the flank angle. A smooth transition will be characterized by $\rho = 1$ mm in both cases. If inspection after fabrication has been carried out and it reveals that the weld toe geometry is much poorer than the given values, one should reduce the life predictions or improve the weld toe by grinding. Weld beads that rise steeply from the plate with a radius close to zero (say 0.1 mm) give rise to very high stresses and result in $K_w$ values between 4 and 7. However, the affected material volume will decrease, as we shall see in the next section.

### 5.4.4. *Conclusions on the various stress approaches*

As we have pointed out in the above discussion, we have three different approaches for fatigue life estimates:

– *Alternative A:* we can use the nominal stress range as the key parameter to fatigue life for a given detail. The approach requires that we have S-N curves available for all possible detail geometries and loading directions. This is required because the stress concentration factors associated with the various geometries do not affect the nominal stress, but are inherent in the S-N curves. Hence, we should have an S-N curve for each, single geometry. This is, in practice, impossible. The procedure is to define one test population where the joints have very similar geometries with almost equal quality from a fatigue point of view. The associated S-N curve for this population is then defined as a class or category in rules and regulations. If the variation in geometry for details assembled in one population is too large, the drawback will be a large scatter. This will penalize the favorable geometries within the actual category. The problem with defining the nominal stress range for a detail when integrated in a complex structure has already been mentioned. Despite the drawback of the method it is still the most used method in rules and regulations.

– *Alternative B:* we can use the geometrical stress range as the key parameter to fatigue life for a given detail. This approach requires only one reference S-N curve, usually a butt weld between two plane plates. This joint has, by definition, $Kg = 1.0$. The curve will also be applicable for all other details when the influence of the global geometry of the actual joint is explicitly accounted for by analytical formulae or FEA. The method is appealing, but one must bear in mind the simplified assumptions on which it is based:

1) The weld bead and local toe profile for the actual welded detail are as for the butt joint used as reference.

2) Surface conditions at the weld toe are the same as for the butt joint used as reference.

3) Welding procedure and heat treatment are the same.

4) The actual FEA model is refined enough to capture the geometrical stress concentration for the detail in question.

The designer should carefully check the various points when using the geometrical stress approach. A fillet weld may in fact often have less favorable toe geometry than the reference butt weld due to the toe angle, as we have just shown. Furthermore, when carrying out FEA, both element type and element mesh will play an important role.

– *Alternative C:* we can use the weld notch stress range as the key parameter to fatigue life for a given detail. We have already pointed out that this is the most logical method from a theoretical point of view. In this case we also need only one S-N curve as a reference curve, i.e. a butt joint between two plane plates where the weld bead is ground flush with the plates. This joint has, by definition, $Kg = Kw = 1.0$. Again, this curve will also be applicable for all other details when the influence of the global geometry and weld toe geometry of an actual joint is explicitly accounted for by analytical formulae of FEA.

The approach has the advantage of eliminating assumption 1) discussed in relation to the geometrical stress approach. We have emphasized the weld notch stress approach in the present chapter due to the fact that it is an excellent pedagogical presentation of the fatigue problem for welded joints. All possible stress concentrations are explicitly treated. However, we must be aware of the fact that this method also is a simplification due to the fact that it assumes that it is only the SCF at the point of the weld toe that governs the fatigue life. The method disregards the stress gradient as one departs from the weld toe through plate thickness. One must be aware of the fact that it is only a small material volume at the weld toe where the maximum stress actually occurs. The stresses will decrease rapidly at a small distance from the surface as shown in Figure 5.15. It is a fact that a small $\rho/T$ ratio will increase SCF significantly, but over a smaller material volume compared to the volume affected by the stress concentration for a larger $\rho/T$ ratio. This is illustrated in Figure 5.15 with a large $\rho/T$ ratio to the left and a small $\rho/T$ ratio to the right. As can be seen, the SCF increases when the $\rho/T$ is decreased, but at the same time the stress gradient becomes steeper. As a result, a smaller material volume will be affected by the high stresses. Consequently, the assumption that the fatigue life is influenced and scaled by the notch SCF only will be overly pessimistic. This has lead to the concept of the fatigue notch factor instead of the linear stress concentration we have discussed. The fatigue notch factor is smaller than the calculated $K_w$. The concept of a fatigue notch factor remains somewhat elusive. We will not pursue this concept any further in the present chapter. The reader is referred to Chapter 10 for further discussion of the subject. In that chapter we propose a weld notch a *stress field* instead of a *stress concentration* as the key to fatigue life. This is more in line with physical realities.

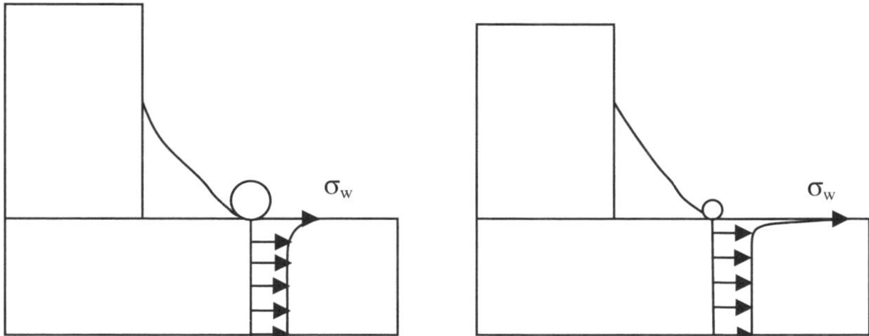


**Figure 5.15.** *Stress concentration and stress gradient for a large $\rho/T$ ratio (left) and a small $\rho/T$ ratio (right). The geometrical stresses are assumed the same for the two cases*

Finally, let us remind ourselves that all the geometrical and notch stress approaches are suitable for welds with loading transversely to the weld seam direction and with cracks that emanate from the weld toe. The stress concentrations we have defined refer to the principal stresses acting normal to the weld seam. If the cracks are likely to emanate from the weld root, these approaches are not applicable. In this case or any case where the weld throat shear stress range is used as the key parameter, the nominal stress approach should be preferred. However, such cases are less frequent compared to the numerous cases of potential toe cracking.


## 5.5. Some comments on finite element analysis

The objective for an FEA, from a fatigue point of view, is to reveal the geometrical stress concentrations at critical hot-spots. The accuracy must be very high so that the fatigue life estimates have sufficient credibility when using the geometrical stress method. The effect of all stress concentrations that are not inherent in the fatigue test data and associated S-N curve must be taken into account in the stress analysis.

Usually, a relatively fine mesh model using thin shell elements or volume elements is applied. The welds are usually not included in the model. As a result, the stresses will approach infinity at the sharp corners where the weld bead is located, if the element size is decreased towards zero. Hence, the analysis involves both considerations for element type, element size, and extrapolation techniques for stresses. A general rule of thumb is that the element size should not be appreciable larger than the plate thickness in the vicinity of the hot-spots. Furthermore, it is important not to have too steep a change in the density of the elements mesh where the geometrical stresses are to be analyzed; Ref [6].

Figure 5.16 shows the most common geometry effect in a structure, i.e. a scallop hole in a plate. The hole has a butt weld at each edge transverse to the applied loading.

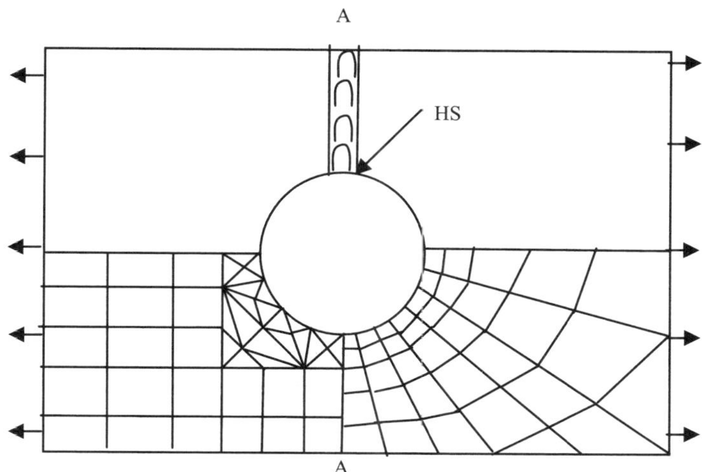


**Figure 5.16.** *Transverse butt weld in a plate with a hole. Two alternative finite element meshes are shown (each of them is doubly symmetrical)*

The geometry is unfortunate in the way that the notch effect of the weld will be superimposed on the geometrical stress concentration because of the hole. This will occur at the hot-spot (HS) as indicated on the figure. The design should, from a fatigue point of view, be avoided, but due to functionality and fabrication requirements it must in some cases be accepted. What is important is that the designer recognizes that the geometrical and the local notch stress concentration are occurring at the same hot-spot and that this must be taken into account. This can be done by using the geometrical stress approach using an S-N curve obtained from test with butt joints in plane plates without a hole, i.e. a D-curve; see section 5.6.2. The key to fatigue life must then be the geometrical stress given by equation (5.21). If there are other geometrical stress risers in the structure they can be dealt with in exactly the same manner. The stress concentrations can be revealed by FEA and the geometrical stress stresses will read:

$$\sigma_g = K_g \sigma_n.\tag{5.26}$$

The weld bead is usually not modeled in the FEA model, as shown in Figure 5.16. This will require a refinement of the model that is not cost-efficient. Two possible element models are shown in the figure. The elements are plane stress elements and the model mesh is symmetrical about two axes. The mesh shown to the left is hardly refined enough to reveal a stress concentration of $K_g = 3.0$ for an infinite plate with a hole. Furthermore, some of the elements are distorted with corner angle less than 45 degrees. The mesh shown to the right is much more appropriate for revealing the stress concentration. (The mesh is an *alternative* and not *compatible* to the one shown on the left.) As can be seen, the elements are more regular and refined in the critical area. Stresses at nodes and mid-lines are easily determined and extrapolated. In some cases fictitious bar elements are located at the free edge of the hole. The tensile stresses in the bars will then determine the stress concentration very accurately.

Let us take another detail as shown in Figure 5.17. This is the same problem as we discussed thoroughly in the previous section, a gusset is welded on to a plane plate subjected to in-plane loading in the direction of the gusset; see Figure 5.9. The nominal stress in the plate is $S = \sigma_n$. Figure 5.17 also shows the stress variation of the principal stress perpendicular to the weld as one approaches the weld toe at the end of the stiffener. The fully-drawn curve ending at $\sigma_w$ is the real stress distribution. This stress distribution is analogous to the strain distribution we measured experimentally in Figure 3.5, Chapter 3, for a tubular joint. The dotted line in Figure 5.17, ending at $\sigma_g$, is the stress distribution one would have if the weld was not present. To determine this fictitious stress level one must carry out a linear extrapolation through the stress points at some distance from the weld toe. This will eliminate the effect of the weld notch. These points are chosen at a distance T/2 and 3T/2 from the figure. This procedure must be followed when the stress points are obtained by measurements on real joints and also when they are obtained by FEA models; for FEA models, we will explain why in what follows. In practice, the stress distribution has to be determined by a FEA model as shown in Figure 5.18. Two alternative models are presented. The first model uses volume (solid) elements, whereas the second model uses thin shell elements. By using the first model it is in fact possible to model the weld itself; in the second model the weld cannot be modeled directly. However, even if the weld is included in the first model it is usually not possible to model the weld toe profile in detail. This can usually only be done in a refined sub-model as was shown in Chapter 3, Figure 3.4. Hence, both models in Figure 5.18 will have sharp notches creating a singularity for the stress fields. The volume element model will have a sharp fillet due to the weld, whereas the shell model will have a sharp corner at the end of the stiffener. In both cases the stress will increase towards infinity at these locations as the element size becomes very small. This is a principal difference from the element model presented in Figure 5.16. The phenomenon will have an impact on how to choose the element size and how to determine the geometrical stresses. As the element size becomes smaller, the FEA obtained stresses at spots at a distance farther than T/2 from the weld toe will converge towards constant values. These stresses are a good approximation for the true stresses. It has been found that the convergence is reached when the element size approximately equals the plate thickness $T$. For the spots closer to the sharp corners, the stresses will, as discussed, continue to increase with smaller element size, i.e. no convergence will be reached. This is the explanation of why the geometrical stresses at the weld toe are determined by extrapolation from the spots lying at a distance T/2 and 3T/2 from the weld toe for both the models in Figure 5.18. Furthermore, it is important that the stress reference points at T/2 and 3T/2 are not inside the same element. In addition to the recommended element size, it is recommended that one uses higher-order elements with 20 nodes for the volume element model and 8 nodes for the thin shell element model. The stresses at 3T/2 and T/2 have to be found first by interpolation from the Gaussian integration points of the elements. Let us underline that the principal stress used in fatigue life prediction is a vector – and not a color. The color plots often used to reveal the level of the geometrical stresses fails to give us the direction of the stresses. The stress direction relative to the weld direction is, as we have seen, particularly important. Hence, the stress analyst must from time get a print out of the stress arrows in addition to the color plots. Young, aspiring designers often seem to forget this.

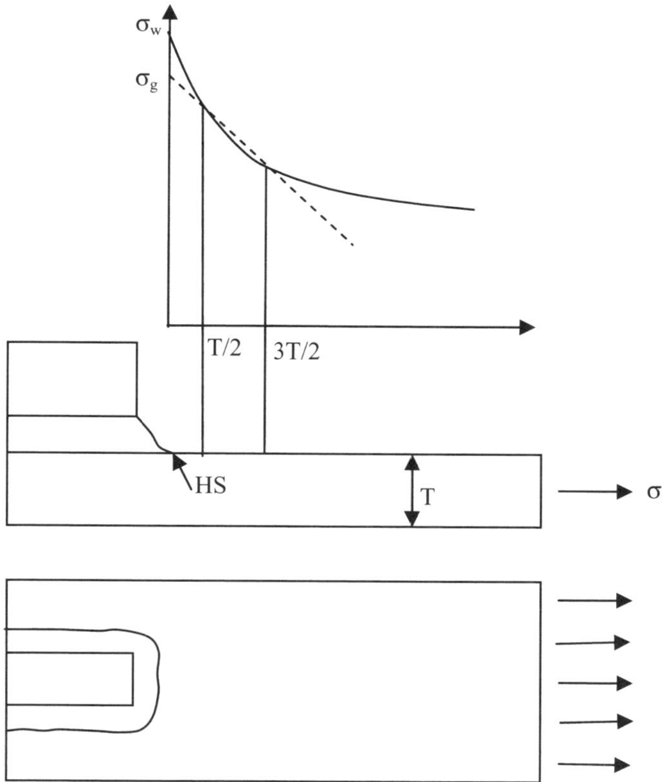

**Figure 5.17.** *Definition of geometrical stress and notch weld stress*

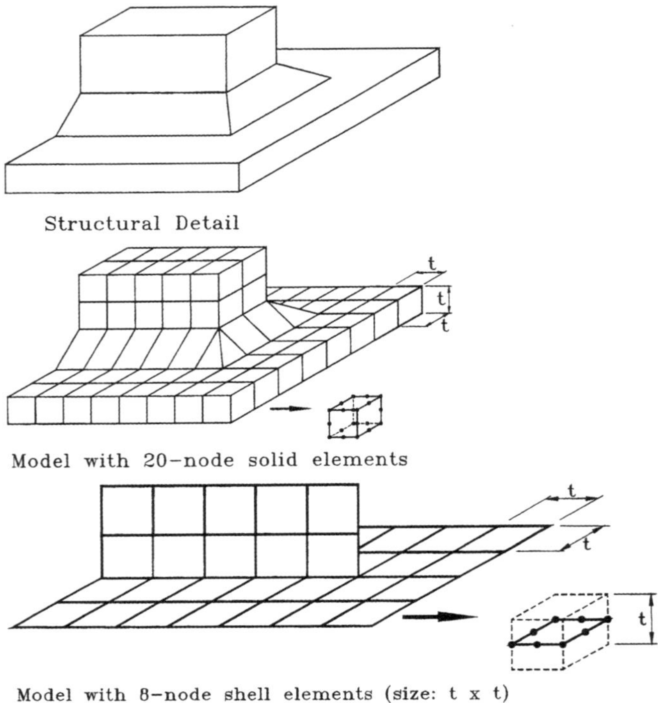



**Figure 5.18.** *Finite element models for plate with welded gusset, Ref [14]*

Before finishing with this example, it should be noted that if the loading in Figure 5.18 is changed to act transversely to the direction of the gusset, the model based on volume elements will still reveal the geometrical stress concentration, whereas the model with the thin shell elements will not. This is because a shell element is a plane without a geometrical thickness as can be seen from Figure 5.18, bottom right. The geometrical stress concentration in the transverse direction to the gusset will increase with the thickness of the gusset; see Figure 2.19, Chapter 2. The thin shell model will not capture this effect. The case is one of the pitfalls when on is determining geometrical stresses based on thin shell models for large structures. Thin shell models are by far the most-used models for large, welded plate structures. Figure 5.19 shows a thin shell element model for the tubular joint test described in Chapter 3. The stress results for in-plane bending are shown in Figure 5.20. As can be seen from the stress distribution, there is a non-linear rise in the stresses close to the intersection of the members as expected; see Figure 3.5, Chapter 3. In the thin shell mode, the location of the weld toe is not directly defined as the thickness of the shell is not physically modeled, as already pointed out. In this case the distances T/2 and 3T/2 are defined from the intersection between the mid-plane of the elements. The geometrical stress is defined to occur at the same intersection. Again, this stress is determined by extrapolation from the spots located T/2 and 3T/2 from the intersection. The approximation has been shown to give results close to strain measurements, as discussed in Chapter 3. More details are found in Ref [6].

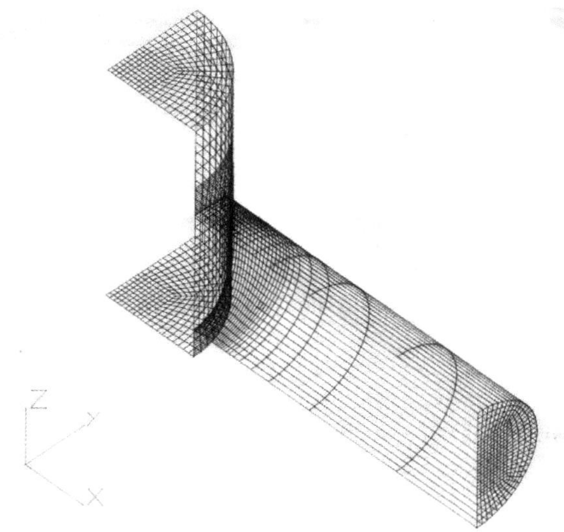

**Figure 5.19.** *Thin shell element model of the tubular joint*

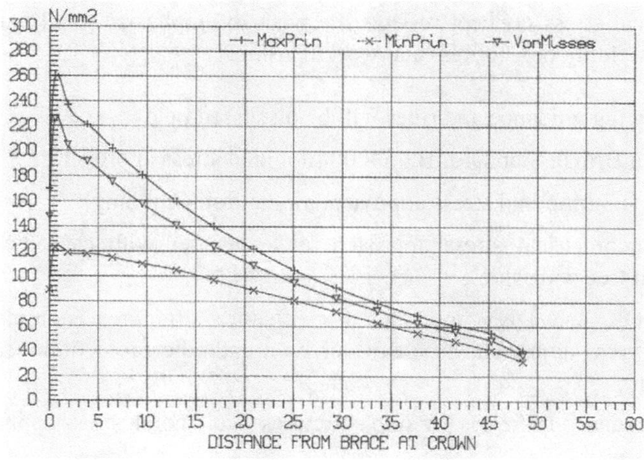



**Figure 5.20.** *Stress (principal stress and Von Moses) distribution at in the chord at given distance from crown intersection*

## 5.6. Current rule and regulations

### 5.6.1. *General considerations*

There currently exist a variety of rules and regulations for fatigue life verification of welded joints. As we shall see in what follows, the nominal stress method is still the most frequently used approach in most of the rules and regulations. Typical examples are the Eurocode, DoE, DNV, and NORSOK Refs [1, 4, 6, and 14]. As we have discussed, the method is quite accurate if large test series are available for *exactly the same detail* as the one to be analyzed. If the test series are available for *similar details only*, the scatter in fatigue life will increase and the method becomes less precise.

The geometrical stress approach is also an approximation, as we have discussed, but has many appealing aspects when used in conjunction with FEA of large complex structures. The method has been adopted in the Health and Safety Executive (HSE), UK; Ref [13].

The weld notch stress approach has gained less popularity than the two other methods. The main reason for this is related to the problem of characterizing the highly irregular and variable weld toe geometry. However, the method has been adopted for ship structures by DNV and BV; Refs [18, 19]. The method is convenient for ship structures because these structures have so many geometrical variations of details that the classification concept becomes difficult. Hence, the concept, with one common reference curve and detailed notch stress calculation, becomes attractive. The reason for choosing the weld notch stress concept instead of the geometrical stress concept reflects the wish to emphasize good manufacturing procedures (welding procedures) and workmanship.

The following guidance and rules will be discussed below:

– DoE and British Standard, BS 54000 (nominal stress approach);

– Eurocode 3 (nominal stress approach, air environment only);

– HSE (geometrical stress approach, air, seawater with cathodic protection, seawater with free corrosion);

– NORSOK and DNV guidance for offshore structures (nominal stress or geometrical stress approach, air, seawater with cathodic protection, seawater with free corrosion);

– DNV guidance BV rules for ship structures (weld notch stress approach).

All the different building codes have more or less the same data basis for their S-N curves and the classification of various details is more or less the same. The differences lie in the method of presentation and application. Variations are found particularly in the location of the knee point of the bilinear S-N curves, the influence of seawater and cathodic protection, and the treatment of the thickness correction. The differences between the British and the Norwegian codes are small, and the classification of the welded details is almost the same. In the Eurocode 3, there is a somewhat different format for the S-N curves, as we shall see. Regardless of which code is used, the designer must go through the following stages for fatigue life prediction:

1) Assess the entire structure and various load conditions. Describe the applied loading as outlined in Chapter 4.

2) Identify joints and details that are vulnerable to fatigue. Understand the load transfer for these joints, and determine the service stress history that the joint will be subjected to during service life. For variable amplitude loading, the load spectrum can be presented by stress blocks, each block defined by a given stress range $\Delta\sigma_i$ and a corresponding number of cycles $n_i$, equation (5.13).

3) Detail classification. Determine to which experimental population the joint belongs, and select the most appropriate S-N curve to determine the fatigue capacity $N_{Ti}$ for each block, equation (5.4).

4) Carry out a fatigue damage calculation by comparing the applied number of cycles $n_i$ at a given stress level with the experimental number of cycles to failure $N_{Ti}$, equation (5.7).

There is not much problem with the calculation scheme; the complications are more in foreseeing and determining a realistic service stress history. A correct classification of the detail with related fatigue strength is usually a simpler task.

The designer must also be aware of the consequences of a possible fatigue failure and the designer should assess whether or not the details under consideration are accessible for inspection and repair during service life. If they are accessible for inspection and possible to repair, a combination of *safe life* and *damage tolerance* philosophy may be used. This means that the FDF can be reduced and scheduled inspection can be planned and carried out. Detected cracks must be repaired to prevent failure. In these cases the required FDF will be set to three for most standards, sometimes lower. If inspection and repair are impossible and the consequences of fracture are severe, a FDF of 10 is often required. These issues must be discussed from the beginning of a rule-based analysis.

We shall illustrate the various rules and regulations in a brief overview that gives some of the characteristic features of the rules and regulations. The reader is referred to the original sources for details.

### 5.6.2. *The original fatigue classes and S-N curves from DoE*

The original DoE curves for plated joints were designated B, C, D, E, F, F2, G, and W, (DoE, Ref [4]). The curves are valid for high-cycle fatigue, i.e. the number of cycles to failure is higher than $10^5$ cycles. B represents the curve with the highest strength, whereas W represents the curve with the poorest. The B and C curves are representative for non-welded material, as well as for welds loaded parallel to the welding direction. The curve for tubular joint is designated T. For welded joints the classification will be a result of:

- joint type;
- joint geometry;
- loading direction;
- welding procedure;
- distortion tolerances;
- post-weld treatment;
- non-destructive inspection (NDI);
- location of the potential crack.

There is a logical explanation for why each point above will have an influence on the fatigue capacity, as we have explained in the foregoing chapters. Joint type, geometry, and loading directions will obviously affect the fatigue capacity. Different welding procedures may create different weld shapes and corresponding local notch stress concentration factors. The choice of welding consumables may also affect the quality of the fusion line and initial defects. Distortion of plates introduces secondary bending stresses and decreases the fatigue capacity. Post-weld improvement methods, such grinding or TIG dressing, will increase the fatigue capacity. The same will be the case after NDI and any necessary repair of initial cracks. Finally, the potential crack locus is important.

We may find it surprising that the material quality does not appear on the list for the joint classification. This is because the fatigue process in welds is dominated by geometrical parameters. The material quality will not have such a strong bearing on the fatigue capacity as it has for unwelded components where the initiation phase plays an important role. However, the material quality may play an important role in very long fatigue lives, when the initiation period is significant; see Chapter 9. In such cases, a high-quality welded joint will also benefit from high-strength base material quality. In any event, the base material, weld, and the heat affected zone (HAZ) will fulfill given requirements regarding Hardness Brinell (HB) and Charpy values for the joints.

Typical figures for various classes are given in Table 5.2. $A_0$ and m are constants for each experimental population that corresponds to a given S-N class. The inverse slope m of the curves is close to 3 for most classes. Hence, the estimated fatigue life is very sensitive to the applied stress range $\Delta\sigma$. Typical values of $s_{logN}$ range from 0.18 to 0.26. This corresponds to coefficients of variation (standard deviation divided by mean value for fatigue life) in the range from 0.4 to 0.7. These coefficients pertain to linear fatigue life, not the logarithmic values. The relation is such that $s^2_{log} = 0.188 \ln(1 + CoV^2)$. Some of the curves are drawn in Figure 5.21 for variable amplitude loading. The curves are valid for high-cycle fatigue and most of the tests have endured more than $10^5$ cycles. As can be seen, the inverse slope m changes from 3 to 5 at $10^7$ cycles. This change was discussed in section 5.3.2. For constant amplitude loading there will be a fatigue limit at $10^7$ cycles. There is a cutoff stress level at $10^8$ cycles. These original curves were based on test results in air, but were also used for seawater with cathodic protection before new information came to hand; see Figure 5.5.

| Class | m | $A_0$ | $s_{logN}$ | CoV |
|---|---|---|---|---|
| B | 4.0 | 2.343E15 | 0.182 | 0.44 |
| C | 3.5 | 1.08 E14 | 0.204 | 0.50 |
| D | 3.0 | 3.99 E12 | 0.210 | 0.51 |
| E | 3.0 | 3.29 E12 | 0.251 | 0.63 |
| F | 3.0 | 1.73 E12 | 0.218 | 0.54 |
| F2 | 3.0 | 1.23 E12 | 0.228 | 0.56 |
| G | 3.0 | 5.70 E11 | 0.179 | 0.43 |
| W | 3.0 | 3.70 E11 | 0.184 | 0.44 |
| T | 3.0 | 4.79 E12 | 0.264 | 0.67 (geometrical stress) |

**Table 5.2.** *Statistical details for various S-N classes, air conditions; Ref [4]*

As we have discussed, the T S-N curve, which is representative for tubular joints, is based on geometrical stresses. It is noted that the T class has the highest coefficient of variation. This is due to the fact that this class is based on an experimental population containing a variety of geometrical configuration and loading modes. Although the scatter is reduced due to the application of the geometrical stress approach, additional scatter is introduced due to a large variation of stresses around the welded circumferential between members, as well as due to irregularities in local toe geometry.

Based on the discussion above, it is not surprising that the S-N curve for tubular joint is close to the one for butt joints. The first curve takes explicit account of the stress concentration $K_g$, whereas the butt joint in plated joints has, by definition, $K_g = 1.0$.

In rules and regulations, the median S-N curve for tubular joints is somewhat better than for butt joints, but due to higher scatter (CoV = 0.67 for tubular joints and CoV = 0.51 for butt joints) the design curves for the two classes coincide.

Some frequently occurring classes for plated joints are shown in Table 5.3. The first types are continuous welds, essentially parallel to the direction of applied stresses (type 2). These details have high fatigue strength due to the fact that the weld bead does not give rise to stresses as for perpendicular loading. Hence, the cases are classified B, C, and D, as shown. When the loading is perpendicular to the welds (type 3), the butt weld with full penetration is classified as a D class when welded from both sides and in a flat position. This is the most common case for high-strength joints. The potential crack locus is at the weld toe. If the butt joint is welded from one side only it will be classified as an F class if it is made with permanent backing strip. The potential crack locus will, in this case, be at the root; see Figure 5.22, to the right. The backing strip must not be tag welded, because the tag weld will give a G class. One solution often used is to bond a ceramic backing bar to the back of the joining plates. These types of one-sided joints can also be made with a TIG root pass. If it is possible to grind the backside with a tool afterwards, it is possible to obtain a C class for this one-sided joint. If the joint configuration is changed from a butt joint with full penetration to a fillet welded attachment (type 4) the classification will drop from a D to an F class. If the weld toe is close to the plate edge, the class will drop further down to a G class. The reason is that both the weld toe and the plate edge are stress risers. Hence, one should try to avoid such configurations. For these cases we also see that increased attachment length will downgrade the joint from F to F2.

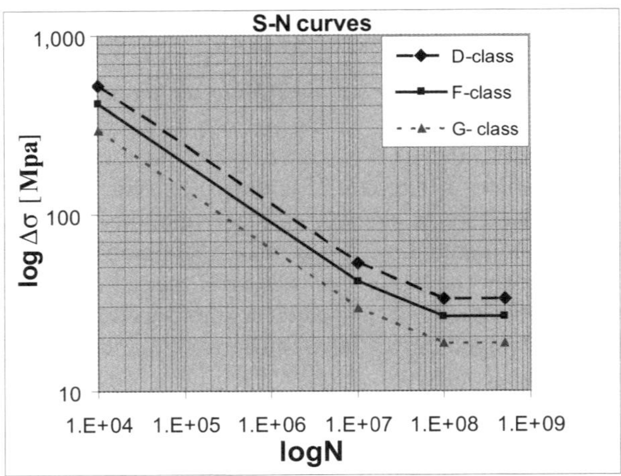


**Figure 5.21.** *Graphs of DoE S-N curves for variable amplitude loading*

Table 5 3, below, gives an overview of certain plated welded details.

*Type 2:* continuous welds essentially parallel to the direction of applied stresses.

Full or partial penetration butt welds or fillet welds built up by plate or section and with no attachments.

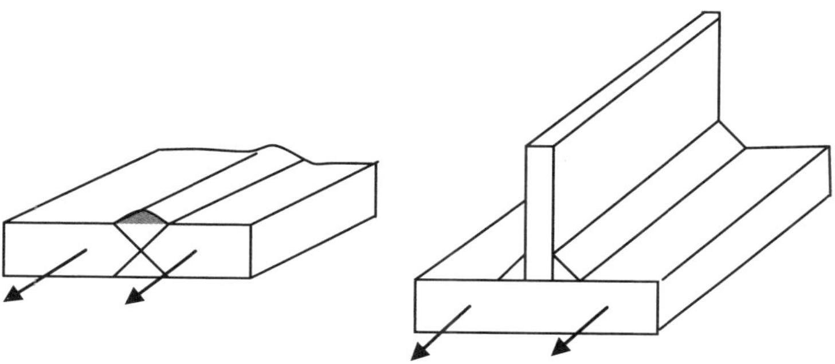

| Conditions | Potential crack locus | Class |
|---|---|---|
| (a) Full penetration butt welds with the weld overfill dressed flush with the surface and finish-machined in direction for stresses. Weld bead proven to be free from defects. | Any defect in weld | B |
| (b) Butt or fillet weld made by submerged arch welding (SAW) or shielded metal arc welding (SMAW) with no stop-start position within the length. | Surface ripples | C |
| (c) As (b), but with welds containing stop-start positions. | Start-stop positions | D |

*Type 3:* transverse butt welds in plate (essentially perpendicular to the direction of applied stresses).

Full penetration butt welds welded from both sides and join plates of equal width and thicknesses. Differences in width or thickness will be machined to a smooth transition not steeper than 1 in 4.

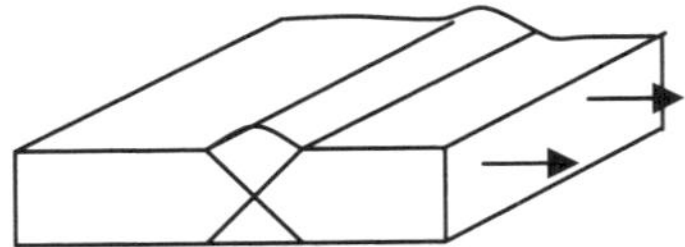

| Conditions | Potential crack locus | Class |
|---|---|---|
| (a) With the weld overfill dressed flush with the surface and finish-machined in direction for stresses. Weld bead proven free from defects. | Any defect in weld | C |
| (b) With the weld made manually or automatically (other than SAW) provided all runs are made down-handed. | Weld toe | D |
| (c) Weld made other than as in (a) and (b). | Weld toe | E |

*Type 4:* welded attachments on the surface or edge of a stressed member.

Parent metal of the stressed member adjacent to toes or ends of bevel-butt or fillet welded attachments.

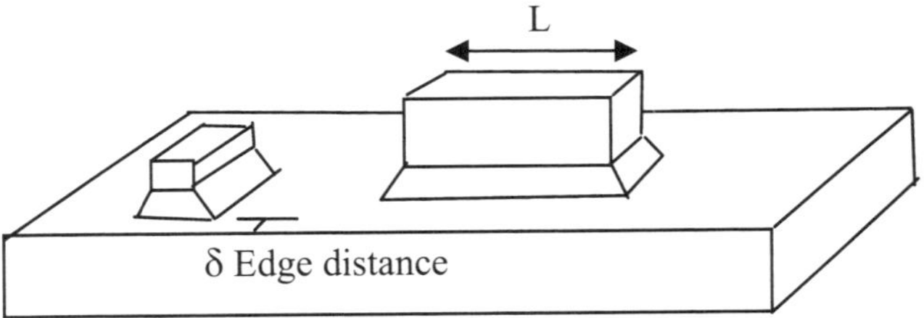


| Conditions | Potential crack locus | Class |
|---|---|---|
| (a) With attachment length parallel to the applied stresses. L<150 mm δ>10 mm | Weld toe | F |
| (b) With attachment length parallel to the applied stresses. L>150 mm δ>10 mm | Weld toe | F2 |
| (c) All types of attachments Edge distance less than 10 mm. | Weld toe/edge of plate | G |

**Table 5.3.** *Overview of some important, plated welded details*

For load bearing welds, the joint classification will be different whether the crack grows from the weld toe or the weld throat, as has already been pointed out. The applied stress range is also different at these two locations. The same joint may therefore be checked against two different S-N curves and with two different stress ranges. The case is shown to the left in Figure 5.22, where fatigue cracking may start at the weld toe or the weld root. Different stress ranges are applied for the two cases. In the first case the joint is classified as F-class (Category 71 Eurocode), while in the latter case the joint must be classified as W-class (Category 35 Eurocode). The ratio between plate thickness and weld leg length will govern which of the failure modes is the critical one, i.e. the one having the shortest predicted fatigue life.

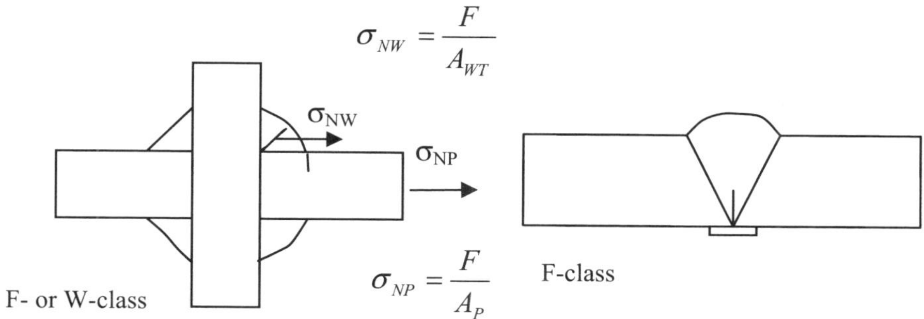


**Figure 5.22.** *Important cases with possible root fatigue cracks*

Life prediction for toe cracking must use the nominal stress range in the plate and the F-class S-N curve. The F-class applies if the weld undercuts near plate edges are dressed. For root cracking, the nominal stress range in the weld throat must be used in conjunction with a W-class S-N curve. Such a case must be regarded as a special one, and cracks that emanate from the weld toe are by far the most common. Figure 5.22, right, shows a one-sided butt weld that is another important case with possible cracks from the weld root. We have already discussed this case.

### 5.6.3. S-N life predictions according to Eurocode 3-Air environment

The Eurocode 3, Ref [1], presents 14 parallel curves that all have an inverse slope of m = 3. The curves belong more or less to the same classification groups (or categories) as the original DoE, but they are designated by figures instead of letters. The characterizing figure for each class corresponds to the stress range at $N = 2 \; 10^6$ cycles. The S-N curves also have a slightly different appearance; see Figure 5.23. As can be seen, the curves change slope at $N = 5 \; 10^6$ and prescribe a cutoff limit at $N = 10^8$. Some details from Eurocode 3 are listed in Table 5.4. The closest comparable classes from DoE are also given. As an example, the E class in DoE will be close to the category 80 in the Eurocode 3.

The calculation scheme in Eurocode may also have a somewhat different format. The Miner rule is used to calculate an equivalent stress range as follows:

$$\Delta\sigma_e^m = \frac{\sum_{i=1}^{k} n_i \Delta\sigma_i^m}{\sum_{i=1}^{k} n_i} \tag{5.27}$$

The fatigue criterion can now be written in the following format:

$$\gamma_f \Delta\sigma_e < \frac{\Delta\sigma_R}{\gamma_m}. \tag{5.28}$$

Where $\Delta\sigma_R$ (fatigue strength) is taken from the S-N curve at the total number of service life stress cycles n, the safety factors $\gamma_f$ and $\gamma_m$ account for uncertainty in fatigue loading and fatigue quality respectively. The difference from the former codes is that the equivalent applied stress is compared to the fatigue strength at given service life, instead of comparing service life with fatigue design life at a given applied stress spectrum. The Eurocode 3 format offers the opportunity to explicitly take into account the load and material partial safety factors. These factors are in many cases set to 1.0. Exemptions are when welds are difficult to inspect and the consequence of fracture is severe.

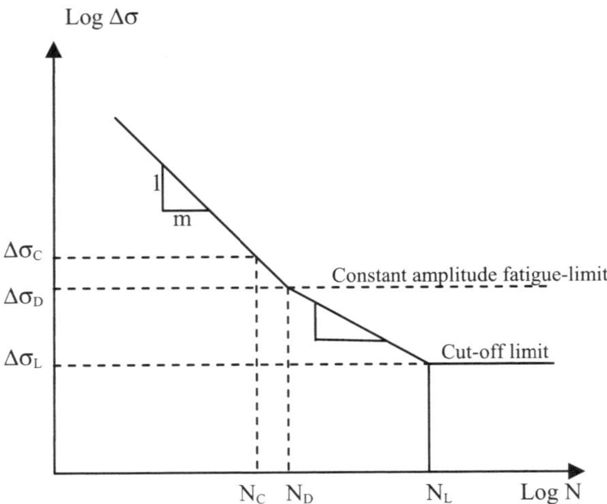


**Figure 5.23.** *S-N curve in Eurocode 3*

| | Log A | | | | |
|---|---|---|---|---|---|
| Category $\Delta\sigma_C$ (MPa) | N < 5 10$^6$ (m = 3) | N > 5 10$^6$ (m = 5) | N = 5 10$^6$ $\Delta\sigma_D$(MPa) | N = 10$^8$ $\Delta\sigma_L$(MPa) | Comparable Class DoE |
| 100 | 12.301 | 16.036 | 74 | 40 | |
| 90 | 12.151 | 15.786 | 66 | 36 | D and T |
| 80 | 12.001 | 15.536 | 59 | 32 | E |
| 71 | 11.851 | 15.286 | 52 | 29 | F |
| 50 | 11.401 | 14.536 | 37 | 20 | G |
| 45 | 11.251 | 14.286 | 33 | 18 | W |

**Table 5.4.** *Eurocode 3 S-N data for various categories in air, Ref [1]*

Spreadsheet for fatigue life predictions according to Eurocode 3 is given in Appendix B.

### 5.6.4. S-N life predictions according to HSE

It was the HSE rules, Ref [13], that first introduced the geometrical stress approach for welded plated joints. The design curves for various plated as-welded joints were normalized to one basic S-N curve, designated the P curve, that is equal to the D curve. The equation for the P curve reads:

$$\log N = \log A_2 - \frac{m}{3}\log(\frac{T}{16}) - m \ log \ \Delta\sigma_g \tag{5.29}$$

where N is number of cycles to failure and $T$ is the plate thickness (mm). The common P-curve is to be used by applying the hot-spot stress range $\Delta\sigma_g$, i.e. the SCF that is related to the overall geometry of the joint must be taken into account. We have explained the concept in detail in section 5.4. The idea is that the various original classification qualities (D, E, F, and so forth, DoE, Ref [4]) for plated joints are due to the differences in the geometric SCF only; hence they will all coincide if the SCF is explicitly taken into account. The assumption is obviously a simplification, as we have discussed thoroughly, but is both useful and acceptable from a practical point of view. The P-curve is divided in to segments where the slope (–1/m) is changing at N = 10$^7$ cycles for joints in air. Appropriate data are given in Table 5.5.

|  | $N < 10^7$ cycles | | $N > 10^7$ cycles | |
|---|---|---|---|---|
| Air | $LogA_{2u}$ | $m_u$ | $LogA_{2l}$ | $m_l$ |
|  | 12.182 | 3 | 15.637 | 5 |
| Seawater with cathodic protection | $N < 10^6$ Cycles | | $N > 10^6$ Cycles | |
|  | $LogA_{2u}$ | $m_{1u}$ | $LogA_{2l}$ | $m_l$ |
|  | 11.784 | 3 | 15.637 | 5 |

**Table 5.5.** *Parameter for the HSE P-curve, Ref [13]*

From Table 5.5 one will recognize that the $logA_2 = 12.182$ is very close to the original value for the D-class, i.e. a butt weld; see Ref [4]. The rationale for this is obviously due to the fact that this specific joint has a geometrical SCF of 1.0. When using the hot-spot stresses, the actual SCF has to be calculated for other classifications. To make the results coincide with the original classification based on the nominal stress range, the SCFs in Table 5.6 have been proposed for the various classes. In specific cases these suggested values may be replaced by results from FEA. It is always useful to keep an eye on these SCFs for comparison when assessing the results from an FEA.

| S-N curve | SCF |
|---|---|
| D | 1.0 |
| E | 1.14 |
| F | 1.34 |
| F2 | 1.52 |
| G | 1.83 |
| W | 2.54 |

**Table 5.6.** *Geometric SCF*

### 5.6.5. S-N life predictions according to NORSOK and DNV

The S-N curves presented in NORSOK and DNV were originally identical; see Refs [6, 14 (2003)]. However, in the latest version of the DNV (2005) some changes were made. The DNV and NORSOK regulations are also very close to the ABS fatigue guidance; see Ref [15]. The classification groups are designated A, B, C, D,

E, F, G, and W for welded flat plates, whereas class T is used for tubular joints. The curves are given in Table 5.7. The reader will recognize the classes found in DoE, but with more refinement in the definition of classes. For example, the former F2 class has been divided into an F1 class and an F3 class. We have limited the presentation to curves for details with effective cathodic protection exposed to seawater. These curves change slope $-1/m$ from $m = 3$ to $m = 5$ at $N = 10^6$ cycles. Under constant amplitude loading a fatigue limit is assumed to appear at $N = 10^7$ cycles. These curves predict shorter fatigue lives than the air curves at high stress ranges, but the curves coincide at $N = 10^7$ cycles; see Figure 5.5. The exponent k for the thickness correction is also given for the various classes. The reference thickness is 25 mm for plated joints and 32 mm for tubular joints. The guidance also offers the possibility to apply the geometrical stress approach with the D curve as the reference curve. The SCF that must be used for each traditional class to obtain the same fatigue life as for the geometrical stress approach is listed in the right-hand column of Table 5.7. We recognize with insignificant changes the figures given in Table 5.6.

| Category | Log $A_2$ | | | | K | SCF |
| $\Delta\sigma_C$ (MPa) | $N < 10^6$ ($m = 3$) | $N > 10^6$ ($m = 5$) | $N = 10^6$ $\Delta\sigma_D$(MPa) | $N = 10^7$ $\Delta\sigma_D$(MPa) | | |
|---|---|---|---|---|---|---|
| C | 12.192 | 16.320 | 115.9 | 73.1 | 0.15 | |
| C1 | 12.049 | 16.081 | 103.8 | 65.5 | 0.15 | |
| C2 | 11.901 | 15.835 | 92.7 | 58.5 | 0.15 | |
| D | 11.764 | 15.606 | 83.4 | 52.6 | 0.20 | 1.0 |
| E | 11.610 | 15.350 | 74.1 | 46.8 | 0.20 | 1.13 |
| F | 11.455 | 15.091 | 65.8 | 41.5 | 0.25 | 1.27 |
| F1 | 11.299 | 14.832 | 58.4 | 36.8 | 0.25 | 1.43 |
| F3 | 11.146 | 14.576 | 51.9 | 32.8 | 0.25 | 1.61 |
| G | 10.998 | 14.330 | 46.3 | 29.2 | 0.25 | 1.80 |
| W1 | 10.861 | 14.101 | 41.7 | 26.3 | 0.25 | 2.00 |
| W2 | 10.707 | 13.845 | 37.1 | 23.4 | 0.25 | 2.25 |
| W3 | 10.570 | 13.617 | 33.4 | 21.1 | 0.25 | 2.50 |
| T | 11.764 | 15.606 | 83.4 | 52.6 | 0.25 | 1.0 |

**Table 5.7.** *NORSOK S-N data for various categories under seawater with cathodic protection, Ref [14]*

Each S-N curve for welded plates is given by the equation:

$$\log N = \log A_2 - m \log \left[ \Delta\sigma \left( \frac{T}{T_{ref}} \right)^k \right] . \tag{5.30}$$

An interesting updating of the DNV (2005) rules is the increase in fatigue life obtained by various improvement techniques; see our discussion in section 5.2.2.6. A short summary is given in Table 5.8.

| Improvement method | Minimum specific yield strength | Multiplication factor for fatigue life |
|---|---|---|
| Grinding | < 350 MPa | $0.01f_y$ |
| | > 350 MPa | 3.5 |
| TIG dressing | < 350 MPa | $0.01f_y$ |
| | > 350 MPa | 3.5 |
| Hammer peening | < 350 MPa | $0.011f_y$ |
| | > 350 MPa | 4 |

**Table 5.8.** *Improvement on fatigue strength obtained by various methods*

The improvements require high-quality workmanship after given procedures; see Ref [5]. For hammer peening, some restrictions on the applied stresses are given; for details see Ref [6].

### 5.6.6. *S-N life predictions for ship structures*

Large seagoing vessels consist almost entirely of welded steel sections. During an operation period of 20 years the structure is subjected to about $10^8$ wave-induced loading cycles. A good overview of the problem is given in Ref [16]. Traditionally the FDF has been set to 1.0 for normal seagoing vessels. During the service life, docking and inspections are regularly planned, typically with a 5 year interval between docking and inspections. It is planned that ship-shaped floating production storage and offloading (FPSO) vessels for oil exploitation are to be kept in service at a given field without any scheduled docking. This makes fatigue damage a matter of concern; see Ref [17]. Details vulnerable to fatigue damage are found in the intersections between longitudinal stiffeners at transverse girders or bulkheads; see Figure 5.24. The figure shows the stiffener as a continuous member through a cutout in the girder web. The intersection is reinforced by a welded bracket connecting the webs and a knee plate on top of the stiffener. The possible cracks are indicated. This connection type has gone through detailed study for optimization to achieve a good fatigue design. The shown solution is not a typically good one due to the presence of sharp corners and edges.

The DNV classification note 30.2 "Fatigue assessment of ship structures", Ref [18], recommends the weld notch stress approach for fatigue life prediction of such details. As already explained, the main reason is the variety of detail configurations found in hull structures and the wish to focus on good workmanship (favorable weld bead geometry). One peculiarity for the ship rules is that the curves are assumed to be the same for air and under cathodic protection in seawater at high stress ranges. This may result in appreciably longer life predictions if the loading is relatively severe.

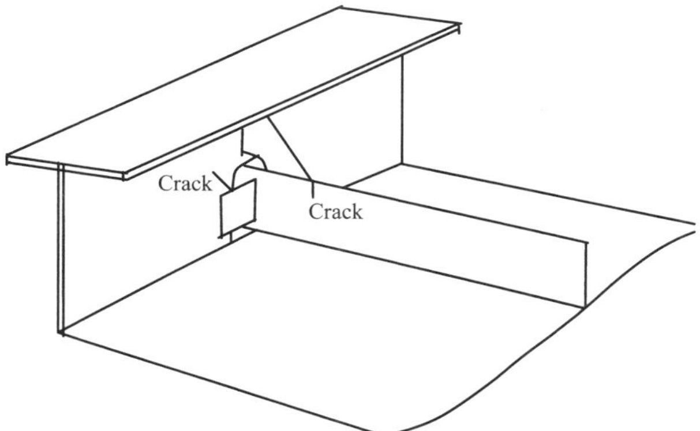


**Figure 5.24.** *Longitudinal stiffener in a bottom panel at the intersection with a transverse girder. Cracks are indicated at both the bracket plate and the knee plate*

The offshore rules modified this assumption according to research results, i.e. the fatigue life in seawater with CP is less than in air at high stress ranges. However, the ship rules give a reduction in predicted fatigue life if the protective paint layer on the structure is deteriorating. The S-N curve reads:

$$\log N = \log A_2 - \frac{m}{4}\log(\frac{T}{25}) - m\log\Delta\sigma_w.$$

(5.31)

The data for the S-N curves are given in Table 5.9. As can be seen, the curve for air and cathodic protection in seawater can be simplified by a one-slope curve. This curve is defined in a way that will give the same annual damage as the two-slope curve for a standard long-term load spectrum on a vessel.

| Environment | $N < 10^7$ | | $N < 10^7$ | | |
|---|---|---|---|---|---|
| | $\log A_2$ | m | $\log A_2$ | m | |
| Air or seawater with cathodic protection | 12.65 | 3 | 16.42 | 5 | Two-slope |
| Air or seawater with cathodic protection | 12.76 | 3 | 12.76 | 3 | One-slope |
| Corrosive environment | 12.38 | 3 | 12.38 | 3 | |

**Table 5.9.** *Notch stress based S-N curves in air
or in seawater with cathodic protection*

The user of the curve must be aware of the fact that the applied stress range will be multiplied with both $K_g$ and $K_w$ before using the curves; see equation (5.23). If the geometric stresses are determined by FEA, the results must be multiplied by a notch stress factor $K_w$ to take account for the local weld notch geometry. If a $K_w$ of 1.5 is used, the curve in Table 5.9 will be identical with the D or P curves that are based on the geometrical stress range concept. Hence, any butt weld with a more severe weld toe stress concentration will be penalized. The reader should look back to our discussion in section 5.4.3 where we used equations (5.24) and (5.25). We obtained $K_w = 1.9$ for a normal weld quality.

The BV regulation for ship structures, Ref [19], recommends almost the same S-N curves as the DNV. The curve reads:

$$\log N = \log A_2 - \frac{m}{3.33}\log(\frac{T}{16}) - m\log\Delta\sigma_w . \tag{5.32}$$

As can be seen by comparison with equation (5.31), the penalty term for greater thickness is somewhat more severe. The constant for the curve is given in Table 5.10. $K_w$ must be set to 1.65 to make the curve coincide with the D curve used for the geometrical stress approach.

| Environment | $N < 10^7$ cycles | | $N > 10^7$ cycles | |
|---|---|---|---|---|
| | $\text{Log}A_2$ | M | $\text{Log}A_2$ | m |
| Air/corrosive protection | 12.76 | 3 | 16.61 | 5 |

**Table 5.10.** *Parameters in BV S-N curve*

To facilitate the calculation scheme, the DNV note contains catalogs for common standard details, as shown in Table 5.11. These figures give the total stress concentration $K_t$ near cope holes in welded plate girders. The effect of plate distortion area is also given in the right-hand column.

| Geometry | K-factor | |
|---|---|---|
| 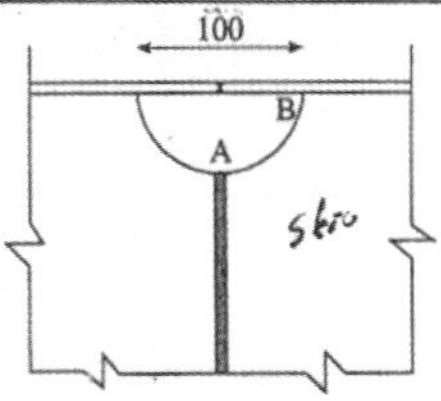 | $K_g \cdot K_w = 3.0$ at point A<br>$K_g \cdot K_w = 1.8$ at point B | $K_g \cdot K_w \cdot K_{te} = 4.5$ at point A<br>$K_g \cdot K_w \cdot K_{te} = 1.8$ at point B<br>An eccentricity of 0.15 t included |
| 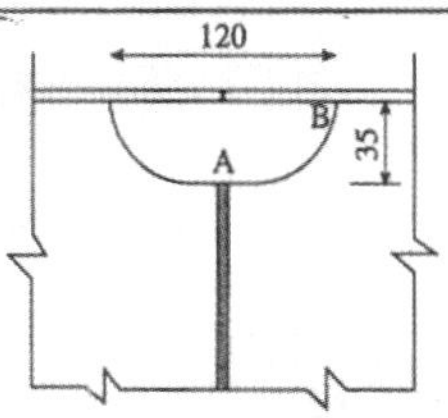 | $K_g \cdot K_w = 1.7$ at point A<br>$K_g \cdot K_w = 1.8$ at point B | $K_g \cdot K_w \cdot K_{te} = 2.6$ at point A<br>$K_g \cdot K_w \cdot K_{te} = 1.8$ at point B<br>An eccentricity of 0.15 t included |
| 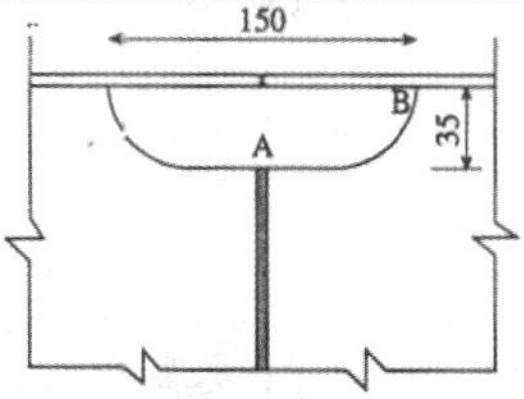 | $K_g \cdot K_w = 1.6$ at point A<br>$K_g \cdot K_w = 1.8$ at point B | $K_g \cdot K_w \cdot K_{te} = 2.4$ at point A<br>$K_g \cdot K_w \cdot K_{te} = 1.8$ at point B<br>An eccentricity of 0.15 t included |

**Table 5.11.** *SCF for common detail in ship structures, Ref [18]*

As can be seen from Table 5.11, it is advantageous to avoid half-circular cope holes as show on the top of the figures in the left-hand column of the table. The disadvantage of this configuration is that the geometrical stress concentration of the hole coincides with the notch effect of the weld, i.e. at point A. If one reduces the curvature of the cope hole at the locus of the butt weld, the geometrical stress concentration $K_g$ reduces appreciably and as a result the total $K_t$ reduces from 3.0 to 1.7. Consequently, it will be more likely that a fatigue crack will appear at spot B than at spot A. The right-hand column shows the influence on $K_g$ if there is a distortion of 15% in the welded plate of the girder. A solution not shown in the table is where the cope hole is omitted. If this solution is possible from a fabrication point of view, it should be preferred for details that will be subjected to fatigue loading. Such solutions are usually sought for offshore structures. The optimum fatigue detailed design for the intersection shown in Figure 5.24 can also be found in the DNV rules.

For the hull structure of ships, the energy-spectrum approach is often applied for the load spectrum for detailed fatigue life verification. The DNV guidance gives the following formula:

$$D = \frac{\upsilon_0 T_d}{A_2}\Gamma(1+\frac{m}{2})\sum_{n=1}^{N_{load}} p_n \overset{\substack{All\ seastates\\ All\ headings}}{\sum_{i=1,j=1}} r_{ij}(2\sqrt{2m_{0ijn}})^m \tag{5.33}$$

where:

$\upsilon_0$ = average frequency [1/s]

$T_d$ = target service life [lives]

$p_n$ = probability for load condition designated n

$r_{ij}$ = the relative number of stress cycles in short-term condition i,j

$m_{0ijn}$ = zero spectral moments (RMS) of stress response process for sea condition $i,j$ load condition $n$.

We recognize that this equation is an elaboration of equation (5.20) with $h = 2$. For a bilinear S-N curve the equation reads:

$$D = \upsilon_0 T_d \sum_{n=1}^{N_{load}} p_n \overset{\substack{All\ seastates\\ All\ headings}}{\sum_{i=1,j=1}} r_{ij} \left[ \frac{\kappa^{m_u}}{A_{2u}}\Gamma\left(1+\frac{m_u}{2};\left(\frac{S_0}{\kappa}\right)^2\right) + \frac{\kappa^{m_l}}{A_{2l}}\gamma\left(1+\frac{m_l}{2};\left(\frac{S_0}{\kappa}\right)^2\right) \right] \tag{5.34}$$

where:

$\Gamma(a,x)$ is the complementary, incomplete Gamma function

$\gamma(a,x)$ is the incomplete Gamma function

and:

$$\kappa = 2\sqrt{2m_{0ijn}} \; . \tag{5.35}$$

The DNV guidance also allows for a simplified fatigue check based on long-term distributions using equations (5.14), (5.15) and (5.16). The corresponding permissible maximum stress ranges are given in tables for various curves, target service life, and FDF; see Ref [18]. The solution for the single-slope S-N curve reads:

$$\Delta\sigma_{0\,permissible} = \left(\frac{DA_2}{v_0 T_d}\right)^{1/m} \frac{\left(\ln n_0\right)^{1/h}}{\left[\Gamma(1+m/h)\right]^{1/m}} \; . \tag{5.36}$$

$\Delta\sigma_0$ is the highest likely stress occurring during $n_0$ load cycles. If the number of cycles is set to $10^8$ cycles, the extreme stress is likely the highest stress range encountered during approximately $T_d = 20$ years of service. This extreme stress range has a probability of exceedance given by $1/n_0 = 10^{-8}$. A graph based on equation (5.36) is given in Figure 5.25 as a function of the shape parameter $h$. We must bear in mind that a low value of the shape parameter $h$ means that the extreme stress range $\Delta\sigma_0$ is followed by many stress cycles of less severity. A high value for the shape parameter h means that the extreme stress cycle is followed by many other stress cycles of relatively greater severity. This is illustrated by the exceedance diagram on the right-hand side of Figure 5.25. Hence, for h = 0.7 the permissible extreme stress will be quite high, whereas for h = 1.0 it will be much smaller. These considerations are based on the fatigue criterion alone. In the first case (low value of $h$) it is likely that it will be the yield criterion that will set a limit on the maximum permissible stress range. However, for the yield criterion one can exclude the notch effect of the weld.

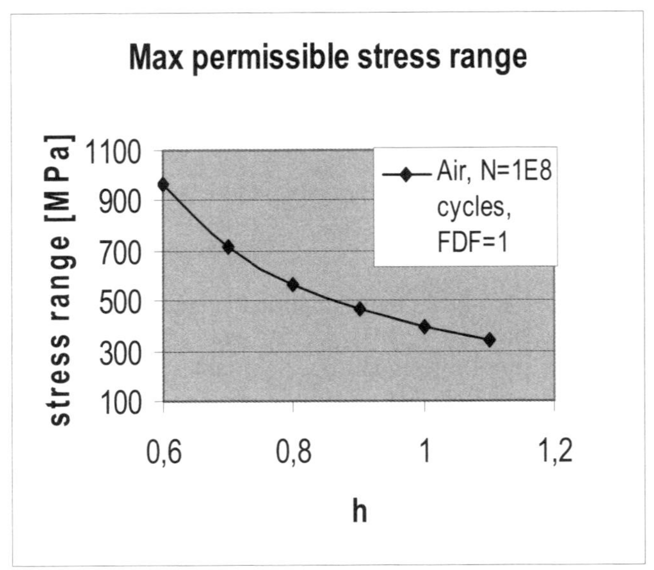


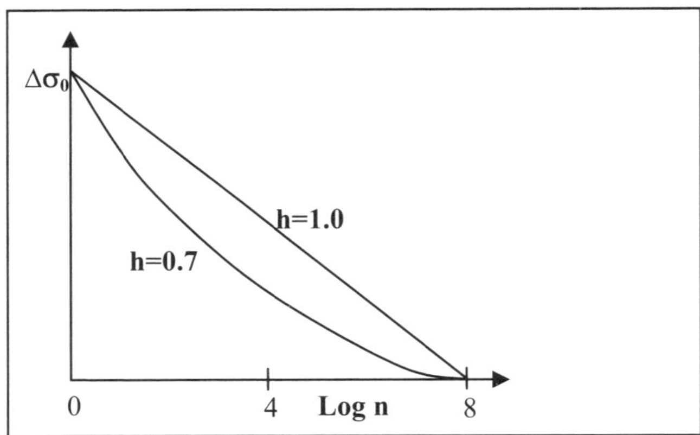


**Figure 5.25.** *Upper: maximum permissible extreme stress range for the fatigue criterion as a function of the Weibull shape parameter. Lower: exceedance diagram for two different shape parameters*

One problem that often arises is the combination stress ranges caused by two time-dependant loading components. If the ranges of the two loading components are regarded as two stochastic variables, the total stress ranges can be combined by the equation:

$$\Delta\sigma_T = \sqrt{\Delta\sigma_F^2 + \Delta\sigma_P^2 + 2\rho_{FP}\Delta\sigma_F\Delta\sigma_F} \qquad (5.37)$$

where $\Delta\sigma_F$ is the stress range due to a load component $F$, whereas $\Delta\sigma_P$ is the stress range due to a load component $P$. The parameter $\rho_{FP}$ is the correlation between the two loading components. As can be seen from equation (5.37), the stress ranges will be added directly if the correlation coefficient equals 1.0. For the hull structures of ships, the two loading components can be the vertical and horizontal bending of the hull beam.

A final comment on the weld notch stress approach will be made before finishing with the subject. The reference curve as given in Table 5.9 is a modified C-curve. The test specimens in this class are smooth, without welds, and the stresses are often constant through the thickness of the specimen. For a welded detail in a ship structure, the maximum notch stress is limited to a small volume at the weld toe only. There is an inconsistency if one relates these two conditions. The surface conditions for the actual welded detail and the reference detail (C-curve) are not the same. The weld leaves a type of fingerprints that reduces the fatigue crack initiation time, as we discussed in Chapter 2. The other inconsistency is the fact that fatigue lives are scaled solely according to maximum notch stress at the weld toe, ignoring differences in stress gradients through the plate. This is overly pessimistic and will result in too short fatigue lives for joints with abrupt weld toe geometry, i.e. with a small toe radius $\rho$; see Figure 5.15. As we have discussed, the inflicted material volume becomes smaller and smaller as $K_w$ increases due to a decreased $\rho/T$ ratio. To compensate for this, a $K_w$ of 1.5 has been suggested by DNV. This corresponds to an angle of 45 degrees and a toe radius 2 mm. However, this is an unrealistically smooth transition for most butt joints. We have seen, from measurements presented in Chapter 3, that the toe radius may often be less than 1 mm, sometimes as low as 0.1 mm. Hence, one may argue that the choice of such an unrealistically smooth transition is a result of disregarding the limited volume actually subjected to the stress rise. A more rational way to deal with this phenomenon is to use a weld notch stress field approach instead of the stress concentration approach. The stress situation at the weld toe is too involved to be characterized by one parameter only. We shall pursue this line of thought in Chapter 10.

## 5.7. The industrial case: an offshore loading buoy

FPSO units are extensively used in offshore oil exploitation. These ship-shaped structures are permanently moored for long-term service. In Figure 5.26 a concept is shown where the floater has a mating cone in the bottom that is connected to a loading buoy by a clamping mechanism. Details from the loading buoys are shown on the lower part of the figure. It consists of an outer cone, a central turret, and an inner pipe for the oil flow. The steel quality is a medium-strength C-Mn steel with yield strength close to 350 MPa. The central turret is connected to the mooring lines at the turret table as shown, whereas the lower part of the oil pipe is connected to the flange of the flexible oil riser. These connections are subjected to large dynamic forces from the mooring line and the riser respectively. Hence, for both connections the fatigue criterion is of major importance. We will, in what follows, see the analysis carried out to verify the fatigue life for the lugs and welded details in the turret table. Each lug eye will in fact be subjected to a load spectrum as was shown in Figure 2.7 in Chapter 2, i.e. a two -frequency tension loading. The load cycles can easily be identified by cycle counting. If we use an energy spectrum approach, we will get a two-peak spectrum. Once the dynamic loading has been determined, the corresponding stresses at various hot-spots can be used as the basis for the fatigue life predictions. Two actual hot-spots are indicated in the upper part Figure 5.27. The first one is the edge of the hole in the turret table lug, the second on is the nose of the horizontal bracket between the lugs. The first hot-spot is a machined one and will be classified as C, whereas the toe of the bracket is welded. As can be seen, the welded brackets are smoothly shaped in order to keep the $K_g$ low at the nose of the brackets. In this case one can either try to determine the geometrical stress and use the hot-spot approach, i.e. a D-class, or use an F-class in combination with the nominal stress range. It is assumed that the fillet weld is ground at each spot where the toe is crossing the edge of the bracket plate. The lower part of Figure 5.27 shows the element mesh used for determining the stresses. As can be seen, the element mesh is quite refined in the lugs in order to reveal the SCF at the edge of the hole. The mesh at the bracket nose is somewhat more coarse and irregular. All elements are volume elements. For a given tension load F in the lug, the stress plots are given in Figure 5.28. As can be seen, the stresses are highest on the bracket side of the weld, not on the lug side. Based on these figures, the mechanical transfer function (MTF) for the bracket nose is given by:

$$MTF = \frac{\sigma_1}{F} = \frac{191.7}{7374} = 0.026 \ MPa/kN \tag{5.38}$$

If we multiply all the load cycles by this factor we get the stress spectrum. The obtained stresses can be regarded as geometrical stresses, but we have not carried out an interpolation according to what was shown in Figure 5.17. Hence, as a first assessment, the largest stresses are treated as nominal stresses. We then use an F-class for this detail instead of a D-class as is recommended for the geometrical stresses. This will lead to a conservative estimate, and if the prediction is acceptable we do not have to do any further refinement. Details from fatigue life estimates are given in Appendix B and are based on the NORSOK S-N curves for cathodic protection. Only some of the load levels are included. The fatigue life prediction is 280 years. For a target service life of 20 years, this corresponds to an FDF of 28. As the hot-spot is impossible to inspect during service, the required FDF is set to 10. The obtained safety margin is so large that no further analysis is required. It must however be clear to the reader that the approach is a shortcut compared to the recommended procedure. It thus requires experience and judgment from the designer. We shall therefore proceed to a more formal verification for another detail. Figure 5.29 shows the results from a three-dimensional FEA sub-model for a welded corner within the structure of the turret. As can be seen, the element size is about half the plate thickness. The stress plot at the bottom indicates the critical hot-spot in the upper corner, and the arrow defines the interpolation line towards this corner. As can be seen to the right on the figure, the MTF obtained by extrapolation is close to 0.0599 MPa/kN. In this case we have carried out the interpolation procedure that permits us to use the D-class curve.

As can bee seen from Figure 5.29 the stress gradient near the hot-spot is very steep. If we had chosen a distance 0.4 T (Recommendation in Eurocode) instead of 0.5T as an interpolation point, the hot-spot stress would have increased appreciable. The example reveals the difficulties in determining the correct hot-spot stress.

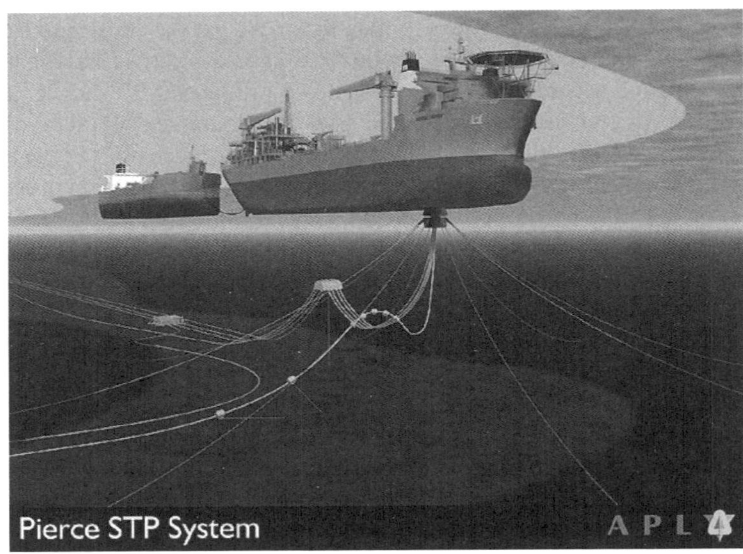


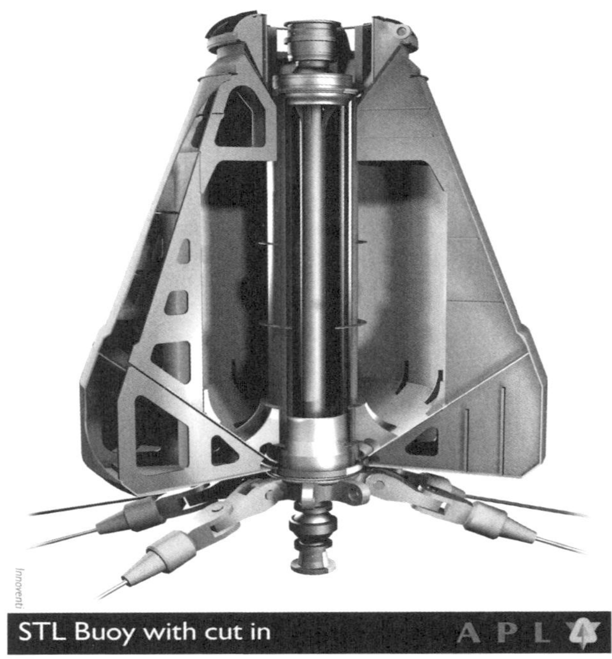


**Figure 5.26.** *Upper: FPSO connected to loading buoy. Lower: details from loading buoy with mooring line connections. Courtesy of Advanced Production and Loading AS, Arendal, Norway*

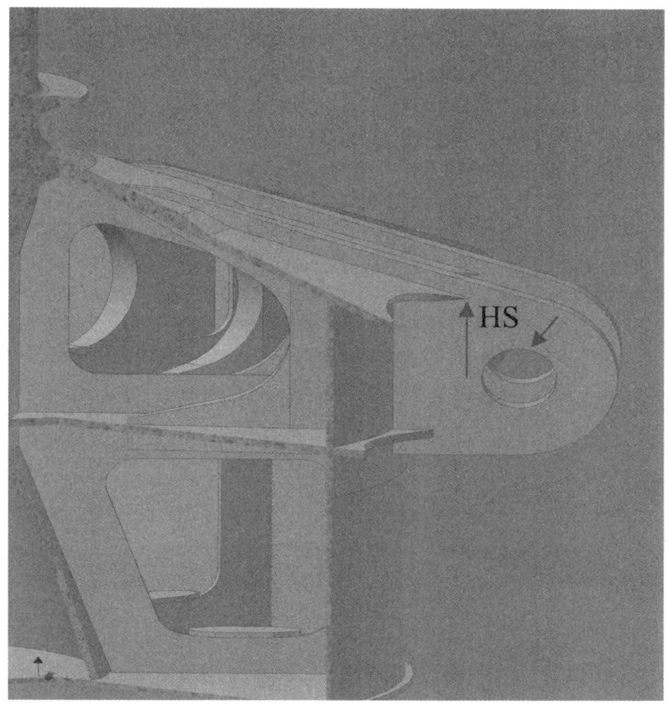


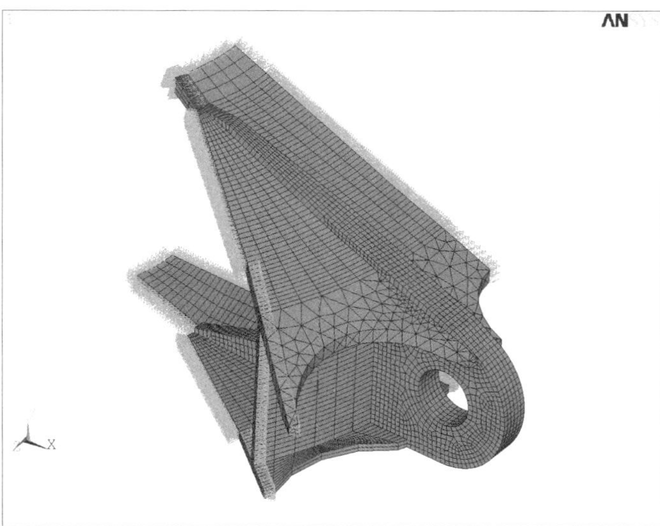


**Figure 5.27.** *Upper: lug in central turret for mooring line connection in an offshore loading buoy. Hot-spot at nose of welded plate. Lower: finite element model to determine geometrical stresses at hot-spots. Courtesy of Advanced Production and Loading AS, Arendal, Norway*

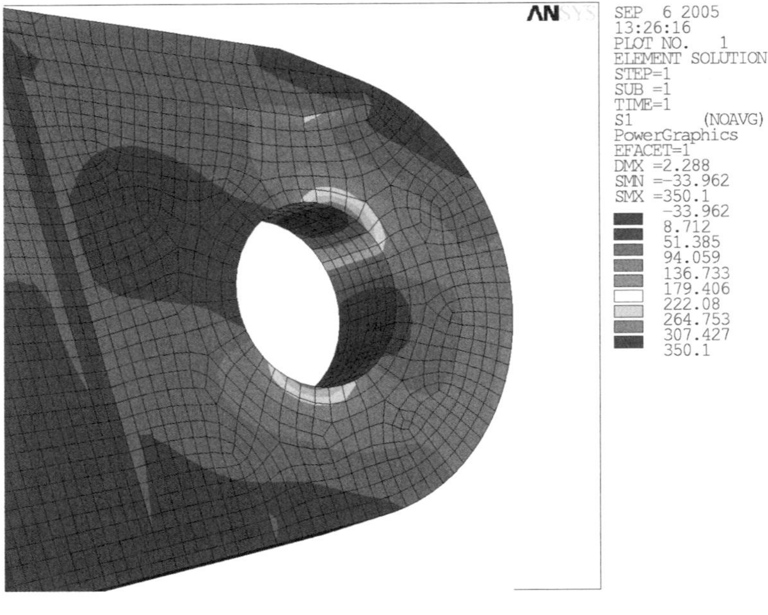

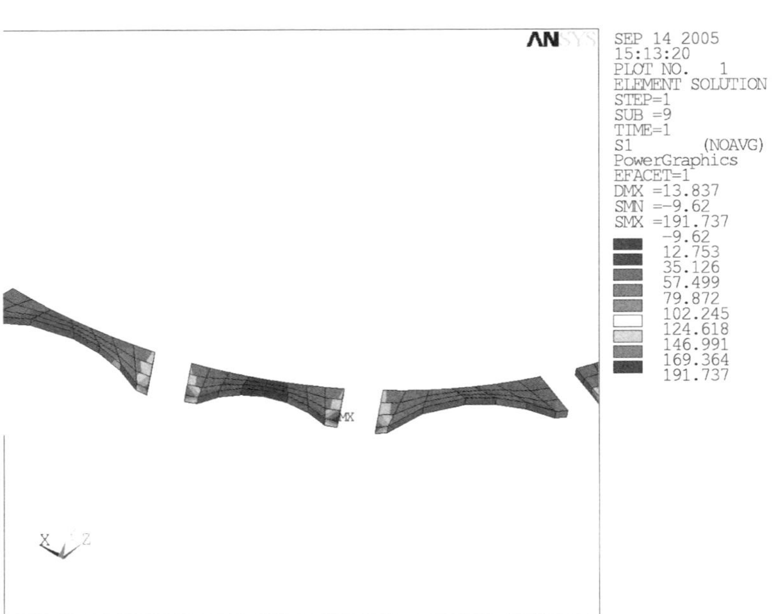

**Figure 5.28.** *Result for stress analysis tension 7374 KN. Upper: lug. Lower: bracket plate*

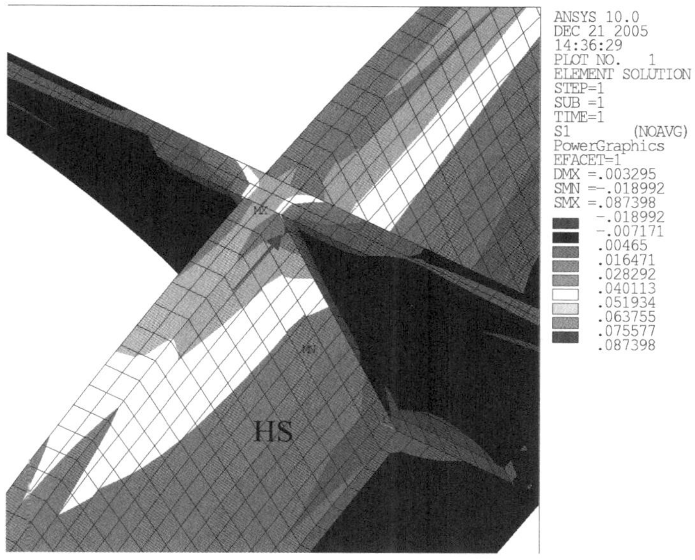


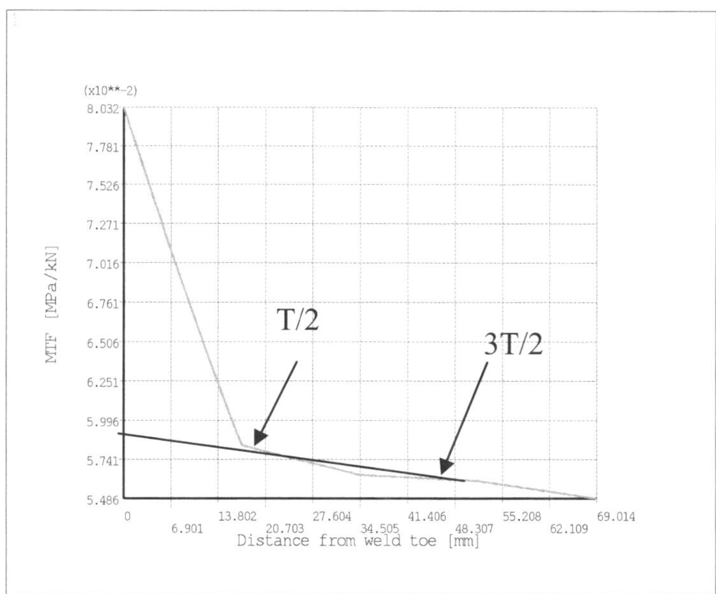


**Figure 5.29.** *Refined sub-model of a welded intersection between two plates within the turret. Upper: hot-spot and interpolation line. Lower: interpolated stress values. Courtesy of APL*

## 5.8. References


1    ENV: EUROCODE 3- Steel Structure-Fatigue

2    A. Almar Næss, *Fatigue Handbook*, Trondheim, Tapir, 1985

3    S.J. Maddox, *Fatigue Strength of Welded Structures*, Abington Publishing, 1991

4    *Offshore Installations: Guidance on design construction and certification*, 4th edn, London, HMSO, 1990, Chapter 21, Steel

5    P.J. Haagensen and S.J. Maddox, "Specification for weld toe improvements by burr grinding, TIG dressing and hammer penning", IIW-XIII-WG2-1995

6    DNV: Fatigue Strength Analysis of Offshore Steel Structures, Recommended practice RP-C203, 2005

7    A. Bignonnet, *Corrosion Fatigue of Steel in Marine Structures, Proceeding from Steel in Marine Structures*, Delft, Elsevier, 1987, pp 119–37 (ISBN 0-444-42805-4)

8    P.S. Pao, "Mechanism of corrosion fatigue" in *ASME Handbook Fatigue and Fracture*, ASM International, 1996, pp 185–92 (ISBN 0-87170-385-8)

9    E. Haibach, Modifizierte lineare Schadeakkumulations-Hypothese, (technical document), Technische Mitteilung des LBF, No TM 50/79, 1970

10    C. Berger *et al*, "Betreibfestigkeit in Germany – An Overview", *International Journal of Fatigue* 2002 (24), pp 603, 627

11    X. Niue and G Glinka, "The weld profile effect on the SIF in weldments" *International Journal of Fracture* 1987 (35), pp 3–20

12    F.V. Lawrence Ho, "Fatigue Test Results and Predictions for Cruciform and Lap Welds" in *Theoretical and Applied Fracture Mechanics* 1984 (1)

13    A. Stacy and J.V. Sharp, *The Revised HSE Fatigue Guidance*, OMAE, 1995, pp 1–16

14    NORSOK Design of Steel Structures, N-004, Annex C, NORSOK standard 1998

15    *ABS: Guide for Fatigue Assessment of Offshore Structures,* ABS 2003

16    I. Lotsberg *et al*, "Fatigue assessments of Floating Production Vessels", Conference proceedings, BOSS'97, Delft University of Technology, July 1997

17    E. Ayala-Uraga and T. Moan, "Reliability-based assessment of welded joints using alternative fatigue failure functions. ICOSSAR 2005 pp 1071-1079, Millpress, Rotterdam, ISBN 90 5966 0404

18  DNV, Fatigue Assessment of Ship Structure Classification Note 30.7, Det Norske Veritas 2003

19  BV, Fatigue Strength of Welded Ship Structures, Bureau Veritas, July 1998

# Chapter 6

# Applied Fracture Mechanics

## 6.1. Introduction

Fatigue life predictions have been carried out using the S-N approach for over a century and long before the physics of the fatigue process was properly understood. As we have seen, the model is based on simplified assumptions and statistical analysis of the entire fatigue life. The method does not try to analyze the fatigue process itself. Its only goal is to estimate the time to failure at a given probability level. One may in fact use the S-N curves without being aware of the fact that it is a crack growing to a critical size that is the cause of the fatigue failure. In the fracture mechanics approach the entire crack growth history is modeled, not only from the final fatigue life, but also from a small, initial crack to the final, critical crack leading to fracture. The formulation is based on physics, not just life statistics as is the case with the S-N model. After a crack has initiated, the standard S-N curves are no longer applicable, whereas a fracture mechanics model can describe the probable crack behavior and propagation towards final failure. Because the S-N model cannot deal with the presence of cracks, the fracture mechanics model is an indispensable tool in situations when a crack is detected and sized. Furthermore, a fracture mechanics model gives valuable support when planning scheduled in-service inspection. In such cases we will have to know the possible damage evolution in advance in order to know what crack sizes to look for and at various time stages during the service life. The principal difference between the S-N model and the fracture mechanics model is shown in Figure 6.1. To the left is shown the conventional S-N plot at a given stress range level; the crack depth $a$ as a function of number of cycles for one of the tests is given to the right.

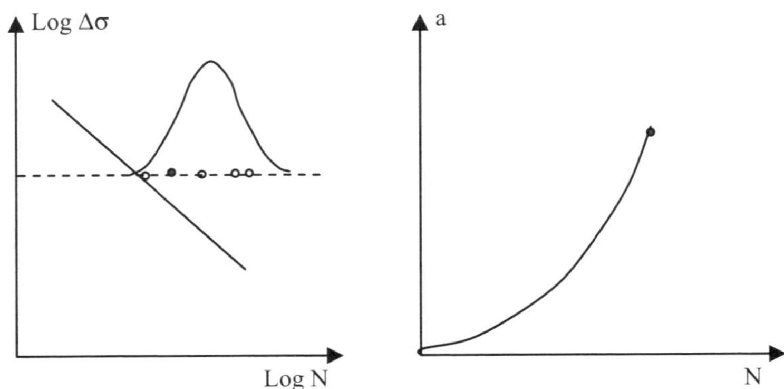


**Figure 6.1.** *Left: S-N curves with several tests. Right: the crack depth history of one test*

The basis for the fracture mechanics approach is that it considers the stress field and not just the stress concentration at the weld notch. Furthermore, it describes the synergism between the weld notch effect ant the presence of a sharp fatigue crack. This synergism is reflected in the concept of the stress intensity factor (SIF). This factor determines the local stress field in front of the crack and is the key parameter along with the material characteristics for predicting the crack behavior.

From a textbook point of view one may add that a fracture mechanics model can be used as a pedagogical aid to help the reader understand the fatigue process, whereas the S-N model is of lesser help in this respect. The fracture mechanics approach provides insight into the interplay of parameters that primarily influence the fatigue process and final life. For a long time there has been debate about the usefulness of the fracture mechanics as a tool when compared to the S-N approach. Generally, the practicing designer prefers the relatively simple and rough estimate that the S-N curves provide, whereas academics have advocated the use of fracture mechanics. This difference in opinion may be explained by the fact that academics are fond of cracks and the theory that is associated with them. The practicing engineer is not fond of cracks or the theory that predicts their behavior. The conclusion of the authors of this book is that the S-N model and the fracture mechanics approach should be used as complementary tools when analyzing fatigue of welded structures. Fracture mechanics has obviously matured to the point where its application is both reliable and useful. As a consequence it has gained acceptance in new rules and regulations as an alternative to the S-N approach. The British standard BS7910 is entirely devoted to the application of fracture mechanics for fusion-welded joints in order to describe the behavior of cracks. Therefore, the reader should certainly have a general knowledge of fracture mechanics issues.

One may argue that as long as the S-N-estimated fatigue life exceeds the target service time with a required safety margin, one does not have to worry about the fatigue process. There are, however, some very important cases were knowledge of the fatigue crack behavior in necessary:

– where S-N curves for the detail and loading mode in question do not exist;

– where a crack has been detected after fabrication or in-service;

– where preventative scheduled inspections are to be planned.

In the first case the best strategy is, of course, to carry out life testing to establish an S-N curve. However, because this is costly and time consuming, one may try to estimate the life by trying to predict the growth of an initial crack that likely to appear in the detail to a final, critical crack. The actual geometry of the joint must be properly modeled and the material has to be characterized. A crucial issue is that one must assume the size of an initial crack which usually is small, not measurable and is random by nature. In addition to these uncertainties it is also a question of whether the fracture mechanics approach is applicable for small cracks. The second case listed above is more straightforward. If a crack has been detected and sized, one can verify whether or not it will grow to a critical size during the target service life. If it will reach a critical crack size, repair should be carried out before the structure enters into service. This so-called engineering critical assessment (ECA) may support decisions that can help to avoid costly repair. If one uses current quality standards as the basis for decision-making, they are usually overly conservative with respect to allowable defects. It is for these cases that the linear elastic fracture mechanics (LEFM) was originally developed and outlined as in BS7910. When it comes to inspection planning it is obvious that when developing a scheduled fatigue-inspection program at the design stage, one has to know the most likely crack geometry and size as time elapses during service life. This will be of great help when selecting the inspection method and the time interval. We will treat this issue in more detail in the following sections. Figure 6.2 may be used as the basic geometry for our problem formulation. This is the fillet weld that we were discussing in Chapter 5, but now we have shown the presence of a semi-elliptical crack emanating from the weld toe. Before addressing this and similar problems we will briefly outline the basic concepts of LEFM.

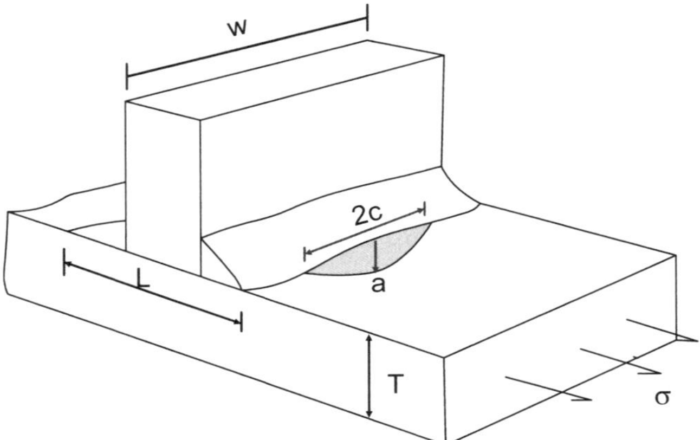


**Figure 6.2.** *Fillet weld with a semi-elliptical crack at the weld toe*

## 6.2. Objectives of this chapter

In this chapter the reader will learn the basic concepts of fracture mechanics. The theoretical part is not extensive, but the reader will be able to calculate the SIF for cracks that emanate from the weld toe and use this parameter for assessing both the danger of unstable fracture and stable crack propagation. The stable crack propagation will be based on Paris's law and it will be shown how this equation can be calibrated to describe the behavior of fatigue cracks in welds. The theory is limited and confined to LEFM. For a more thorough presentation the reader is referred to the summary given in Appendix A and Refs [1, 2 and 3].

## 6.3. Basic concepts of linear elastic fracture mechanics

### 6.3.1. *The local stress field ahead of the crack front*

Before applying the LEFM theory to details with weld notches, as shown in Figure 6.2, we shall outline the general theory for simple plane components containing sharp fatigue cracks. Figure 6.3 shows an infinite plate with a central crack subjected to in-plane loading. The crack is slitting the plate through the thickness and has a rectangular shape. The loading mode is denoted as stress mode I and is characterized by stresses acting normal to the crack plane so that the crack surfaces are moving directly apart. Other possible loading modes are shown in Figure 6.4 and are referred to as stress mode II-shear mode and stress mode III-tearing mode (out-of-plane shear). Mode I is by far the most important for welded joints and we shall confine our analysis to this mode in what follows. Furthermore, we shall assume linear elastic behavior of the material. We shall have a closer look at the mixed stress mode in Chapter 11.

The local stress field at the crack front can be found by using the Airy stress function with complex harmonic functions. According to Westergaard, Ref [4], the solution for the case in Figure 6.3 reads:

$$\sigma_x = \frac{K_I}{\sqrt{2\pi r}}\cos\frac{\theta}{2}(1-\sin\frac{\theta}{2}\sin\frac{3\theta}{2})$$

$$\sigma_y = \frac{K_I}{\sqrt{2\pi r}}\cos\frac{\theta}{2}(1+\sin\frac{\theta}{2}\sin\frac{3\theta}{2}) \tag{6.1}$$

$$\tau_{xy} = \frac{K_I}{\sqrt{2\pi r}}\cos\frac{\theta}{2}\sin\frac{\theta}{2}\cos\frac{3\theta}{2})$$

where $r$ and $\theta$ are cylinder coordinates with origin at the crack tip. $K_I$ is the referred to as the SIF given by:

$$K_1 = \sigma_0\sqrt{\pi a} \tag{6.2}$$

where $\sigma_0$ is the nominal uniform stress field as it appears in the plate without the crack. Half the crack length is designated a for a through-thickness crack; see Figure 6.3. As can be seen from equation (6.1), the stress situation is two-dimensional. Ahead of the crack front for $\theta = 0$ (r-axis coincides with x-axis) we will have the normal stress in x and y direction that are equal in magnitude. Furthermore, the theoretical values will approach infinity as $r$ approaches 0. If we have a plane-strain condition at the crack front we will have a third component which reads:

$$\sigma_z = \upsilon(\sigma_x + \sigma_y) \tag{6.3}$$

where $\upsilon$ is the Poisson ratio. Plane-strain condition will appear at the crack front in the mid-area of thick plates. The associated three-axial stress condition can sustain very high stresses before the yield sets in. Hence, we have a three-dimensional local severe stress field ahead of the crack with a singularity at the crack tip.

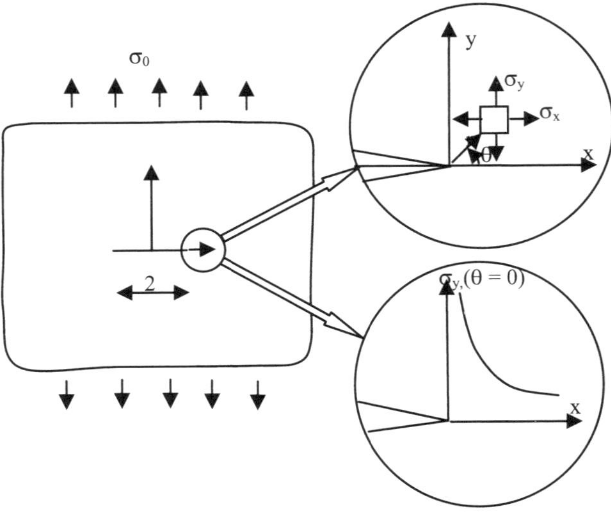

**Figure 6.3.** *An infinite plate with a central through-thickness rectangular crack subjected to loading mode I*

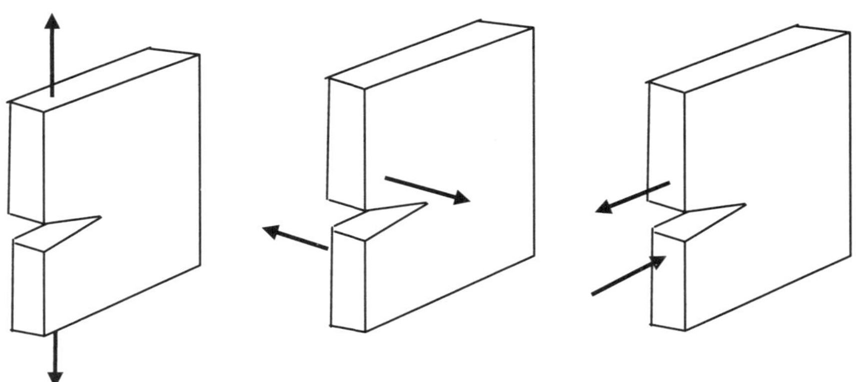

**Figure 6.4.** *Various loading modes. Mode I: normal to crack plane loading. Mode II: shear loading. Mode III: tearing loading*

As can be seen from equation (6.1), the SIF alone determines the magnitude of the local stress field ahead of the crack front. The concept of the SIF leads to both a fracture criterion under an extreme load case and a crack-propagation model under repetitive loading. It is therefore the controlling factor for assessing the criticality and behavior of a crack. As can be seen from equation (6.2), it is proportional to the applied uniform stress $\sigma_0$ and the square root of the crack size. This means that if the applied stress is doubled, the local stress field ahead of the crack front is doubled as well; likewise, if the square root of the crack size is doubled. The SIF must not be confused with the stress concentration factor (SCF). The latter is the stress magnification at a given spot only, whereas the SIF controls the entire stress field ahead of the crack front. It is to be noted that the dimension for the SIF is MPa, $m^{0.5}$ in SI units. The SIF concept is of a general nature and may be extended to a variety of geometries and loading modes, including semi-elliptical surface cracks that propagate from the weld toe. For items and crack geometries other than the one shown in Figure 6.3 the SIF will read:

$$K_I = \sigma_0 \sqrt{\pi a} F(a) \tag{6.4}$$

where $F(a)$ is a geometry function taking into account the geometrical deviations from the central through-thickness crack in an infinite plate. The through-thickness crack case (see Figure 6.3) can be treated as a reference case with $F(a) = 1.0$. $F(a)$ is dependent on both the geometry of the crack and the geometry of the component, and its various deviations from the reference case can be as follows:

– finite dimensions of the plate (the material ahead of the crack front is reduced);

– surface cracks (the crack has one front in the material and a "mouth" out to the surfaces);

– elliptical cracks (curved crack fronts);

– cracks in the vicinity of a notch (the applied stress $\sigma_0$ has an SCF and a gradient).

These deviations are treated by introducing correction factors that are multiplied to obtain the $F(a)$. If these correction factors are less than 1.0, it means that the local stress field at the crack front is less severe than the field occurring for the reference case. If they are larger than 1.0 the stress field is more severe. Some cases are shown in Figure 6.5. The plates now have finite width w and finite thickness $T$. To the left on the figure is shown a central through-thickness crack and two symmetrical edge cracks in the plate. Both types of cracks still have straight crack fronts, as does the reference crack in Figure 6.3. These cases will have an increase in $F(a)$ compared to the reference crack. For the central crack in Figure 6.5, this is due to the fact that the ligaments ahead of the fronts of the crack are reduced. This will increase $F(a)$ and the following correction term for $F(a)$ is given:

$$f_w = \frac{1}{\sqrt{\cos\dfrac{\pi a}{w}}} \qquad (6.5)$$

In the case of the symmetrical edge cracks to the left in Figure 6.5, the ligament on one side has disappeared compared with the reference case. This will increase the geometry function by a multiplication factor of:

$$M_1 = 1.1215 . \qquad (6.6)$$

The functions given in equations (6.5) and (6.6) are correction factors due to the finite width of the plate and the presence of a crack mouth a the surface respectively. If the edge crack is located with its crack mouth at a fillet as shown to the right in Figure 6.5, the material in the crack area will be subjected to a stress concentration. The SCF caused by the fillet will decrease as the distance from the surface increases. This notch stress field will of course have an impact on the stress field ahead of the crack front. For very small cracks an approximation for this effect will read:

$$F(a) \approx M_1 \cdot \text{SCF} . \qquad (6.7)$$

As can be seen, we have compounded the effect of a free surface ($M_1$) and the effect of the fillet notch (SCF). The expression takes into account the maximum SCF, but ignores the stress gradient. As we have seen for welded joints, the SCF can typically have a value between 2 and 5. Hence, this correction factor has a very strong influence on $F(a)$ compared to the other correction factors. As we shall see below, for larger cracks this gradient must be taken into account and the geometry function will decrease below the SCF value with increasing crack depth. In equation (6.6) the crack depth is assumed small compared to the width of the plate.

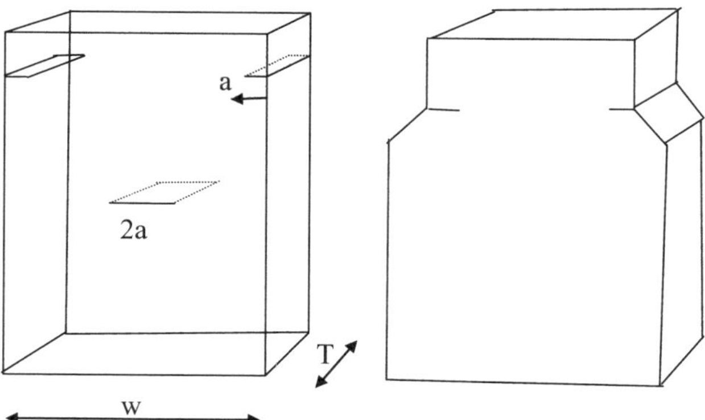

**Figure 6.5.** *Left: central through-thickness crack and symmetrical edge cracks with straight fronts. Right: symmetrical edge cracks at a fillet*

So far we have just dealt with straight crack fronts. Curved crack fronts are shown in Figure 6.6. To the left is shown an embedded elliptical crack with long axis 2c and short axis 2a. If we assume that the crack dimensions are small compared with the plate dimensions, the only correction compared to the reference case will be due to the curved crack front. The correction factor reads:

$$f_\varphi = \frac{1}{\Phi}\left[\sin^2\varphi + \left(\frac{a}{c}\right)^2 \cos^2\varphi\right]^{1/4}. \tag{6.8}$$

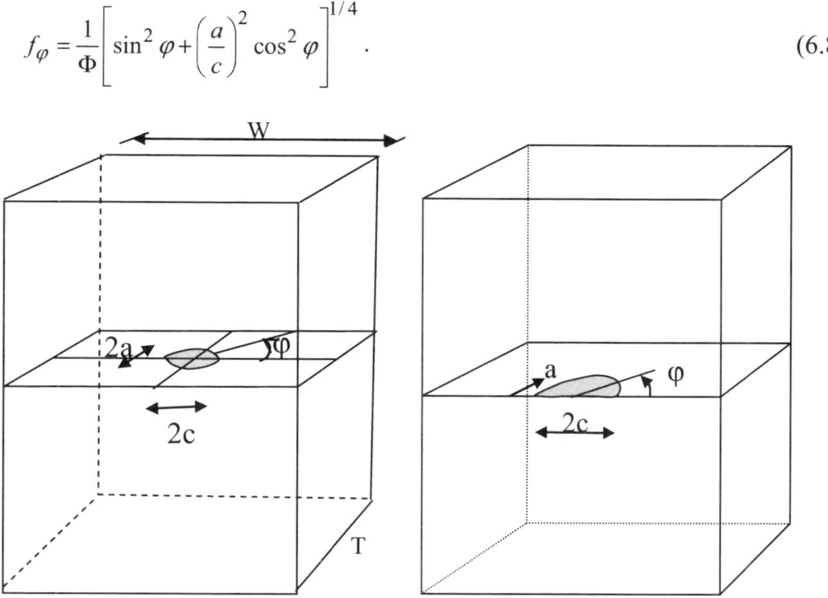

**Figure 6.6.** *Embedded elliptical and surface semi-elliptical cracks in a plate*

In this case, $F(a) = f_\varphi$ will vary along the crack front and the angle $\varphi$ gives the actual position for where the geometry function is calculated; see Figure 6.6. The end points of the short axis and the end points of the long axis are the most interesting points. The function $\Phi$ can be found by a complete elliptical integral which in turn can be approximated by:

$$\Phi = \left[1 + 1.464(a/c)^{1.65}\right]^{1/2}.$$

(6.9)

For circular cracks the expression in equation (6.8) reduces to:

$$f_\phi = \frac{2}{\pi} = 0.64.$$

(6.10)

In this case the correction factor is constant along the circular crack front. As can be seen from equation (6.10), an embedded circular crack has a smaller geometry factor compared to a through-thickness crack with a straight front when the crack size is the same. To the right in Figure 6.6 is shown a surface crack with a semi-elliptical shape. Again, the $F(a)$ will increase compared to the embedded crack because the crack now has a mouth at the plate surface at one side and also because the crack front is approaching the other side of the plate. Furthermore, we have to make a correction for the curved crack front. This problem is solved by Newman and Raju, Ref [5]. The equation for uniform tensile loading reads as follows:

$$F(a/c, a/T, c/w, \varphi) = \left[M_1 + M_2(a/T)^2 + M_3(a/T)^3 \cdot\right] \frac{g \cdot f_\varphi \cdot f_w}{\Phi} .$$

(6.11)

The sub-functions read:

$$M_1 = 1.13 - 0.09(a/c)$$

$$M_2 = -0.54 + 0.89/(0.2 - a/c)$$

$$M_3 = 0.5 - 1/(0.65 + a/c) + 14(1 - a/c)^{24}$$

$$g = 1 + \left[0.1 + 0.35(a/T)^2\right](1 - \sin\varphi)^2$$

(6.12)

$$f_\varphi = \left[(a/c)^2 \cos^2\varphi + \sin^2\varphi\right]^{1/4}$$

$$f_w = \left[\sec(\pi c/w) \cdot (a/T)^{1/2}\right]^{1/2}$$

As can be seen from the formulae in equation (6.12), all the correction factors are dependent on the ratios $a/c$, $a/T$ or $c/w$. They are correction factors for the curved shape of the crack, and the finite thickness and finite width of the plate. Hence, the function includes all the types of corrections except the possible stress gradient factor. The expressions are valid for shallow flaws with $a/c < 1.0$. A typical solution for various locations along the crack front is given in Figure 6.7.

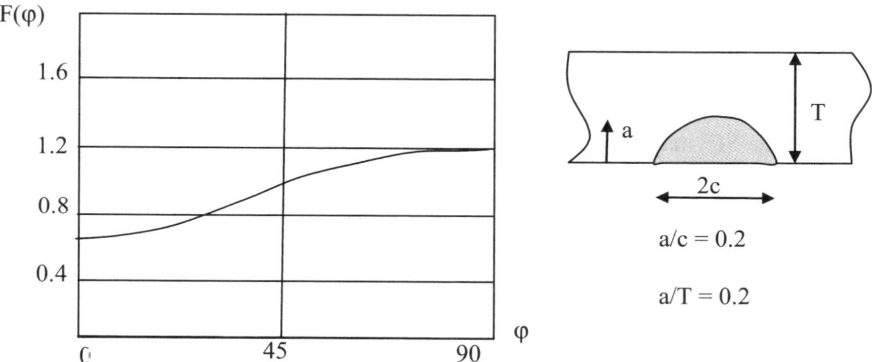


**Figure 6.7.** *Geometry function along the crack front for a semi-elliptical crack*

The values at the point at the surface ($\varphi = 0$) and at the deepest point ($\varphi = \pi/2$) are particularly interesting. At the deepest point of the crack front we will have:

$$g = 1.0$$
$$f_\varphi = 1.0$$

$$(6.13)$$

and at the ends of the crack at the surface:

$$g = 1.1 + 0.35(a/T)^2$$
$$f_\varphi = (a/c)^{0.5}$$

$$(6.14)$$

The semi-elliptical surface crack in finite plates is particularly interesting to us because it has the same configuration as the fatigue surface crack that was shown in Figure 6.2. The only correction that has to be added is the important effect of the weld notch. In this case the expression in equation (6.11) has to be multiplied by a stress gradient factor $M_k$. For very small cracks $M_k$ will in fact coincide with the SCF of the weld notch as was shown for the case in Figure 6.5 and in equation (6.7). For fatigue crack growth calculations, we need to have a more refined solution for larger cracks. Several investigations have been done to determine $M_k$. The factor can be defined as the ratio between the $F(a)$ determined with notch, and $F(a)$ determined without a notch for a given crack:

$$M_k(a) = \frac{F(a)_{with\ notch}}{F(a)_{without\ notch}}.$$

(6.15)

This ratio can be determined for a straight surface crack ($a/c = 0$) using a two-dimensional FEM analysis or by a three-dimensional analysis where the curved crack shape is modeled together with the footprint of the welded attachment. The 2-D analysis is a good approximation, but the 3-D analysis is considered to be more accurate. The SIF may be obtained by a FEM analysis of the joint, either directly by including the crack geometry in the FEM model or indirectly by an analysis of the body that does not have a crack. In the direct approach, the SIF is obtained from the stress distribution in the vicinity of the crack tip or the singular displacement field at the crack tip. The indirect method must be used in conjunction with the weight function method (WFM). Due to the fact that the WFM needs only one FEM analysis for the body without cracks, it requires much less computational effort than the direct FEM analysis which must be carried out where there are more that one cracks. Although less accurate, the WFM is therefore attractive and competitive. The method is based on the fact that there exists a function $w(x, a)$ from which the SIF can be obtained by integration (see Ref [6]):

$$K_I = \int_0^a \sigma(x)w(a,x)dx$$

(6.16)

where $\sigma(x)$ is the stress distribution over the crack depth center line for a body without cracks. The weight function $w(x, a)$ is a unique property for a given body geometry. Equation (6.16) reflects the fact that when a crack slits over a highly-stressed area ($\sigma(x)$ is high), the material ahead of the crack front will act as an alternative stress path, i.e. it will be highly-stressed locally. The results from a 3-D analysis (see Refs [7, 8]) are shown in Figure 6.8. $F(a)$ is shown for a plane plate and a T-butt joint. $M_k(a)$ can be found from equation (6.15). The crack has an aspect ratio of $a/c = 0.2$ and the solution pertains to the deepest point along the crack front ($\varphi = 90$ degrees). The T-butt joint has a weld toe shape characterized by an angle of 45 degrees and a very sharp toe radius. As can be seen from Figure 6.8, the geometry function for a shallow, surface crack ($a/T < 0.2$) in a plane plate is close to 1.15. This value is obtained from equation (6.11) and is the same as was shown in Figure 6.7. The geometry function for the same crack at a weld toe in a T-butt joint is as high as 3 for very small cracks ($a/T \cong 0$) and decreases rapidly towards 1.0 as $a/T$ approaches 0.2. Hence, the $M_k(a)$ function in the T-butt joint is close to the SCF at the weld toe for very small cracks, but drops rapidly with the crack depth. For deeper cracks ($a/T > 0.2$) it can be seen that cracks in a plane plate have higher values for the geometry function than in the T-butt joint. Hence, the welded attachment has two different effects on the geometry function, depending on the crack depth. For very small cracks there is a significant increase in $F(a)$ compared to the plane-plate solution, whereas for larger cracks the geometry function will be somewhat lower than for plane plates. The first effect is explained by the severe stress concentration of the attachment and the weld bead, whereas the second effect is explained by the fact that the attachment gives some additional ligament (alterative load path) for larger cracks that have extended the crack front out of the stress concentration area. The stress concentration effect from the weld notch is by far the most important, whereas the additional load path effect is benign and is usually ignored. Hence, although the alternative load path effect may give an $M_k$ slightly less than 1.0, the minimum value for $M_k$ is set to 1.0.

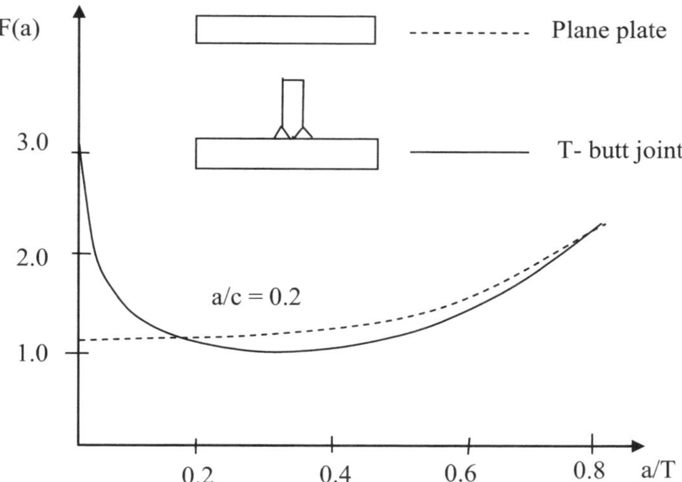


**Figure 6.8.** *Geometry functions for semi-elliptical surface cracks in plates and in T-butt joints with L/T = 1.25, θ = 45 degrees subjected to membrane loading. Reproduced from Refs [7, 8]*

For as-welded joints under membrane loading Ref [7] suggests the following equation:

$$M_k = f_1(\frac{a}{T},\frac{a}{c}) + f_2(\frac{a}{T},\theta) + f_3(\frac{a}{T},\theta,\frac{L}{T}). \tag{6.17}$$

For practical solutions, the following equations are given (see Ref [9]):

$$M_k = v\left( a/T \right)^w \tag{6.18}$$

and for $L/T < 2.0$ and $a/T < 0.05(L/T)^{0.55}$ we get the following:

$$v = 0.51(L/T)^{0.27}$$
$$w = -0.31 \tag{6.19}$$

The results are obtained from 2-D analysis. Based on 3-D analysis, for $\theta = 45$ degrees we will get (see Ref [9]):

$$M_k = f_1(a/T, a/c) + f_2(a/T) + f_3(a/T, L/T) \tag{6.20}$$

which is a simplification of equation (6.17). Details for calculating the sub-function $f_i(\cdot)$ are found in Ref [7]. The equations for the bending loading mode are also found in Ref [7]. We will return to the application of these equations, based on BS7919 recommendations, in section 6.6.


## 6.4. Fracture criterion due to extreme load

When a cracked item is subjected to an extreme load, an unstable, brittle fracture may be inflicted due to the presence of the crack. The material ahead of the crack front is separated by cleavage because the local stress field in this region has reached the limit of what can be sustained by the actual material. As we have seen, this local severe stress field that causes the onset of fracture is determined by the SIF, in this case designated $K_{Imax}$. This upper limit of the SIF is designated $K_{IC}$ and is referred to as the fracture toughness of the material. Hence, the brittle fracture criterion can be written as follows:

$$\sigma_{0\max} \sqrt{\pi a} F(a) = K_{IC} . \tag{6.21}$$

From this simple equation one can calculate the permissible crack length for a given maximal stress when the fracture toughness is known. This gives a tool for damage tolerance assessments. As the simple equation is based on LEFM, it has to be modified for practical materials such as steel because of local yielding at the crack tip. However, for central through-thickness cracks in thick plates, $K_{IC}$ can be regarded as a material constant. This is because the stress situation at the crack front for this case is associated with a plane stain condition. This gives a three-axial stress situation as was shown in equation (6.3). This stress condition will, to a large extent, be a hindrance to the yielding mechanism and will make an engineering practical material behave almost according to linear elastic theory at the crack tip. The plane strain condition is caused by restraints from the large material volume surrounding the crack. The same condition will also appear in a case with a semi-elliptical surface crack at a weld toe such as the one we showed in Figure 6.2. In this case the primary stresses are found in a plane normal to the plane of the crack surfaces. Plane strain condition will obviously be achieved by restraints from the large material volumes on each side of the elliptical crack front. Hence, yielding is to a large extent avoided. Based on these considerations this book will not address the subject of elastic-plastic fracture mechanics. We will confine ourselves to present mixed-mode failure based on the so-called R6 criterion.

### 6.4.1. *Mixed mode rupture*

There are two failure modes that can lead to unstable rupture:

– global criterion: net section yielding;

– local criterion: brittle fracture starting from a crack tip.

The first criterion is checked in a straightforward manner by calculating the nominal section stresses of the uncracked area and comparing it with the yield stress (or ultimate stress). The failure will occur when:

$$\sigma_N = \frac{F}{A} \geq f_y \tag{6.22}$$

The failure mode in this case is global overload of the entire section. Even if this criterion is not critical it may happen that the material in the vicinity of a crack front is stressed locally to a level that the material can not sustain. The crack will then develop rapidly and result in a more or less brittle rupture. The two criteria below may be used to verify the resistance. Fracture will occur when:

$$K_{I\,\max} \geq K_{IC} \tag{6.23}$$

$$CTOD_{\max} \geq CTOD_C$$

The first equation is based on the stress intensity factor that characterizes the stress field in the vicinity of the crack tip, whereas the latter equation is based on the crack tip opening at the crack tip characterizing the strain in the vicinity of the crack tip. The first criterion is used for pure, brittle fractures as we just have discussed, whereas the latter may be used for mixed-mode local fractures. A more thorough discussion is given in Chapter 11. In the present chapter we will use the global yield criterion together with the local stress intensity approach. This will lead to the so-called R6 criterion often used in rules and regulations.

### 6.4.2. *The R6 criterion and critical crack size*

To illustrate the R6 criterion we will take a central crack in a plate as an example; see Figure 6.9 and Figure 6.10. In Figure 6.9 we recognize the check against global net section yielding. In Figure 6.10 we recognize the check against local brittle fracture at the crack front.

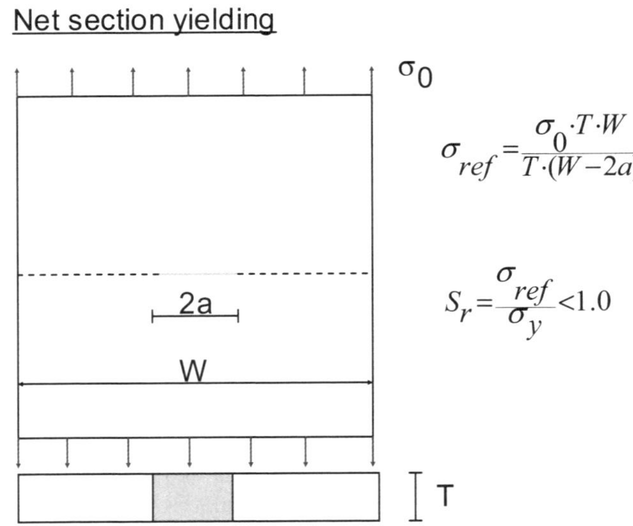


**Figure 6.9.** *Wide plate with central through-thickness crack, net section yielding check*

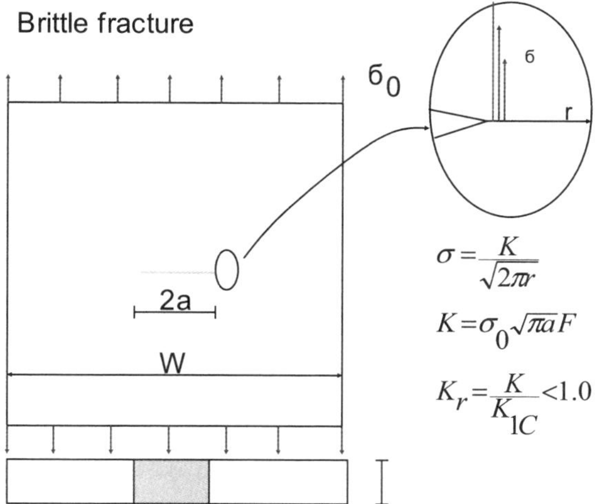



**Figure 6.10.** *Wide plate with central through-thickness crack, brittle fracture check*

The interaction between the two failure modes is accounted for in the R6 criterion. It is obvious that each of the two criteria given in Figures 6.9 and 6.10 must be fulfilled in order to avoid net section yielding and brittle fracture when these two failure modes are considered separately. If there is a danger of a mixed-mode failure, a proper limit on the sum of the two criteria must be set. This is shown in Figure 6.11. The limit line of the diagram is given by the following equation:

$$K_r = \left[ \frac{8}{\pi^2 S_R^2} \ln\ \sec\left( \frac{\pi}{2} \cdot S_r \right) \right]^{1/2} . \tag{6.24}$$

From this equation the critical (biggest admissible) crack size $a_c$ can be determined. Although we have shown the method for a through-thickness central crack in a wide plate, the verification procedure remains the same for the semi-elliptical surface cracks found near the weld toes. Opposite to the influence of the initial crack depth, the fatigue life is not so sensitive to the values of $a_c$. For medium-strength steel with yield stress up to 400 Mpa, critical crack size $a_c$ can, for practical purposes, be set equal to the plate thickness.

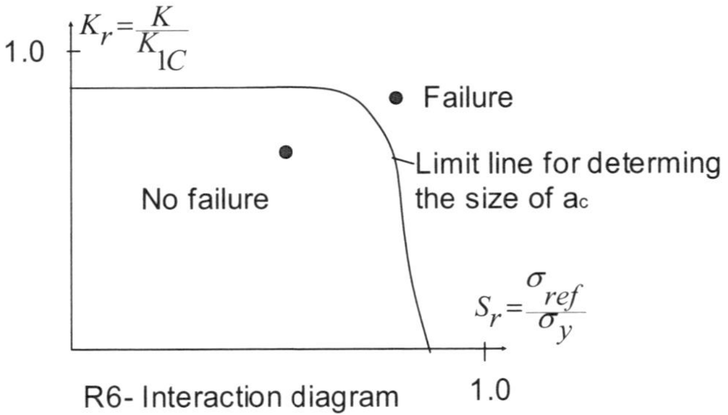


**Figure 6.11.** *Illustration of the R6 criterion*

## 6.5. Fatigue threshold and fatigue crack growth

### 6.5.1. *Crack growth models*

For a welded joint made of medium strength fatigue crack propagation from the weld toe region is a rather complex problem, involving stress gradients due to the weld toe notch effect, residual stresses, and micro-structural non-homogeneities in the heat-affected zone (HAZ). In the present chapter we shall limit the presentation to loading model I. Hence, we will simply write $K$ instead of $K_I$ for a short denotation. Multi-axial fatigue will be treated in Chapter 11 with a more formal denotation. The fatigue propagation is often predicted using a LEFM theory. For semi-elliptical cracks at the weld toe, the deepest point at the crack front will propagate under plane strain conditions, as we have discussed. During one external load cycle the crack front will advance by a micro-mechanism involving crack blunting and re-sharpening as the stresses increase and decrease respectively; see Refs [1, 2]. During each load cycle, the formation of a striation may be observed on the fatigue crack surface. It is assumed that the crack growth rate at a macroscopic average level related to the stress intensity factor range (SIFR). A semi-empirical relation is given by the Paris-Erdogan law:

$$\frac{da}{dN} = C(\Delta K)^m, \quad \Delta K > \Delta K_0 \tag{6.25}$$

where $C$ and $m$ are treated as material parameters for a given $R$ ratio and environmental condition. $\Delta K$ is the SIFR at the crack tip, corresponding to the applied stress nominal range $\Delta\sigma$. The crack depth $a$ is measured from the plate surface to the crack tip. The crack tip is defined as the deepest point along the crack front for a semi-elliptical surface crack as was shown in Figure 6.2. Equation (6.25) is coupled to similar equations in other crack front directions, e.g. the end points of the crack at the surface. However, in the present analysis the growth model will be simplified by the one-directional approach. The shape evolution of the crack given by the aspect ration $a/2c$ will be derived from experimental data, and introduced in the calculations as a forcing function on the aspect ratio $a/2c$. In what follows, only the crack opening mode I is discussed, i.e. the case where the crack surfaces move directly apart. Although many problems for welded joints are of the mixed-mode type, mode 1 is considered to be the dominant mode for fatigue propagation and fracture. The SIFR can be written in the following form:

$$\Delta K = K_{max} - K_{min}$$

$$\Delta K = \Delta\sigma \sqrt{\pi a}\, F(a) . \tag{6.26}$$

The validity of equation (6.25) is given in region B in the diagram in Figure 6.12. Region B is the regime for stable crack growth and the linear curve corresponds to equation (6.25) for a log-log scale. Region A is the near-threshold regime where the crack growth rate tends to drop towards very low values. It is often assumed that $da/dN$ is zero if $\Delta K$ is less than a certain "threshold" value $\Delta K_0$. Another way of taking account of this behavior is to introduce a modified crack law in the threshold regime (regime A in Figure 6.12):

$$\frac{da}{dN} = C(\Delta K - \Delta K_0)^m . \tag{6.27}$$

The threshold values for $\Delta K_0$ for normal welded steel qualities are found in the range 3-7 MPa m$^{0.5}$. The concept of $\Delta K_0$ is usually based on experiments with long cracks, typically several mm in a compact tension (CT) specimen. For shallow surface cracks at weld toes the crack growth rate may be considerably higher. These shallow cracks do not exhibit the same retardation as long cracks for low values of the SIFR. This becomes even more pronounced if they grow in a tensile residual stresses caused by the welding process. The conditions are illustrated schematically in Figure 6.12. It is difficult to quantify how much faster small surface cracks at weld notch grow compared to long cracks. Because of this uncertainty it is recommended that one extrapolate the curve from region B into region A when the small values of SIFR are due to small crack depths. For the same reason we shall use equation (6.25) and not equation (6.27) for fatigue-growth calculations.

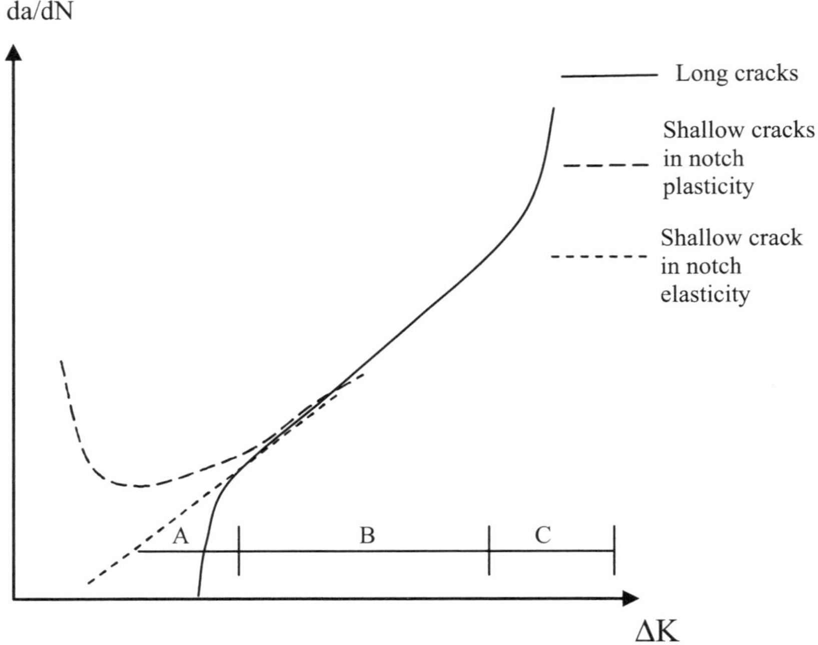


**Figure 6.12.** *Likely growth curves for shallow cracks at weld notches compared with long crack behavior*

In region C of the diagram, the crack growth rates are greater than what a simple extrapolation of the linear curve from region B would predict. The reason for this behavior is the onset of unstable fracture due the fact that $K_{Imax}$ approaches the fracture toughness of the material, $K_{IC}$, see equation (6.21) In this regime the Foreman equation has been suggested (see Ref [10]):

$$\frac{da}{dN} = \frac{C(\Delta K)^m}{(1-R)K_C - \Delta K}. \tag{6.28}$$

The reader must be aware of the fact that the parameters C and m are only valid for the equation to which they are fitted. Parameters for equations (6.25), (6.27) and (6.28) are not interchangeable. In this book we shall base all of our discussion on equation (6.25). This is the most common equation for welded joints and also the one for which the most material data are available. The number of cycles to reach given crack depths is calculated by numerical integration of equation (6.25) as follows:

$$N_P = \int_{a_0}^{a_c} \frac{da}{C\left(\Delta\sigma\sqrt{\pi a}\,F(a)\right)^m}$$

(6.29)

where $a_0$ is the initial crack and $a_c$ is the final crack depth giving some sort of malfunction or fracture. A large number of values for the SIFR is needed during numerical integration of equation (6.29). Clearly, the computational efforts and accuracy are crucial when selecting calculation method.

### 6.5.2. *Parameters C and m*

The material characteristics of the HAZ and the base material will influence the crack growth rate through the parameters C and m in the Paris equation. Although treated as material parameters, both the environment and the applied stress ratio R may have a strong bearing on C and m. The parameters are valid for a given environment and R-factor. The dependence on the R-ratio may be partly eliminated by introducing the crack-closure concept. It is assumed that during a lower part of the loading cycle the crack remains closed. This part of the cycle does not open the crack and will not contribute to the crack growth. The effect is more pronounced for lower values of the applied R-ratio. This observation leads on to consider only the effective SIF range:

$$\Delta K_{eff} = K_{max} - K_{op} = U \cdot \Delta K$$

(6.30)

where $K_{op}$ is the value at which the crack opens (or closes). In fact, the crack may close before the nominal stresses approaches zero. Under constant amplitude loading this is mainly due to a small, plastic crack tip zone developing at top of the loading cycle. As the crack grows through a succession of these zones, plastically deformed material is left within its wake. This plastic wake is constrained by the elastic surroundings during unloading and causes the crack to close while still subjected to tensile stresses. Typical values at a stress ration R = 0.0 are $K_{op} = 0.3\,K_{max}$; see Ref [3]. A further discussion of these topics will take place in Chapter 12 where we model the effect of an overload. If $\Delta K_{eff}$ is used in equation (6.25), the C and m factors will remain almost constant for different R-values; the $\Delta K_0$ likewise. The great advantage of the concept is that C and m values determined from one test series at specific R-factors can be used under other loading conditions, provided that the curve fitting is based on an effective stress range concept. The problem is to determine the effective range of the SIFR from case to case. Elber (Ref [11]) proposed the following formula for aluminum:

$$U = A + B R$$

(6.31)

where A and B are constants for a given material. There is at present a lack of experimental evidence for this equation for welded steel joints. Based on a literature study carried out by Verreman (Ref [12]) it may be assumed that A and B are close to 0.75 and 0.35 respectively. This means that under full alternating loading (stress ratio R = -1) we will have U = 0.4 which corresponds to $K_{op}/K_{max}$ = 0.2, whereas for a stress ration R = 0 we will have U = 0.75 with $K_{op}/K_{max}$ = 0.25. Further experiments and theoretical studies are needed to verify these values. We will return to some of these topics in Chapter 12.

### 6.5.3. *Residual stresses*

The presence of residual stresses has for a long time been a source uncertainty for fatigue crack growth in welded joints, even under laboratory conditions. During a test series, these stresses are cumbersome to measure and difficult to control. Probably the best solution is to carry out tests with stress-relieved specimens and then apply the results to as-welded conditions by explicitly taking account of the presence of residual stresses when calculating the SIF factors. If the residual stress distribution $\sigma_r(x)$ along the crack line is known, the corresponding SIF can be found by the weight function method, equation (6.16). By the principle of superposition, the maximum and minimum SIF during the external loading cycle is found:

$$K_{max} = K_r(\sigma_r, a) + K_{max}(\sigma_n, a)$$

$$K_{min} = K_r(\sigma_r, a) + K_{min}(\sigma_n, a) \tag{6.32}$$

$$\Delta K = K_{max} - K_{min}, \quad R_{eff}(a) = \frac{K_{min}}{K_{max}}.$$

$R_{eff}$ may be very different from the nominal applied stress ratio. There are then two alternative ways of taking account of the residual stresses in the calculations. The first is to select C and m values fitted to equation (6.25) based on nominal $\Delta K$ and various R values corresponding to $R_{eff}$. The other is to calculate $\Delta K_{eff}$ as described earlier and provide data based on this effective SIFR.

As seen from equation (6.32), $R_{eff}$ will be a function of crack depth because the self-equilibrating residual stresses may change from large tensile stresses to compression. If this is the case, then $R_{eff}$ and $\Delta K_{eff}$ will decrease during crack propagation. This fully explains why as-welded test specimens may give relatively higher growth rates for smaller cracks than for the larger ones, compared with results obtained from stress relieved specimens. If these results are plotted against nominal SIF, the result will be a low exponent m in Paris equation. Small surface-breaking cracks do not obey the retardation found for longer cracks at low values of SIFR. This becomes even more pronounced if the cracks are propagating in residual tensile stress field as we have shown.

A final comment should be that if $\sigma_r(x)$ is not known, as usually is the case, constant tensile residual stresses of magnitudes near the yield stress may be assumed for a conservative crack growth estimate.

### 6.5.4. *Some notes on the size of the initial cracks*

Regarding welded joints, there has been a discussion as to whether $N_P$ in equation (6.29) could be treated as the entire fatigue life, ignoring the number of cycles $N_I$ to crack initiation. The approach is usually to assume initial flaws of such small size that $N_P$ equals $N_T$. However, these small, initial cracks must be regarded as fictitious cracks chosen to make LEFM describe the entire fatigue process. They are not real, physical cracks (flaws, intrusions) created by the welding procedures. It is therefore difficult to relate the derived model to actual detected flaws in the weld. Furthermore, the initial cracks are often so small that LEFM is not applicable, typical values being from a = 0.01 to 0.05 mm. This crack size is close to the micro-structural features of the material. Ordinary structural steel qualities often have a grains size near 0.01 mm. Engesvik (Ref [13]) concluded that it may be dubious to apply LEFM at crack depths of less than 0.1 mm. Before this stage the crack initiation phenomenon is probably better modeled by the Coffin and Manson equation which is based on a local stress-strain approach; see Chapter 2.

### 6.6. Geometry function and growth parameters given in BS7910

The British standard BS7910 (Ref [9]) is issued as a guide to engineering critical assessments for flaws and imperfections in fusion-welded joints. Both unstable fracture and fatigue crack growth are treated for planar crack-like defects. When using equation (6.25) for welded plated joints, data are available for most of the parameters. For the case shown in Figure 6.13, fatigue crack growth is modeled by the simple version of the Paris law:

$$\frac{da}{dN} = C(\Delta K)^m = C(\Delta S \sqrt{\pi a} \, F(a))^m, \qquad \Delta K > \Delta K_0$$

$$N = \frac{1}{C} \int_{a_0}^{a_C} \frac{da}{\left(\Delta S \sqrt{\pi a} \, F(a)\right)^m}$$

$$(6.33)$$

In simple cases where $F(a)$ is assumed constant we have the following:

$$N = \frac{1}{(1 - m/2)(C(\pi\Delta\sigma F)^m} \left[ a_f^{1-m/2} - a_i^{1-m/2} \right].$$
(6.34)

The equation may give reasonably good estimates for the final fatigue life for welded plated joints if $F(a/T)$ is set constant to a value corresponding to F(0.01).

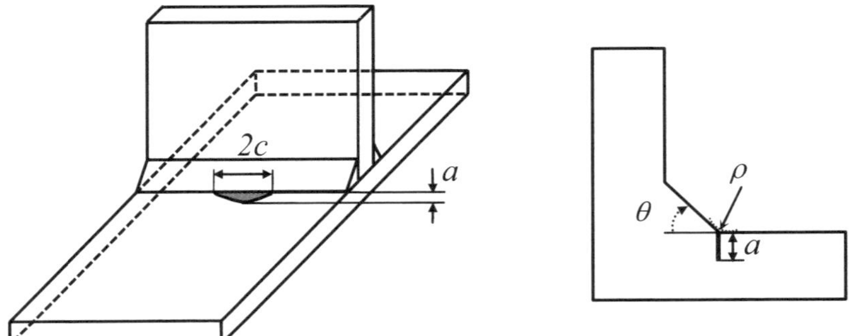

**Figure 6.13.** *Typical geometry and crack in a fillet welded joint*

### 6.6.1. *The geometry function*

The geometry function is determined by equations (6.11) and (6.12). The correction factor $M_k$ due to the stress gradient is given by equation (6.17). Figure 6.14 gives the gradient correction $M_k$ and geometry function $F(a/T)$ for fillet welded joints for an edge crack.

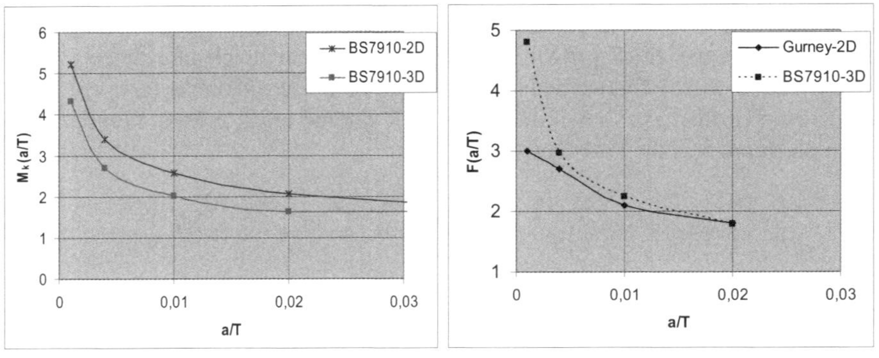

**Figure 6.14.** *Left: Mk correction based on 2-D and 3-D analysis of an edge crack. Right: geometry functions F(a/T) for an edge crack*

The solutions for $M_k$ to the left in Figure 6.14 are obtained by the method that we explained when we presented equations (6.19) and (6.20). The geometry function $F(a/T)$ to the right is valid for a surface crack with a straight front. The solution designated BS7910-3-D has been obtained simply by multiplying $M_k$ to the left by the free surface factor $M_1 = 1.12$. As can be seen, the function designated BS7910-3-D is very close to the solution presented by Gurney (see Ref [14]) for cracks with depth greater than $a/T = 0.005$ ($a = 0.1$mm for $a$ 20-mm-thick plate). The solution given by Gurney is based on a two-dimensional FEA analysis using a weld toe angle of 45 degrees and ignoring the weld toe radius. Hence, the stress field used had to be extrapolated toward this singular point at the weld toe. The analysis was carried out for a joint without a crack and the SIF was subsequently obtained by the weight function method. The BS7910-3-D solution is also determined for a flank angle of 45 degrees and very small values for the toe radius. Both functions are valid for straight-edge cracks. As can be seen from the figure, the Gurney solution is approximately 10% smaller than the BS7910-3-D solution for cracks with $a/T > 0.005$. If a correction is carried out for the elliptical shape of the crack with $a/c$ close to 0.20, the BS7910-3-D solution for a curved crack front will approach the Gurney straight crack solution.

The BS7910 guide also gives values for geometry functions at other points along the crack front. However, in the present analysis the growth is simplified by the one-directional approach. The shape evolution of the crack is derived from experimental data, and introduced into the calculations as a forcing function on the aspect ratio $a/2c$, where $2c$ is the crack length at the plate surface (see Figure 6.13):

$$2c = 2.92a + 3.83 . \tag{6.35}$$

This expression is based on extensive amount of crack-path data obtained by ink staining of the fatigue crack at various stages during the tests; see Ref [15]. Based on the above equation, a crack depth of 1 mm will typically have a length at the surface that is close to 6 mm. In our model used in the spreadsheet FLAWS-CG (found in Appendix C) we have used the Gurney edge solution directly as an approximation for all types of crack geometries.

### 6.6.2. Parameters C and m

In BS7910, two alternatives are suggested for the relationship between the growth rate $da/dN$ and the SIFR for a log-log scale. Mean values and scatter are given for the parameters m and C. The first alternative is based on a single linear relationship, whereas the second alternative proposes a bilinear relationship. The main difference between the two models is that the bilinear one models the gradual decrease in the growth rate for low values of the SIFR before the threshold $\Delta K_0$ is reached. For the simple linear relationship m is set to 3.0 and only the upper bound for the $C$ is given. The earlier document PD6493 (see Ref [16]) recommended $C = 3.0 \times 10^{-13}$ (units MPa, mm), whereas BS7910 recommends as high as $5.21 \times 10^{-13}$ (units MPa, mm): see Table 6.1. Hence, the value is increased as much as by 80% from the first document. The mean values are not given, and we have listed the mean value found in Ref [17] in Table 6.1. This is done because the figure for the mean plus two standard deviations (mean + 2SD) given in Ref [17] coincides with the upper bound PD6493. For conversion of units for the parameter $C$, see section 6.10.

Data for the bilinear relationship are given in Table 6.2. The stage A/stage B transition point is 363 N/mm$^{3/2}$ for the mean curve and 315 N/mm$^{3/2}$ for the mean plus two standard deviations curve. For shallow surface cracks the threshold value of SIFR is given as 63 N/mm$^{3/2}$ as a lower limit regardless of the applied stress ratio.

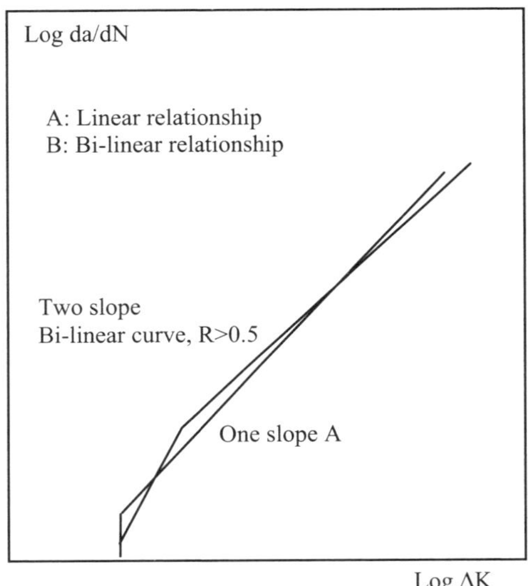


**Figure 6.15.** *Sketch of one-slope and bilinear relationship*

| Growth curve: | Ref: | m | C |
|---|---|---|---|
| Mean | 17 | 3 | $1.85 \times 10^{-13}$ |
| Mean + 2SD | 9 | 3 | $3.00 \times 10^{-13}$ |
| Upper Bound | 16 | 3 | $5,21 \times 10^{-13}$ |

For *da/dN* in mm/cycle and *ΔK* in N/mm$^{3/2}$

**Table 6.1.** *Growth parameters in air one-slope curve m = 3.0*

| Stage A | | | |
|---|---|---|---|
| Mean curve | | Mean + 2SD | |
| $C$ | $m$ | $C$ | $m$ |
| $1.21 \times 10^{-26}$ | 8.16 | $4.37 \times 10^{-26}$ | 8.16 |
| Stage B | | | |
| Mean curve | | Mean + 2SD | |
| $C$ | $m$ | $C$ | $M$ |
| $3.98 \times 10^{-13}$ | 2.88 | $6.77 \times 10^{-13}$ | 2.88 |

For $da/dN$ in mm/cycle and $\Delta K$ in N/mm$^{3/2}$

**Table 6.2.** *Growth parameter in air*


## 6.7. Fracture mechanics model for a fillet welded plate joint

### 6.7.1. *Basic assumptions and criteria for the model*

Based on the above presentation of applied fracture mechanics and the recommendations given in BS7910, we shall demonstrate the LEFM model's abilities and shortcomings for fillet welded joints. The objective is to establish a model that is consistent with rules and regulations both based on the S-N approach (Eurocode 3) as presented in Chapter 5 and applied fracture mechanics (BS7910) as presented in this chapter. Emphasis is laid on how to choose growth parameters in conjunction with a fictitious, initial crack size to obtain both reliable crack growth paths and predictions of the entire fatigue life. There is no doubt that the model outlined above and the parameters given by BS7910 are well suited to describing the behavior of large cracks detected and sized during inspection. It is however more uncertain as to whether the model is capable of describing the entire fatigue process from initiation to final fatigue failure for high-quality welded joints. This will obviously be an approximation and we will pinpoint the merits and shortcomings of such a model with reference to two data bases presented below. Based on these considerations, a model will be established to meet the following criteria:

– the model should be corroborated by S-N data for the joint in question when these data are available (database 1 and 2);

– the model should predict a crack evolution that coincides with measured crack growth histories before failure (database 1).

With this background we will endeavor to model the fatigue process in fillet-welded joints. Before proceeding let us have a closer look at the experimental data.

### 6.7.2. *Data for crack growth measurements (database 1)*

Database 1 contains crack growth measurements made directly on fillet-welded joints. The results have been presented in Refs [15, 18]. 34 non-load-carrying cruciform and T-joint test specimens were tested under constant amplitude axial loading. The test joints were described in Chapter 3. All the test joints were fabricated from C-Mn steel plate with a 25-mm thickness. The nominal yield stress was 345 MPa. The welding procedures were taken from normal offshore fabrication practice. The joints were proven free from cracks and undercuts. The specimens were tested under constant amplitude axial loading at $\Delta S = 150$ MPa with a loading ratio of $R = 0.3$. Experimental details are to be found in Ref [15]. Typical crack size histories as a function of time are shown in Figure 6.16.

The total fatigue lives for the 34 specimens have been plotted in Figure 6.17. The main characteristics of the measurements are given in Table 6.3. As can be seen, the time to reach a crack depth of 0.1 mm is close to 30% of the total fatigue life, whereas the total life is only 10% less than the prediction of the F-class S-N curve ($N = 513,000$ cycles). Hence, the test series is of normal quality and comparable with the population pertaining to the F-class and Category 71.

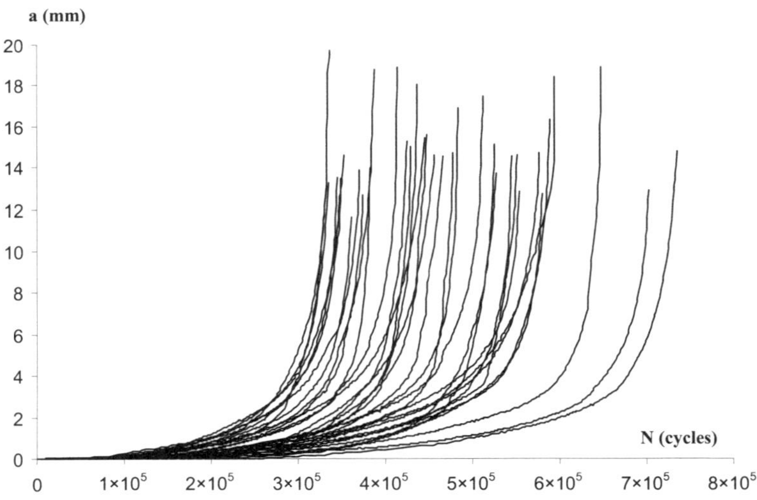


**Figure 6.16.** *Crack growth histories in database 1*

| $N_i$ | $N_p$ | $N_t$ | $N_t$ **F-class** |
|---|---|---|---|
| 145 (0.34) | 323 (0.20) | 468 (0.22) | 513 (0.54) |

**Table 6.3.** *Statistics for database 1 (in 1,000 cycles, COV in parenthesis)*

### 6.7.3. *Data for fatigue lives at low stress levels (database 2)*

Database 1 is the most important one for model adjustment and calibration because the entire crack growth histories are known. When it comes to assessment of the fatigue life, the S-N curve statistics themselves represent what we will call database 2. However, the data point pertaining to this curve has its center of gravity at a stress range in the region of 120 to150 MPa. In addition, we have in this chapter compiled results from fatigue life tests in several large experimental investigations carried out in Europe (see Ref [19]). The applied stress ranges are in the region of 80 to 105 MPa and the thickness of the plates is from 16 to 38 mm. All the selected specimens are as-welded and the loading ratio $R$ is between 0 and 0.3. We have chosen this database because the data points are near the "knee point" of the S-N curve where the data are scarce and where service stresses often occur. These data points are also plotted in Figure 6.17. As can be seen, they have a mean value substantially longer than the prediction of the median F-class curve.

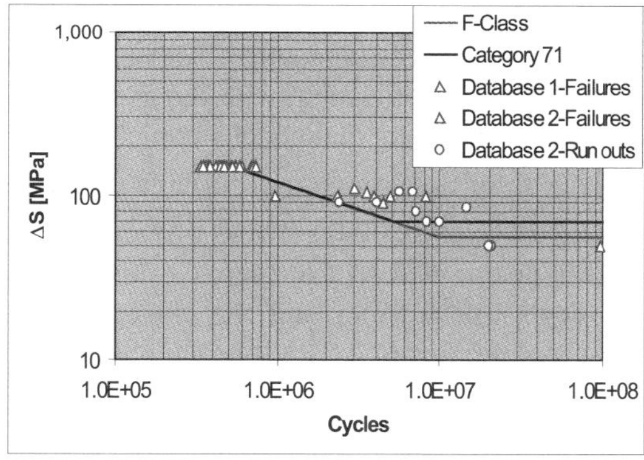


**Figure 6.17.** *Plot of S-N points for databases 1 and 2, together with median curves in current rules*

### 6.7.4. *Procedure and curve fitting*

The numbers of cycles to reach given crack depth are calculated by numerical integration of equation (6.25) as follows:

$$N = \frac{1}{C} \int_{a_0}^{a_c} \frac{da}{\left( \Delta\sigma \sqrt{\pi a}\, F(a) \right)^m} \tag{6.36}$$

It is noted that equation (6.36) can be reduced to the same form as the S-N equation given in Chapter 5 for constant amplitude loading. The number of cycles is inversely proportional to the stress range raised to a power of m in both equations. The problem that arises when using equation (6.36) is to choose an appropriate combination of the parameters C, m, and $a_0$. For the parameters C and m, suggestions based on growth tests for C-Mn steels are found in BS7910 (Tables 6.1 and 6.2), whereas the depth $a_0$ is non-measurable for high-quality welds and no guidance is found. In this book the initial depth is found by backwards calculations pursuing the origin of the growth histories measured for larger cracks as shown in Figure 6.16 for database 1.

In contrast to the influence of the initial crack depth, the fatigue life is not as sensitive to the values of the critical crack depth $a_c$. In the present work, $a_c$ is set to half of the plate thickness of the joints. The $F(a)$ geometrical function is based on the work by Gurney (Ref [14]), who derived numerical values for $F(a)$ for an edge crack with an average weld toe profile; see Figure 6.14 to the right. In the present analysis we will examine both the linear and the bilinear relationships in BS7910 (Figure 6.15) with respect to the growth rates observed in database 1. We will adopt the m values given and see if the corresponding C values derived from each of the growth curves in Figure 6.16 are compatible with the statistics given in Tables 6.1 and 6.2. The procedure for determining the model parameters are as follows:

1) the slope parameter m of the growth rate curve is chosen in accordance with BS7910 (Tables 6.1 and 6.2) for the actual value of $\Delta K$;

2) the parameter C is determined so that the measured number of cycles spent between these crack depths of 0.1 mm and the final crack depth of 12.5 mm coincides with the measured number of cycles for each joint in database 1;

3) with the C and m values obtained above for a given test, $a_0$ is determined so that the number of cycles from $a_0$ to the first measured crack depth of 0.1 mm coincides with the measured number of cycles. Hence, for each of the 34 tests, a set of the variables m, C and $a_0$ is obtained.

Figure 6.18 shows experimental curves taken from Figure 6.16 and curves obtained from fracture mechanics using the above procedure and equation (6.36). As can be seen from the figures, the $a$-$N$ curves simulated by the two alternative relationships are quite close to the experimental ones.

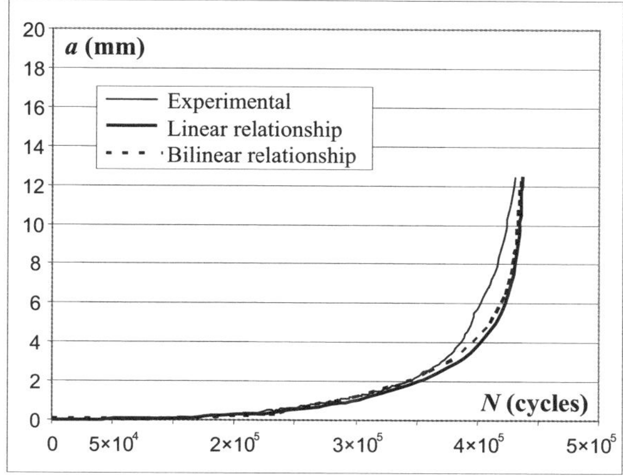

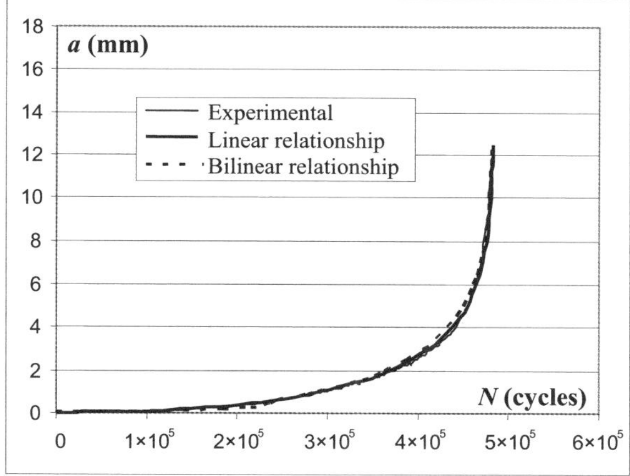

**Figure 6.18.** *Measured and fitted crack growth histories*

The measured curves are somewhat more irregular, but they are very "Paris-like" and the fit must be regarded as excellent considering the fact that we have kept m fixed. If we had optimized the fit by letting both m and C vary, the fit would even been even closer. However, the scope of the present work is to use the recommendation in BS7910.

### 6.7.5. *Growth parameters C and m*

A histogram for the natural logarithm of C (MPa, m) obtained by simulation is shown in Figure 6.19. The values are obtained under the assumption of a single linear relationship between crack growth rate and the SIFR for a log-log scale with m = 3. The variable ln C has a rather even distribution and not a normal distribution as might have been expected.

The mean value for ln C is –29.4 (units MPa, mm) with a standard deviation of 0.20. The statistics for C are shown in Table 6.4.

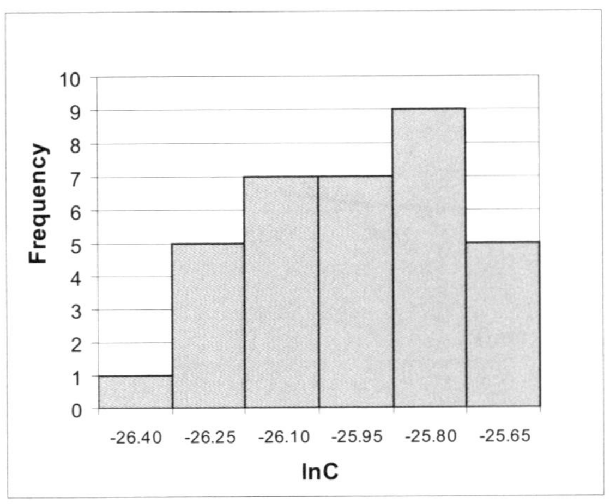

**Figure 6.19.** *Statistics of ln C (MPa, m) obtained from curve fitting to experimental a-N curves*

| Median | Median + 2SD |
|---|---|
| $1.67 \times 10^{-13}$ | $2.48 \times 10^{-13}$ |

For $da/dN$ in mm/cycle and $\Delta K$ in N/mm$^{3/2}$

**Table 6.4.** *Statistics for the parameter C obtained from curve fitting*

The mean value in Table 6.4 is only 10% less than the BS7910 mean value given in Table 6.1. The mean value plus two standard deviations is only 20% less than the value given in PD6493. Figure 6.20, to the left-hand side, shows the experimental results plotted for an arbitrary value of the SIFR together with the BS7910 mean and upper bound curves. None of the results obtained were in the vicinity of the upper bound.

A similar analysis was carried out for a bilinear relationship. The obtained results do not fit the lower line as shown to the right in Figure 6.20. The growth rates are close to 4 times higher than the values given in BS7910. If the geometry function given by BS7910 had been used instead of the Gurney function, on the right-hand side of Figure 6.14, the growth rates would have been reduced to twice the values given in BS7910. The reason for these high growth rates is likely due to the fact that the lower part of the bilinear curve is derived from tests with relatively long cracks (several millimeters) in compact tension specimens. The fatigue process in welded joints comprises crack growth of small surface-breaking elliptical cracks with depths of less than 0.1 mm. These cracks may grow considerably faster than the lower part of the bilinear growth curve in BS7910 prescribes. The results are consistent with the sketch in Figure 6.12. Hence, one should take care when using the bilinear relationship given in BS7910 to calculate the early fatigue-crack growth in welded joints. These curves should only be used when larger cracks are detected and sized.

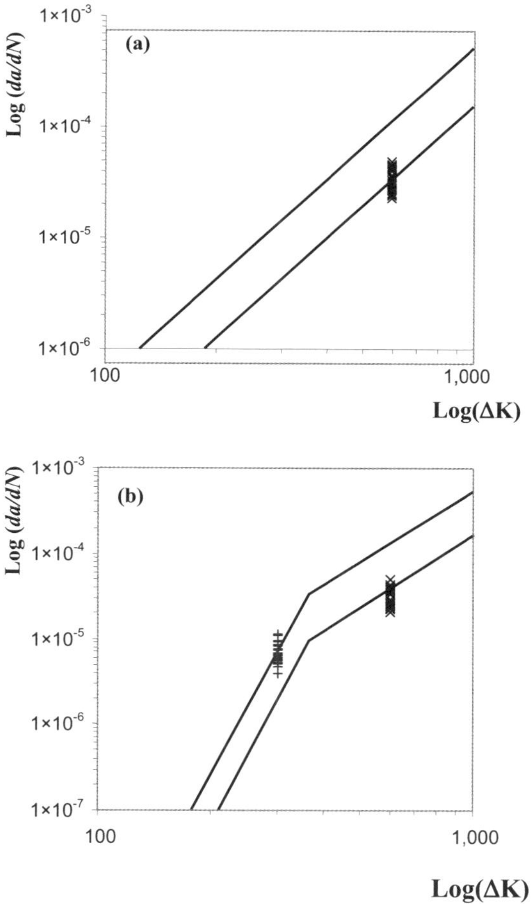



**Figure 6.20.** *Experimental results plotted within the BS7910 scatter band:*
*(a) linear relationship; (b) bilinear relationship*

### 6.7.6. *The initial crack depth $a_0$*

Based on the curve fitting procedure above, the initial crack depth for each test was also obtained. The statistics for the derived initial crack depths are given in Table 6.5. The PDF of $a_0$ is shown to the left in Figure 6.21. As can be seen from Table 6.5, the mean value for the initial crack depth is 0.015 mm and the upper bound is close to 0.03 mm. It should be kept in mind that the model does not take into account the variability of the local toe geometry when determining the initial crack depth, i.e. we have held the geometry function $F(a)$ constant at its mean value. It should also be emphasized that this initial crack depth distribution is a purely theoretical concept, i.e. it cannot be proven that the crack depths are related to initial flaws created by the welding process.

A corresponding analysis was carried out for the bilinear relationship and gave a mean value for the initial crack depth of close to 0.06 mm.

| Mean [mm] | Standard Deviation [mm] | COV |
|:---:|:---:|:---:|
| 0.0151 | 0.0045 | 0.30 |

**Table 6.5.** *Statistics for $a_0$ for the linear relationship between log da/dN and log SIFR*

The curve in Figure 6.21(b) shows the part of the growth-rate curve used when calculating the growth rate from $a_0$ to $a = 0.1$mm. As can be seen, we are very close to the threshold value of 63 N/mm$^2$mm$^{0.5}$ (2 MPam$^{0.5}$).

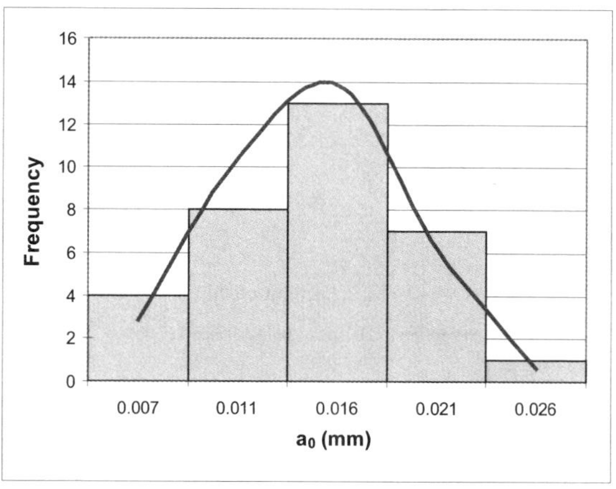

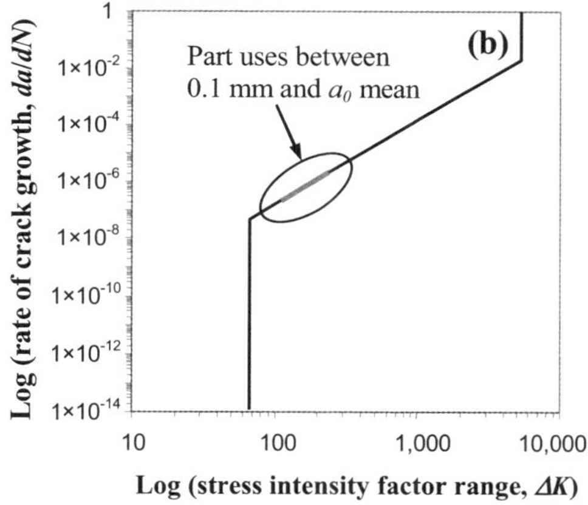


**Figure 6.21.** *(a) PDF for $a_0$; (b) part of the relationship
da/dN versus ΔK uses (MPa,mm)*

### 6.7.7. *Prediction of crack growth histories and construction of S-N curves*

After the curve fitting was carried out for the experimental stress range of
150 MPa, the obtained mean values for $a_0$ and C was substituted into equation (6.36)
to calculate both crack evolution and the fatigue life at various constant-amplitude
stress levels. Both the linear and bilinear results were used. Crack evolutions at
stress ranges equal to 150 MPa and 100 MPa are shown in Figure 6.22.

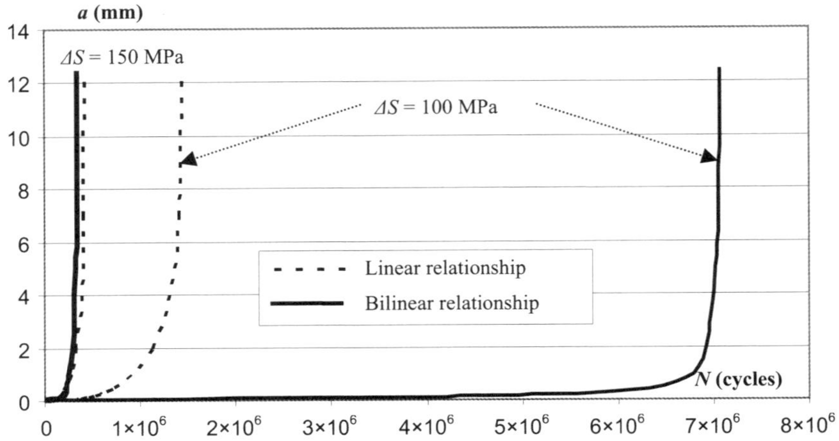


**Figure 6.22.** *Crack evolution from $a_0$ to $a_c$ at stress levels 150 and 100 MPa*

As can be seen from Figure 6.22, the two relationships follow slightly different paths to arrive at the same end point for $\Delta S = 150$ MPa. The fatigue lives predicted at $\Delta S = 100$ MPa are, however, very different. The linear relationship gives a fatigue life close to $1.5 \times 10^6$ cycles, which is again 10% less than the F-class S-N prediction, whereas the bilinear relationship gives a fatigue life close to $7 \times 10^6$ cycles, which is far too long compared to the S-N estimate. The latter scaling of fatigue life with the applied stress range will not be consistent with the slope of the S-N curve. This is because close to half the fatigue life (from the initial crack depth to a depth of 0.6 mm) scales relative to the stress level with a power of m = 8.16 (Table 6.2), whereas the F-class S-N curve has an inverse slope corresponding to m = 3. The predictions made by the linear relationship will correspond to the predictions made by the S-N curve class F, whereas the predictions made by the bilinear curve do not. It is also to be noted that both the F-class curve and the fracture mechanics model (FMM) fail to predict the long, experimental lives given by the data points in database 2 at low stress levels; see Figure 6.17. As for the fatigue-limit $\Delta S_0$, this is determined from the threshold value $\Delta K_0$ by the following equation:

$$\Delta S_0 \sqrt{\pi \, a_0}\, F(a_0) = \Delta K_0 . \qquad (6.37)$$

If we enter a mean value of $a_0 = 0.015$ mm into this equation we get $\Delta S_0 = 95$ MPa, which is far too high compared with the given F-class value of 56 MPa. The fatigue limit is between 40 MPa and 56 MPa as can be seen from Figure 6.17. Even if we choose an initial crack depth equal to the mean value plus two standard deviations (0.025 mm), we will still get an endurance limit as high as 75 MPa. The reason is the same as for the discrepancy found between the experimental growth rates and the growth rates given in the lower part of the bilinear relationship in BS7910. Both the prescribed crack growth rate decrease and the final stop at low SIFR are valid for larger cracks only. They are not applicable to shallow elliptical cracks at the weld toe. The investigation supports the choice of a one-slope curve for the *da/dN* as a function of SIFR. Furthermore, the fatigue limit in the S-N curve cannot be explained by a crack growth threshold phenomenon.

### 6.7.8. *Conclusions for fillet joints with cracks at the weld toe*

The entire fatigue process in fillet-welded joints has been modeled by a pure fracture mechanics approach. The model has been calibrated to describe the entire fatigue process. The simple version of the Paris law has been adopted. The initial crack depth and growth rate parameters have been determined to fit each experimental *a-N* curve in database 1. The following conclusion can be drawn:

The single linear relationship between *da/dN* and SIFR for a log-log scale with slope m = 3.0 gives a good fit between the calculated *a-N* curves and the curves measured at a stress range of 150 MPa. The growth rates found have a mean value close to the mean value given in BS7910. Furthermore, all the rates are well below the upper bound given in BS7910. In fact, the mean plus two standard deviations is quite close to the upper bound given in the document PD6493. The predicted fatigue life is very close to the predictions given by the F-class S-N curve at any stress level above the fatigue limit.

The initial cracks are in a range between 0.005 and 0.03 mm with a mean value of 0.015 mm. The initial crack distribution applied for the model cannot be verified by experimental findings. The concept of a threshold value does not apply for these shallow cracks. Hence, the fatigue-limit stress cannot be determined from the model.

The bilinear relationship between *da/dN* and SIFR for a log-log scale also gives good agreement with experimental *a-N* curves. However, the derived growth rates are higher than the upper bound given in BS7910 for low values of the SIFR. Compared with the F-class curve, the model fails to predict the slope of the curve and overestimates significantly the fatigue life as the stresses approach service stresses.

In conclusion, the FMM should not be used to model the entire fatigue process in high-quality welds that have been proven to be free from detectable initial cracks. Although the model is capable of describing the crack evolution at any one given stress level, it will fail to predict the change in slope of the S-N curve as the stress range decreases. Furthermore, the fatigue limit will be overly optimistic. The FMM should primarily be applied in cases where cracks are found and sized. In other cases, a crack initiation phase should be modeled before the crack-propagation phase is added on. This will be further discussed in Chapter 9.

## 6.8. Fatigue crack growth in tubular joints

Fatigue cracking and fracture is one of the principal modes of failure in fixed offshore platforms that are constructed as framework structures with tubular members. The welded tubular joints are susceptible to fatigue damage, due to repeated environmental loading and high stress SCF at tube intersection areas. These stress concentrations are related to the secondary bending of the tube walls to fulfill the compatibility criteria at the intersection between the branches and the chords. Cracks normally occur at the weld toe at hot-spot regions either on the chord or the branch members. This was discussed in Chapters 2, 3, and 5. The number of stress cycles that may be endured by a joint depends upon the loading mode, the magnitude of the nominal stress ranges, the global joint configuration, and the weld quality, as we have discussed in Chapter 5. The weld fatigue quality of the weld is characterized by a smooth transition geometry between the plate material and the weld toe, and a surface free from flaws.

If the safety margins are low and fatigue failure of a joint results in a total loss of structural integrity it is common practice to plan a periodic in-service inspection program. Usually, the strategy is to repair detected cracks. The initial inspection plan must be derived by a fracture-mechanics approach which predicts future crack depth as a function of time. The time period during which the crack is large enough to be easily detected by a given non-destructive inspection (NDI) method can be determined. Based on this information, an appropriate inspection method and time interval can be selected.

The fatigue damage modeling of tubular joints is a difficult task compared to plated joints for several reasons:

– the geometry of the joint can be complex with a variety of possible configurations and loading modes. High stresses may occur at several locations along the welded intersections between members of the joint. Stress gradients will appear both through the wall thickness of the members and along the welded intersections;

– cracks may initiate and grow at several hot-spots along the weld seam as a function of the variation in the geometrical SCF and the local weld-notch quality;

– as the cracks propagate, a redistribution of the stresses will take place along the welded intersections. The areas where cracks have developed will be subjected to decreasing stresses, whereas intact areas will carry higher loads and increasing stresses.

Stress calculations are indeed a difficult task even with an undamaged joint. As discussed in Chapter 5, stresses can be determined by parametric formulae or finite element analysis (FEA) that are based on thin shell elements. The fact that cracks may initiate at several locations makes a multi-crack model the most realistic approach. The stress redistribution along member intersections can be modeled by FEA models with the presence cracks. FEA software is available for the modeling task. Nevertheless, it is time consuming and requires numerous analyses and much expertise for the interpretation of the results. The results should preferably be calibrated against strain measurements carried out during the test as the cracks propagate. All these topics make fracture mechanics modeling of the crack growth in tubular joints a difficult challenge. How to cope with the problem the crack growth can be divided into five phases (see Figure 6.23):

1) crack initiation at the weld toe;

2) crack growth of surface semi-elliptical shallow cracks;

3) crack growth of larger semi-elliptical cracks approaching the through-thickness crack;

4) crack growth of a through-thickness crack along the circumference of members;

5) ductile or mixed-mode fracture as the cracks reduce the remaining ligament between joint members to a critical size.

The first phase is the crack initiation and early crack growth, during which micro-cracks emanate from various locations along the weld toe at various times. As these cracks grow, they coalesce to form larger semi-elliptical cracks. The second phase is the crack propagation for shallow surface cracks (part-through) which expands along the weld and in the depth until they approach about 25% of the wall thickness. After this stage, phase 3 takes over. The cracks then propagate until they have penetrated the thickness of the tube wall. After the crack has reached the wall thickness, the fourth phase starts, during which a through-thickness crack continues to grow along the weld intersection. The last phase is the ductile rupture or brittle fracture that results in material separation of the whole tube cross section. This case has already been discussed in relation to the R6 criterion.

Most of the LEFM modeling that has been carried out describes phases 2 and 3 only. Usually single-crack models have been employed. The argument for this limitation has been that the crack initiation period is short and that phase 1 can be ignored. Furthermore, phase 4 can be ignored by defining a through-thickness crack as intolerable. Furthermore, phase 1 can be included in phase 2 by using a short, fictitious, initial crack depth in the fracture mechanics model, as we have discussed for plated joints. We concluded that this approximation has some shortcomings. The remaining fatigue life after a through-thickness crack has appeared has been estimated to be as high as 30% of the total life. This is a relatively long period of time, and it is important from an in-service inspection point of view because the cracks are easy to detect during this last period. Hence, there will be a significant difference in results from risk-based inspection analysis in models that include phase 4, compared to models that ignore this phase. However, in the present case we shall confine our presentation to phases 2 and 3, and the division between them is usually set to a crack depth of $a/T$ in the range between 0.2 and 0.3. This division is made from the influence of the $M_k$ correction due to the weld notch; see Figure 6.8. Phase 2 will be strongly influenced by the local weld notch and the model derived for plated joints can be modified and used. In both phases the geometrical stress concentration will be important. The loading mode will mainly be shell bending of the actual member's cylinder wall. This is because the geometrical stress concentration is dominated by local shell bending at intersections. Stress redistribution will not be so important in phase 2. In phase 3 the effect of the weld bead notch will be less important (see Figure 6.8), whereas the stress redistribution as a function of crack depth and length becomes more and more pronounced. In the following discussion, crack opening mode I, where applied stresses are normal to the crack surfaces, is assumed to predominate.

Regarding the first crack nucleation phase, it is possible to use a local, cyclic stress-strain approach at the weld notch to calculate the time to crack initiation. However, we shall assume that the same practical result can be obtained by a fracture mechanics approach using a fictitious, small, initial crack size.

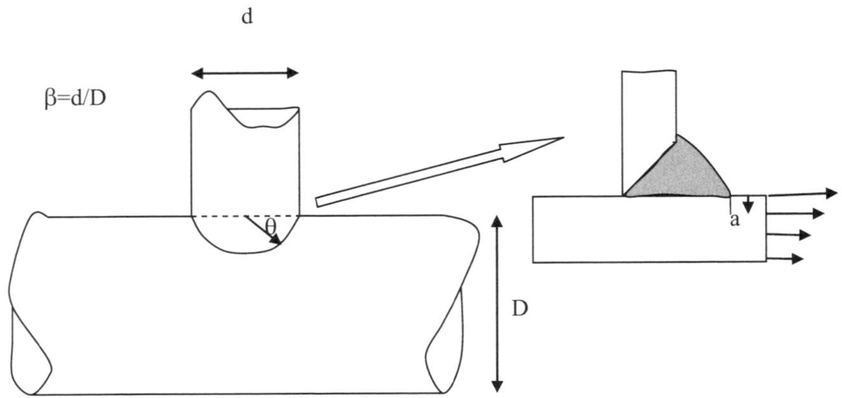


**Figure 6.23.** *T-joint with fatigue cracking from the weld toe at the crown point*

### 6.8.1. *Discussion of current models*

The first model to take account of the load-shedding effect was suggested by Radencovic; see Ref [20]. It is a single-crack model where crack propagation is calculated at different locations along the weld based on known solutions for edge cracks in flat plates under plane-strain conditions. At certain time stages the stresses at each location are modified in accordance with changes in relative stiffness along the cracked weld toe. A more sophisticated and explicit multiple-crack simulation model is presented in Ref [21]. The crack initiation is based on statistical data from welded plated joints, whereas further crack propagation is based on the use of the compliance coefficients derived from a FEM analysis applying thin shell elements. The coefficients are calculated at different stages of the crack propagation and the corresponding SIF is deduced. This method is now presented in the program system SISIF, Ref [21]. Both methods have a theoretically-based calculation of the SIF.

In the present book we shall confine ourselves to presenting a more direct empirical approach. This empirical fracture mechanics model (EFMM) was suggested by Dover *et al*, Refs [22, 23]. The calculations of the SIF are based on empirical calibrations from growth measurements carried out during the UK test program of tubular joints. The SIF is presented in the following form:

$$F(a) = (\frac{0.25T}{a})^{p} B(\frac{T}{a})^{k} \ , \quad a < 0.25T$$

$$F(a) = B(\frac{T}{a})^{k} \ , \quad a > 0.25T \tag{6.38}$$

The geometrical function will be used in conjunction with the geometrical stress range and not the nominal stress range as was the case for plated joints. The SIF calculation follows two different expressions depending on the crack depth range. The first expression is used for phase 2 discussed above, whereas the second expression is used for phase 3. This model has some very appealing features, and one of them is that crack depth at given times can be given on a closed analytical form found by integration of the Paris law. Both the maximum geometrical SCF and the distribution of SCF around the welded intersection between the brace and the chord have a strong influence on the crack growth in phase 3. The governing parameters are chosen to be the diameter ratio $\beta$ between brace and chord, and the stress average factor (SAF). The latter is defined as the maximum SCF divided by the average stress concentration factor $SCF_{AV}$ around the welded intersection. The $SCF_{AV}$ reads as follows:

$$SCF_{AV} = \frac{1}{\pi}\int_{0}^{\pi} SCF(\theta)d\theta \quad \text{axial and out-of-plane bending} \tag{6.39}$$

$$\mathrm{SCF_{AV}} = \frac{1}{\pi}\int_{-\pi/2}^{+\pi/2}\mathrm{SCF}(\theta)d\theta \quad \text{in-plane bending} \tag{6.40}$$

and:

$$\mathrm{SAF} = \frac{\mathrm{SCF}}{\mathrm{SCF_{AV}}}. \tag{6.41}$$

The set of empirical parameters $p$, $k$, and $B$ in equation (6.38) can now be determined as follows:

$$B = (0.669 - 0.1625 \cdot \mathrm{SAF})(T/T_{\mathrm{Re}f})^{0.11}$$

$$k = (0.353 + 0.057 \cdot \mathrm{SAF})(T/T_{\mathrm{Re}f})^{-0.099} \tag{6.42}$$

$$p = 0.231 \cdot (T/T_{\mathrm{Re}f})^{-1.71}\beta^{0.31}\mathrm{SCF}^{0.18}.$$

As can be seen, the model does not explicitly consider the notch effect of the weld and the through-thickness stress distribution when calculating the SIF. One may argue that the fitting parameters $p$, $q$, and $k$ are only applicable to describe the results from the tests that are carried out. The ability of the model to predict crack growth for other joint geometries and loading modes are not substantiated. To validate its generality, the model was used in this book on a series of large-scale tubular joint tests carried out by IRSID, France (see Ref [24]). During these tests, crack growth paths were determined by beach marking. A typical joint is shown in Figure 6.23. The measured crack depths were plotted against the applied number of cycles and compared with the calculated number of cycles. The data used in the EFMM are given in Table 6.6. As seen from Figure 6.24, the results are remarkably good compared with measurements for larger crack depths. One dominant feature for the results is the almost linear trajectory of the curves. The strong acceleration found for the plated joint before the crack propagates through the thickness is not present. We have explained why: it is the load-shedding effect that makes this crack evolution possible. A further comparison for one particular joint thoroughly analyzed by the SISIF program, the SIF factors are listed in Table 6.7 together with results obtained using the ABAQUS software. This program calculates the SIF based on a virtual-crack extension and the associated J-integral (see Appendix A). This latter calculation is based on refined FEM analysis by first calculating the J-integral, then the associated SIF. As can be seen, the empirical $\Delta K$ from the EFMM is about 15% less than the values obtained by SISIF and 20% less than the figured obtained by a full FEA with a cracked member.

| Node | Loading mode | Branch diameter | Chord thickness | $\beta$ | SCF | Geometrical Stress range | SAF |
|------|------|------|------|------|------|------|------|
| C | Axial | 1280 | 75 | 0.27 | 1.91 | 102 | 1.08 |
| B' | IPB | 685 | 40 | 0.50 | 1.89 | 154 | 1.60 |
| D' | IPB | 950 | 40 | 0.72 | 3.91 | 148 | 1.50 |
| E' | IPB | 1280 | 75 | 0.54 | 1.98 | 131 | 1.47 |

**Table 6.6.** *Data used for the EFMM to simulate experimental results*

Growth parameter:

C = 6E-12 (Mpa, m)

m = 3

| Crack depth | EFMM | SISIF | ABAQUS |
|------|------|------|------|
| 11 | 660 | 797 | 828 |
| 14.6 | 732 | 893 | 917 |
| 21.8 | 779 | 896 | 1059 |

**Table 6.7.** *Comparison with SIF calculations by EFMM and more sophisticated models (units MPa,m)*

Finally, the EFFM was tested for an X joint tested in IPB at Agder University College.

The experimental set-up was briefly presented in Chapter 3. The joint was subjected to IPB with the data given in Table 6.8

| Nominal stress range S (MPa) in brace | Hot-spot stress range (brace) | Cycles to reach crack a = 2 mm | Cycles to through-thickness crack | According to T-class median |
|------|------|------|------|------|
| 88 | 230 | 2.0E5 | 4.1E5 | 3.8E5 |
| S | SCF | $SCF_{AV}$ | SAF | |
| | 2.6 | 1.8 | 1.45 | |

**Table 6.8.** *Experimental data for X-joint*

The following figures were used in the analysis:

SCF = 2.6, $SCF_{AV}$ = 1.8, SAF = 1.45.

The results are shown in Figure 6.25. In this case we have chosen an initial crack depth of a = 0.32 mm to make the calculated fatigue life coincide with the experimental life. As can be seen the four measured points indicate a somewhat more curved *a-N* relation than the calculated one. The difference in the number of cycles to reach given crack depths is up to 30%. However, other reasonable choices of the growth parameters *C* and m and m can reduce the gap between measured and calculated crack histories.

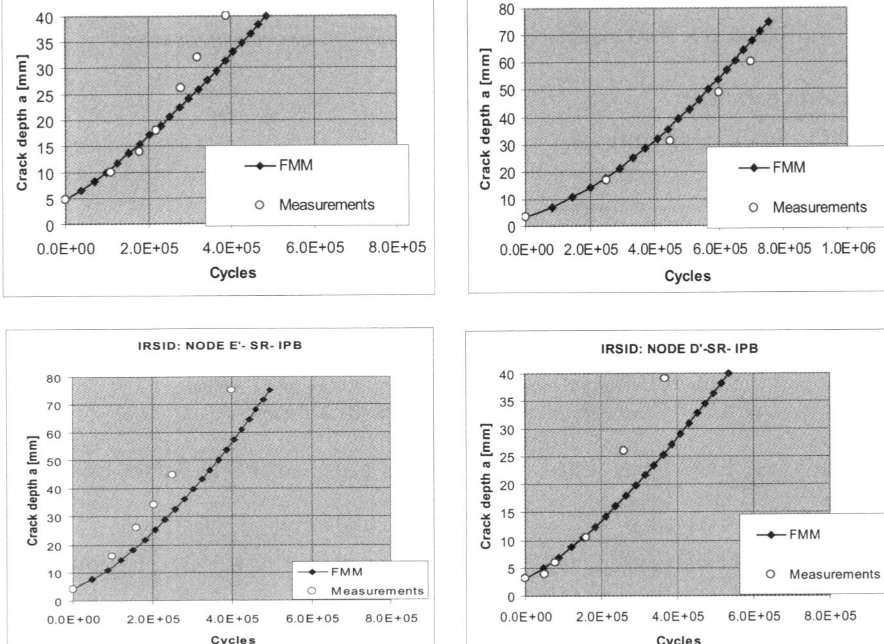

**Figure 6.24.** *EFMM model predictions and experimental measurements*

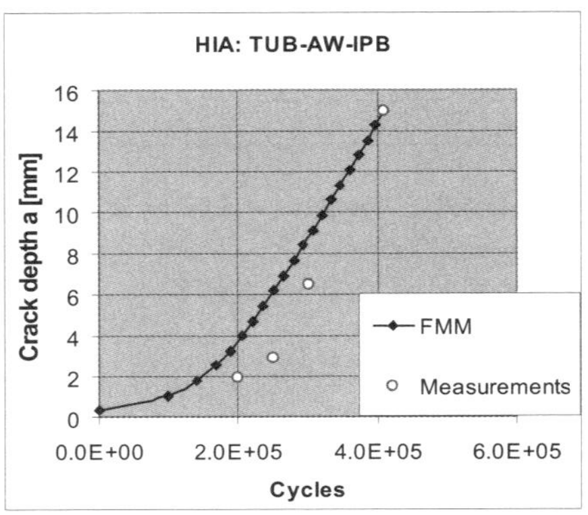

**Figure 6.25.** *EFMM model prediction and experimental measurements*

### 6.8.2. *Conclusion on the empirical fracture mechanics model*

As discussed above, one may ask if the EFMM has sufficient theoretical basis to be capable of predicting crack growth when the loading mode and geometry are totally changed compared to the fatigue tests from which the model parameters were obtained. As we have seen, it gives acceptable results for the behavior of larger cracks in similar joints. However, an obvious objection to the method is that it does not explicitly take into account the weld bead geometry or the geometrical bending stress distribution through the member wall. The model has been included in this book for pedagogical reasons because it can provide the reader with insight into the problem of crack growth in tubular joints. The growth model is included in the spreadsheet FLAWS-CG (found in Appendix C) and the reader can carry out some analysis to see how various parameters influence crack growth and final fatigue life.

### 6.8.3. *Proposal for model improvements*

For a full fracture mechanical analysis of welded tubular joints the reader should follow the guidance from Ref [9]. According to this guidance, the following topics should be addressed:

a) Global structural analysis to determine the nominal stresses in brace members.

b) Usually the analysis is carried out by frame analysis using beam elements.

c) Local stress analysis of the joints to determine the geometrical SCF and the proportion of membrane and bending stresses through the tube walls. Usually the analysis is carried out by a FEA using thin shell elements.

e) Generation of the hot-spot stress range histogram for the joint, based on load history.

f) Crack growth analysis, according to the Paris law.

g) Determination of critical crack depth using the R6 criterion.

The SIFR is calculated using the Newman-Raju solution, which we have shown for plates in equations (6.11) and (6.12). The weld notch magnification factor has to be taken into account. Alternatively, the first phase of crack propagation (up to $a/T = 0.25$) can be treated by the detailed fracture mechanics using the weight function method. This will require a global thin shell FEM analysis of the joint, in combination with a two-dimensional plane-strain stress analysis of a sub-model containing the actual member wall thickness and weld toe profile. A reasonable assumption during this phase is that crack redistribution along the weld has not yet become important, and cumbersome corrections of the stress field through the thickness are not necessary. This phase will explicitly take into account the weld geometry through the calculated stress gradient. For larger cracks it is essential to include the load-shedding effect due to the reduction of stiffness of areas with large crack sizes. A "moment release model" is suggested. This model predicts the reduction of the moment acting over the cross section of the cracked member; see Ref [9]. The model involves the reduction in the bending stress components in the following form:

$$\sigma_{bc} = \sigma_b (1 - \frac{a}{T})$$

(6.43)

where $\sigma_{bc}$ and $\sigma_b$ are the bending hot-spot stress components for cracked and uncracked joints, respectively. The bending part of the SIFR is reduced proportional to the decrease in bending stresses. For further details the reader is referred to Ref [25]. A similar approach was used in Ref [26]. Instead of using the approximation given by equation (6.43), one should continuously update the stresses along the tube circumference by a global FEA model of the joint. The results should then be used in conjunction with a refined sub-model that contains the crack. This model is known as the crack-box method and will be elaborated in Chapter 11.

## 6.9. A brief overview of stiffened panels

Thus far we have considered the behavior of relatively small cracks that emanate from the weld toe. The SIF has been calculated for cracks that have a depth less than the plate thickness and for which the weld notch is an important influence. In some structures that are damage tolerant, it is important to model larger cracks that are through the thickness and are slitting over a large part of the stiffened panel. Typical examples are bottom and deck structures in large tankers; see Figure 6.26. In these cases cracks may start from a weld in the vicinity of a scallop hole in the stiffener and grow through the thickness of the plate. As the crack continues to grow as a trough-thickness crack, the crack fronts will approach the neighboring stiffeners as shown in Figure 6.26. The geometry function for such a crack is given in Figure 6.27, both for the case where the stiffener in the center of the crack is intact and where it is assumed to be slit by the crack together with the plate. The geometry function for a wide plate without stiffeners is drawn for comparison. As expected, the geometry function for the stiffened panel is lower than that for the wide plate when the stiffeners are intact. This is particularly true when the crack fronts are near the position of the stiffeners. The reason is that there is a load transfer from the plate to the stiffeners that results in a stress decrease at the crack fronts in the plate. When using these geometry functions for predicting crack growth, the obtained curves will have a somewhat slower acceleration for larger cracks than the curves shown in Figure 6.18 and Figure 6.22. The curves will, however, be less linear than was the case for the tubular joints shown in Figure 6.24.

If an unstable fracture sets in when the crack fronts are located between the stiffeners, the crack may be arrested when the crack fronts propagate to the stiffeners. This crack arrest may take place provided that the stiffeners are capable of sustaining the increased loading. A good overview is given in Ref [1]. Matters may become complicated because the crack fronts will run into a residual stress field caused by the fillet weld between the stiffeners and the plate. Details are found in Ref [27].

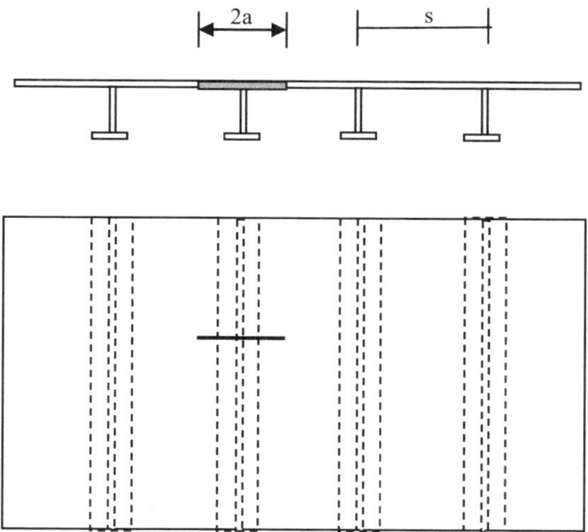


**Figure 6.26.** *Large fatigue crack in a stiffened panel*

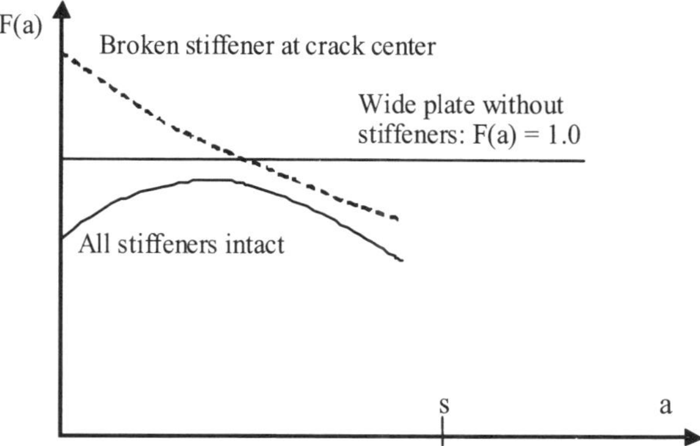


**Figure 6.27.** *Sketch of geometry functions for stiffened panels*

## 6.10. Units and conversion for fracture mechanics parameters

Due to the mathematical expression for the SIF (equation (6.4)) where the square root of the crack length occurs, we obtain slightly peculiar units for this parameter, e.g. MPa m$^{0.5}$. As a consequence, the conversion from one unit system to another is not straightforward. To assist the reader, the most common conversion for the SIF and the crack growth parameter C in the Paris law are given below:

– units system 1: stress in MPa, crack length in m;

– unit system 2: stress in N/mm$^2$, crack length in mm.

(The stress will have the same numerical value for systems 1 and 2.)

The following conversions are given:

$$K_{I1} = \frac{K_{I2}}{\sqrt{1,000}} = \frac{K_{I2}}{31.62} \tag{6.44}$$

$$C_1 = C_2 \frac{1,000^{m/2}}{1,000} \tag{6.45}$$

where subscript 1 refers to unit system 1 and subscript 2 refers to unit system 2. If the exponent m in Paris law equals 3.0, the last equation gives $C_1 = 31.62\ C_2$.

## 6.11. Industrial case: fatigue re-assessment of a welded pipe

### 6.11.1. *Introduction*

We shall return to the offshore loading buoy presented at the end of Chapter 5, in Figure 5.26, where we carried out fatigue life verification for the central turret structure. The item to be analyzed in the present case is the vertical steel pipe in the center of the same turret. The steel pipe has welded attachments for hang-off and support within the turret. The pipe in question was retired from the loading buoy at one field (Field 1) after 5 years of service and was re-installed in a similar loading buoy at another field installation (Field 2) for 10 more years of service. A substantial remaining fatigue life of the steel pipe from Field 1 had to be verified before using it at Field 2. Rules and regulations give some guidance on how to tackle the problem; see Refs [9, 28, and 29]. There are usually two main approaches to demonstrate the fitness for purpose of aging items. The first is to verify a remaining fatigue life that is satisfactory, based on the S-N curve method; see Ref [29]. Alternatively, a remaining fatigue life can be verified with the presence of a crack-like defect based on inspection results and fracture mechanics; see Ref [9]. If no cracks are detected during an inspection, the initial crack equals the crack size that can be missed during the inspection. If a crack is revealed during inspection, one has to either prove that remaining life is satisfactory, using the presence of the crack based on fracture mechanics or by carrying out grinding to get rid of the crack. In the latter situation one can carry out an S-N verification, treating the welded details as good as new. The main objective of the present example is to illustrate how these analyses can be carried out and decisions made with the support of the current rules and regulations. The practical engineering aspects of the problem are emphasized. However, as the rules open for a stochastic analysis and risk-based inspection planning, we can apply a generic fatigue reliability model for welded joints as a supplement to the verification scheme given in the rules and regulations. We shall return to these issues in Chapter 7.

### 6.11.2. *Description of the loading buoy with steel pipe*

The offshore loading buoy was shown in Figure 5.26, Chapter 5. It consists of an outer buoyancy cone with an internal turret. Shuttle tankers that have a mating cone in the bottom hull structure use the loading buoy. The buoy can then be adapted and connected to the tankers. The steel pipe in question is situated in the center of the turret and allows the tanker to connect to the sub-sea riser. At its lower end the pipe is connected to the flexible riser and at the top end to a crude oil swivel: see Figure 5.26. The steel pipe with welded attachments is a C-Mn steel with yield stress of 345 Mpa. It has cathodic protection from sacrificial anodes. The pipe has a welded ring support in the lower end and a cast item in the top support; see Figure 6.28. The lower ring plate is stiffened by vertical welded brackets as shown; the top support is not shown in the figure. The lower ring support is simply sustained horizontally by the inner turret wall. The stresses in the pipe are dominated by the dynamic bending moments from the riser. The critical hot-spots (HS) are at the noses of the welded brackets that support the lower ring. As can be seen from Figure 6.28, the brackets have a concave shape with a gradual transition to the outer surface of the pipe wall. This is a favorable fatigue design. Each toe of the brackets is a hot-spot area categorized as an F2 detail for fatigue life verification, according to Ref [29]. The nominal stresses can then be calculated by simple beam theory for the pipe. Alternatively, an FEA can be carried out and the geometrical stresses can be determined; see Figure 6.28, right. In this case a D-class S-N curve can be used as has been shown for the turret in Chapter 5. It is also possible to use a b-linear S-N curve instead of one linear curve. Further details are given in Ref [30].

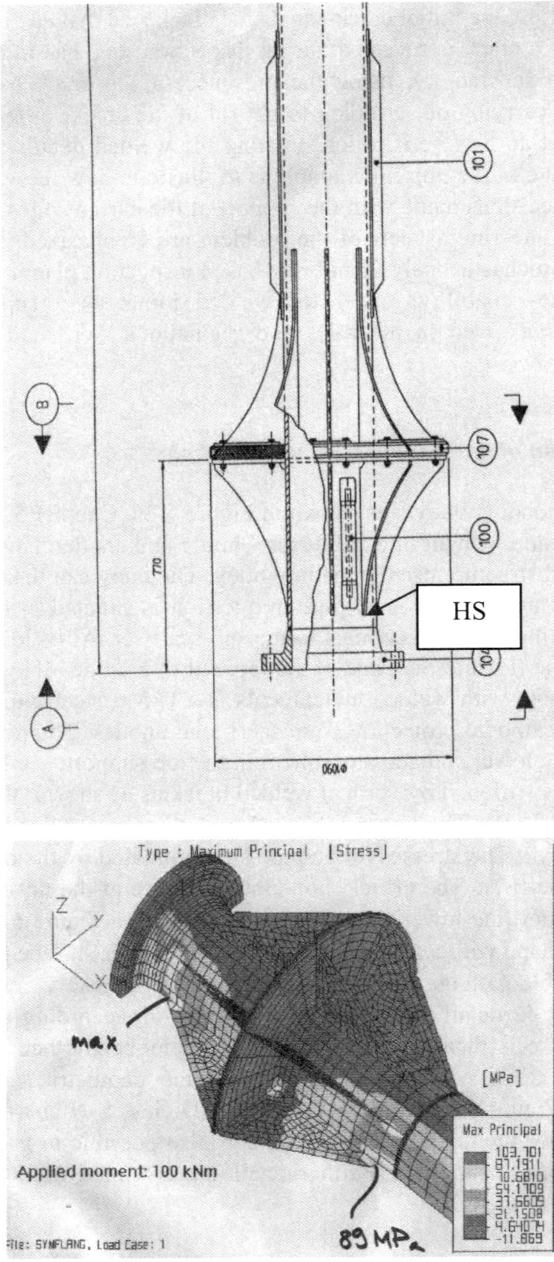



**Figure 6.28.** *Drawing of flowline (above) and stress plot from FEA (below). Courtesy of APL*

### 6.11.3. *Replacement and inspection strategy*

The steel pipes are interchangeable from one buoy to another. After a given period of service, the pipes can be retrieved from the buoy and inspected before being entered into service in another loading buoy at other fields. In such cases a fitness for purpose analysis has to be carried out. When extending the fatigue life of a structural item that has passed phase 1 and entered into a phase 2, there are two alternative ways of carrying out the necessary verification; see Ref [29]. The first is to prove that the fatigue life calculated by the S-N method for phase 1 and phase 2 is longer than the total extended service life multiplied with the required fatigue design factor (FDF). If this is not the case, an inspection is usually carried out after phase 1. The inspection can be carried out *in-situ* and often sub-sea. It is much rarer that the item is retrieved from service for examination in a shop environment. If such an examination is undertaken, the performance of the inspection technique will be far better, i.e. the crack size that may be missed will decrease significantly. Based on inspection results, a fracture mechanics analysis has to be carried out to verify a satisfactory fatigue life remains for phase 2. The S-N based verification will simply read:

$$\mathrm{PFL}_2 = \frac{(1 - D_1)L}{D_2} \tag{6.46}$$

where PFL denotes the predicted fatigue life. $D_1$ is the Miner's sum pertaining to the service time that has elapsed during phase 1, whereas $D_2$ is the Miner's sum during a part $L$ of the remaining service life. $L$ may typically be 1 year. When $D_1$ is calculated for phase 1, one must make a correction for the required FDF as follows:

$$D_1 = \frac{\mathrm{TSL}_1 \cdot \mathrm{FDF}}{\mathrm{PFL}_1} \tag{6.47}$$

where $\mathrm{TSL}_1$ is the elapsed service life and $\mathrm{PFL}_1$ is the original PFL for the phase 1 condition. The FDF is normally set to 3 for welded details that it is possible to inspect and repair, whereas it is set to 10 when inspection is not scheduled and the consequences of fracture are severe. In the present case with a welded pipe and an internal flow of hydrocarbon under pressure, the target is to obtain FDF = 10 for the entire service phase. The pipe cannot be inspected *in-situ* sub-sea and is retrieved for examination in a shop environment when replaced. Details are found in Ref [30]. The history and fatigue life predictions for the two fields are given in Table 6.9. As can be seen, the pipe under consideration was retrieved from Field 1 in 2001 after 5 years of service and was installed at Field 2 in 2003 for 10 more years of service. As was the case in former rules, the S-N curves were assumed to be the same in seawater with cathodic protection as in air; see Ref [29]. The data for the single-slope F2 curve are logA = 11.63 and m = 3.

| Field | Installation | Retirement | Comments | Annual damage | Originally PFL | TSL years |
|---|---|---|---|---|---|---|
| 1 | Summer 1996 | Summer 2001 | Inspection, no findings | 0.0167 | 60 years | 5 |
| 2 | Summer 1995 | - | Replacement 2003 | 0.0081 | 123 years | 10 |

**Table 6.9.** *History of the steel pipes and fatigue life prediction*

Due to difference of the riser load spectrum at Field 1 and Field 2, the predicted fatigue life at the critical hotspots is 60 and 123 years respectively for a new pipe. The difference is partly due to different riser configurations and wave loading, and partly due to the time the shuttle tanker is connected to the buoy. In idle position, the buoy is lowered to a sub-sea position and no loading and damage accumulation are foreseen.

### 6.11.4. Re-assessment based on the S-N approach

After the 5 years of service at Field 1, the accumulated damage is, according to Table 6.9:

$$D_{1t} = \frac{TSL_1 \cdot FDF}{PFL_1} = \frac{5 \cdot 10}{60} = 0.83$$

Hence, most of the fatigue life has already expired when an FDF of 10 is required. At Field 2, a fatigue life estimate of $6.8 \ 10^7$ cycles was found based on an equivalent stress range of 18.5 Mpa. The equivalent stress was found from the long-term stress spectrum. The tanker is assumed to be connected 16% of the time only, and that the annual number of loading cycles will be $5.5 \ 10^5$ cycles. Hence, the annual damage and predicted fatigue life at Field 2 will read:

$$D_{2a} = \frac{5.5 \cdot 10^5}{6.8 \cdot 10^7} = 0.0081 \qquad PFL_2 = \frac{1}{0.0081} = 123 \text{ years}.$$

If we regard the pipe as "as good as new", this will be the fatigue life estimate. It will, as we have discussed, be valid if toe grinding of the welds has been carried out; see Ref [29]. However, if we add the damage accumulated at Field 1 with a safety margin of FDF = 10 we will get:

$$PFL_2 = \frac{\left(1-D_{1t}\right)}{D_{2a}}\frac{1}{} = \frac{0.17}{0.0081} = 21 \text{ years}$$

As the required FDF is set to 10 to avoid any inspection during the remaining planned service life, the prediction is not acceptable. We should have obtained a predicted fatigue life of at least 100 years. The total FDF for both service phases is 6. Consequently, we must use an approach based on inspection and fracture mechanics modeling to verify that a satisfactory fatigue life remains.

### 6.11.5. *Re-assessment based on fracture mechanics*

The hot-spots in this area were examined in detail by both MPI (magnetic particle inspection) and liquid penetrant injection (LPI) in order to achieve a high probability of detection. The inspection was carried out in a shop environment for all hot-spots vulnerable to fatigue cracks, see Figure 6.29. No cracks were found. One may now argue that the S-N result (23 years), with a full reduction due to elapsed service time, is overly pessimistic because is does not take into account the inspection results. On the other hand, the estimate (123 years) based on the "as good as new" assumption is too optimistic for two reasons:

1) at such low stresses the S-N life may contain an initiation period that has elapsed due to the usage at the Field 1. Hence, even if no cracks exist, the S-N life prediction is too long due to already-accumulated damage;

2) the NDI carried out may miss cracks of small sizes.

Based on the NDI performance it is assumed that a fatigue surface crack with a depth of 0.5 mm may exist in a hot-spot after the inspection. This depth usually corresponds to a surface length of 3-5 mm based on normal surface crack appearance with a semi-elliptical shape; see equation (6.35). Larger cracks will be detected by detailed inspections. The probability can be found from a probability of detection (POD) curve for the method in question.

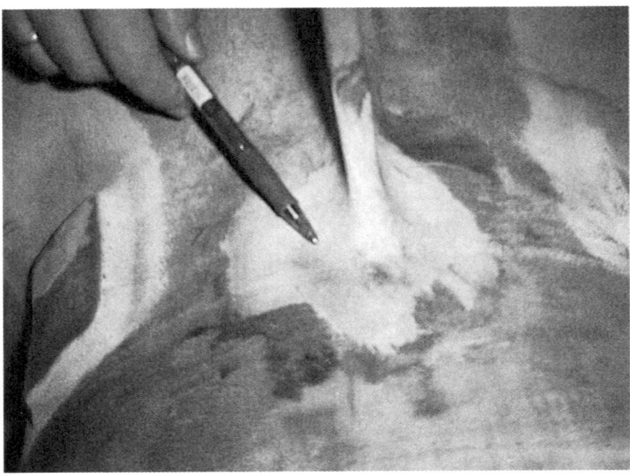

**Figure 6.29.** *Detailed examination of the hot-spots
at bracket toe with liquid penetrant*

A fatigue crack growth analysis is carried out by applied fracture mechanics and adopting the Paris law, equation (6.36). The following data are used:

   – initial crack size $a_i$ = 0.5 mm (may escape an MPI, LPI inspection);

   – final crack size $a_f$ = 8 mm (half the wall thickness);

   – crack growth parameters: C = 1.2E-11, m = 3 (units MPa, m – according to BS7910, Ref [9], median +2 standard deviations);

   – the geometry function $F(a/T)$ for fillet weld F-class is used; see Figure 6.14;

   – applied stress range is $\Delta\sigma$ = 1.13 × 18.5 = 21 MPa (1.13 is the difference between F- and F2-class stress concentration factor; see Ref [29].

The data correspond to a number of cycles close to N = 3.5E7 cycles. The crack evolution is shown in Figure 6.30. The analysis is carried out by the spreadsheet FLAWS-CG (see Appendix C). The number of cycles to failure in Figure 6.30 equals 10 years of continuous use. Again, based on the assumption that the tanker is connected 16% of the time only, we obtain:

$$\text{PFL}_{2,FM} = \frac{3.5 \cdot 10^7}{5.5E5} = 63 \text{ years}$$

As can be seen, our fatigue life estimate is between the two results obtained from the S-N model analyses.

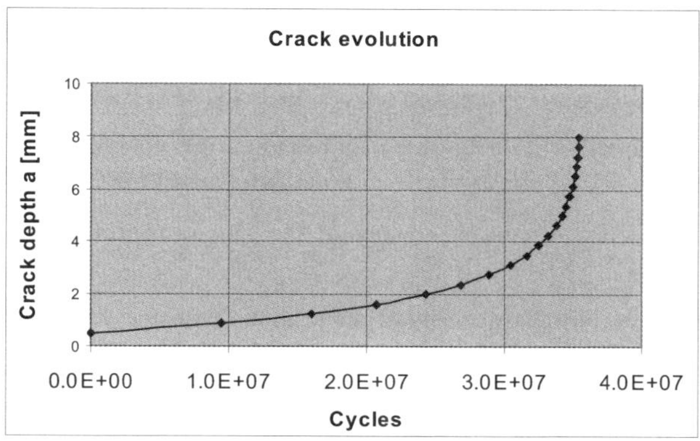



**Figure 6.30.** *Likely crack evolution during phase 2
after inspection at the end of phase 1*

In conclusion, the pipe can be installed in the buoy at Field 2 and remain in service for 6 years without any inspection. Based on reliability and risk analysis, one can try to argue for leaving the pipe in service to the end of phase 2; see Chapter 7.

## 6.12. References


1    D. Broek, *The Practical Use of Fracture Mechanics*, Kluwer Academic Publishing, 1989

2    J. Schive, *Fatigue of Structures and Materials*, Kluwer Academic Publishing, 2001

3    N. Recho, *Ruptures par fissuration des structures*, Hermes, 1995

4    H.M.W. Westergaard, "Bearing pressures and cracks" *Trans. ASME J. Appl. Mechanics*, 1939

5    J.C. Newman and I.S. Raju, "An empirical stress-intensity factor equation for the surface crack". *Engng. Fract. Mech.* 1981, pp 185–92

6    P. Albrecht and K. Yamada, "Rapid calculation of Stress Intensity Factors", *Journal of Structural Engineering* 1977, pp 377–89

7    D. Bowness and M.M.K. Lee, "Prediction of weld toe magnification factors for semi-elliptical cracks in T-butt joints", *International Journal of Fatigue* 1999, 22, pp 369–87

8    D. Bowness and M.M.K. Lee, "Prediction of weld toe magnification factors for semi-elliptical cracks in T-butt joints – comparison with existing solutions", *International Journal of Fatigue* 1999, 22 pp 389–96

9    B.S. 7910, 2005. *Guidance on Methods for Assessing the Acceptability of Flaws in Fusion Welded Structures*, London, British Standards Institution (BSI)

10  R.G. Foreman, *et al*, "Numerical analysis of crack propagation in cyclic loaded structures", *J. of Basic Engn. Trans*, ASME D89, 1967 pp 459–64

11  W. Elber, "The significance of fatigue crack closure", *ASTM STP 486*, pp 230–42

12  Y. Verreman and B. Nie, "Early development of fatigue cracking at manual fillet welds", *Fatigue & Fracture of Engineering Materials and Structures* 19 (6), 1996, pp 669–81

13  K. Engesvik, "Analysis of the uncertainty of the fatigue capacity of welded joints", NTNU, Trondheim, UR-82-17

14  T.R. Gurney, "Finite element analysis of some joints with the welds transverse to the direction of stress", *Welding Research International 6*, 1976, pp 40–72

15  T. Lassen, "The effect of the welding process on the fatigue crack growth in welded joint", *Welding Journal* 69 (2), 1990, 75-s – 85-s.

16  P.D. 6493, 1991. *Guidance on Methods for Assessing the Acceptability of Flaws in Fusion Welded Structures*, London, British Standard Institution (BSI)

17  G.O. Johnston, "Statistical scatter in fracture toughness and fatigue crack growth rates", ASTM STP 798, Probabilistic fracture mechanics and fatigue methods: application for structural design and maintenance, 1983, pp 42–66

18  P. Darcis, *et al*, A Fracture Mechanics Approach for the crack growth in welded joints with reference to BS7910, ECF15, Stockholm, 2004

19  G. Lebas and J.C. Fauve, *Collection de données de fatigue*, Pau, Elf Aquitaine, 1988

20  D. Radencovic, "Calcul de durée de vie des noeuds tubulaires Ecole Polytechnique", Laboratoire de mécanique des solides, CNRS, 1982

21  M. Serror, *et al*, "Simulation de fissuration, Logiciel SISIF", Journée du CLAROM, France, 12 November 1992

22  J.C.P. Kam and W.D. Dover, "Structural integrity of welded tubular joints in random load fatigue combined with size effect", International Conference of Offshore Structures, Glasgow, September 1987

23  S. Darmavasan and W.D. Dover, "Non-destructive evaluation of offshore structures using fracture mechanics", Applied Mechanics Reviews (14), 1988, pp 36–49

24  A. Bignonnet, *et al*, "Experimental study of fatigue crack propagation in welded tubular joints", IRSID rapport no IIS-XIII-1166-85, 1984

25  H.C. Rhee, *et al*, "Reliability of solution methods and empirical formulae of stress intensity factors for weld toe cracks in tubular joints", OMAE 1991, 3B, pp 441–52

26  E. Niemi (Editor) and N. Recho; " Probabilistic approach to fatigue life in T-welded tubular joints Tubular Structures, Third International Symposium, Finland 1989

27  E. Ayala-Uraga and T. Moan, "Time-variant reliability assessment of FPSO hull girders with long cracks". Proceedings of OMAE 2005, June 12-16, Halkidiki, Greece, pp 1-10.

28  NORSOK, Standard N-004, Rev 2, October 2004, Chapter 10, "Re-assessment of structures"

29  DNV RP-C203 Fatigue design of offshore structures – recommended practice DNV 2005

30  APL report, "Fatigue Analysis of Inner Flowline", report no 6330-94573-DE, Advanced Production and Loading

# PART II

# Stochastic Modeling

# Chapter 7

# Stochastic Modeling

## 7.1. Introduction and objectives

We have already encountered the problem of uncertainty several times during the course of the present text. In Chapter 3 we discussed and modeled the scatter in S-N data, in Chapter 4 we treated the fatigue loading as a stochastic process. In Chapter 5 we discussed the uncertainty in the Miner's summation rule and the problem of accuracy when determining the geometrical stresses by finite element analysis (FEA). The highly variable and irregular local weld toe geometry was also discussed. In Chapter 6 we presented the probability distributions for the growth parameter C in the Paris law, as well as for the initial crack depth. Finally, we briefly touched the probability of detection (POD) concept for characterizing the performance of a given inspection technique. All these issues underline the need for stochastic modeling when describing the fatigue behavior of welded joints. Stochastic modeling means taking account the various sources of uncertainty in a logical, rational, and consistent manner. Our tools are statistical methods and probability calculations. In this chapter we shall go into more details about the methods of stochastic modeling. If the analysis is concerned with the probability of survival (or failure) only, the calculation scheme is called *reliability analysis*. If the probability of failure is weighed against the consequences of a fracture it is called *risk analysis*. The latter concept is now extensively used for risk-based inspection planning.

The objective of the present chapter is to give the reader insight and knowledge of the models and methodology that are useful when dealing with the issue of uncertainty in relation to the fatigue problem for welded joints. First, the general concepts of a stochastic analysis are outlined. The limit state functions and safety margins are explained. Then a short review and repetition of elementary life models that are often employed for fatigue reliability assessments are given. Finally, a more advanced random variable model in combination with the Monte Carlo simulation is presented. The practical application of these methods in conjunction with applied fracture mechanics is emphasized. An alternative model using a Markov chain model for the fatigue damage evolution is demonstrated and discussed for decision-making regarding design and in-service scheduled inspections. Finally, the chapter presents a generic reliability model that can be used as a tool for risk-based inspection planning for most welded joints.


## 7.2. Overview of models and methodology

### 7.2.1. *Sources of uncertainty*

Stochastic modeling is a rational attempt to cope with the problem of uncertainty. Various sources of uncertainty are often categorized as (see Ref [1], (DNV 1996)):

– physical uncertainty;

– measurement uncertainty;

– statistical uncertainty;

– model uncertainty.

The physical uncertainty is the natural randomness of a variable, as demonstrated in Chapter 3, for the local weld toe angle and radius. Measurement uncertainty can, for example, be related to strain measurement carried out by strain gauges or the crack depth measurement during the course of one fatigue test. The statistical uncertainty is related to the limited number of observations available. This uncertainty is often dealt with by specifying a confidence interval for the estimated mean and variance of a variable. We gave an example for this at the end of Chapter 3 where we calculated a 75% confidence level for the standard deviation of fatigue life. Model uncertainty is due to imperfections and idealizations in the model formulation. In the present text, both the S-N model for fatigue life prediction and the Paris law used to describe the crack propagation are semi-empirical approximations to a complex problem. From this point of view, one may argue that the scatter in the material parameters C and m is partly due to physical uncertainty, but the scatter also reflects the model uncertainty in the Paris law. A sharp distinction between the different sources of uncertainty is not always possible.

### 7.2.2. *Introduction to the random variable model and related methods*

In this book, two main models for the stochastic modeling of the fatigue problem are reviewed and discussed. These models are:

– the random variable model; and

– the stochastic process model.

In a *random variable model*, the total physical variability in the fatigue process is explained and accounted for by considering the involved parameters as stochastic variables. For the stochastic process model, the variability is lumped together and is inherent in the evolution of the fatigue process. A principal sketch of the problem of uncertainty is shown in Figure 7.1. The source or basic variables are shown to the left in the figure (only two variables $z_1$ and $z_2$ are shown), whereas the result variable y is shown to the right. If we use the model of random variables we have to specify the mean value, standard deviation and frequency functions for the basic stochastic variables on the left-hand side of the sketch. After having carried out a calculation, the result variable on the right-hand side can be determined. This result variable y is usually related to crack propagation and related fatigue life. As the basic variables $z_1$ and $z_2$ entering the calculations are stochastic variables, the results variable y will also become a stochastic variable. The variable y can then also be characterized by its mean, scatter and frequency function. The variable y may be the number of cycles to failure for a given load spectrum or crack depth after a given time period with the same given load spectrum. In the first case, both an S-N model or a fracture mechanics model can be used for the calculation in Figure 7.1. In the latter case, it is only a fracture mechanics model that can be used as basis for the calculations. The frequency function $f_y(y)$ is obtained from many simulations where each simulation generates a set of the variables $z_1$ and $z_2$. Once the frequency function $f_y(y)$ has been determined, the probability that the y exceeds a given value can be calculated. If y is the fatigue life, one can determine the probability that the fatigue life is less than a given limit. If y denotes the crack size, one can determine the probability that the crack size is greater than a given critical value. These events lead to the definition of a limit state function which will be explained later. The calculation scheme demonstrates that the fatigue life is a stochastic variable, as was discussed in the beginning of Chapter 5.

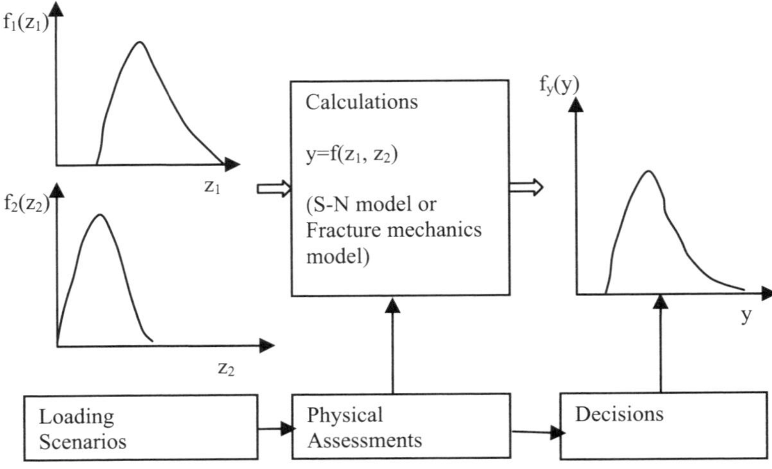


**Figure 7.1.** *Illustration of the problem of uncertainty for fatigue behavior of welded joints*

The bottom of Figure 7.1 also shows the role of the engineer during the modeling. The engineer must initially choose between various loading scenarios and carry out physical assessments during the course of the calculations. The final probability distribution of y will be used as a support for good decision-making based on probability of failure as a function of time. If these probability levels are not acceptable, then changes have to be made regarding design of the joint and/or inspection planning. These types of non-numerical assessments must be carried out by experienced engineers or by using guidance given by, for example, classification societies.

If we do not attack the problem of uncertainty by using the basic variables as explained, but try to characterize the stochastic behavior of the damage evolution directly, the approach is called a *stochastic process* method. In the family of stochastic process models, the Markov chain technique (MCT) has proven to be well suited for the purpose of modeling the fatigue process. It is a Markov chain for which crack depths are related to the model states. The MCT model was suggested by Kozin and Bogdanoff (1985) (see Ref [2]) for general crack growth problems. In Ref [3] the model's ability to deal with the fatigue reliability of welded joints was developed and demonstrated. The final aim was to recommend the most appropriate model for the reliability analysis of welded joints under in-service conditions with scheduled inspections.

### *7.2.3. Requirements for a stochastic model*

As already stated, stochastic modeling involves statistical methods and probability calculations. One important issue is that the applied stochastic model does not conceal the underlying mechanical problem that should be well understood from the start of an analysis. Hence, simplicity is the essence to avoid skepticism or criticism from the practicing engineer. With this pedagogical point of view as the basis, the following model requirements should be met:

1) The model should have a general structure that includes the major sources of uncertainty.

2) It should predict correctly at least the first- and second-order statistical moments in time to reach given crack depths. Likewise, the probability density function (PDF) for the crack depth at given times should be correctly predicted.

3) The sample functions (*a-N* curves) generated by the model should exhibit the major characteristics of experimental measurements.

4) The model should be able to take into account the influence of in-service inspection, both in relation to predicting the most likely influences of future inspections and updating, based on results from inspections already carried out.

The first of these requirements is often satisfied by the use of a fracture mechanics concept, accounting for different geometries, boundary conditions, fabrication procedures, materials, and loading conditions. The uncertainties may either be taken into account through each basic variable in the model or by a bulk approach as discussed for the stochastic process. Requirement 2) is a must if any confidence is to be placed in the obtained probability of failure. The third requirement often ends up with a discussion on the shape of the sample functions as they were shown in Figure 3.8, Chapter 3. If the curves are irregular and are intermingling, this indicates a spatially distributed random crack growth due to material inhomogenities. This will often be the case for small crack growth in the microstructure of the heat-affected zone (HAZ). From an experimental point of view, this is an intra-specimen variability. This will favor a stochastic process model, as a random variable model cannot directly simulate the irregularities found within each sample curve. On the other hand, if the *a-N* curves are smooth and well separated, their paths are more likely to be dominated by variability in joint geometry or weld toe profile. This is described as inter-specimen variability, and is best modeled by a random variable model. A pure Markov chain model is not suitable for this type of scatter.

When inspecting the experimental crack growth curves in Figure 3.8, Chapter 3, it is seen that both types of characteristics are present. Both the intra-specimen and the inter-specimen variabilities are present. However, requirement 3) is generally considered less important than the former ones, and only qualitative considerations will be given to this issue.

The last requirement is a question of how the uncertainties in the performance of scheduled non-destructive inspections (NDIs) are incorporated into the stochastic model, and how easily the model is updated when additional information is available after an inspection has been carried out. As we shall see, the Markov Chain model that works directly with damage states related to crack depth has some very attractive features with regard to these issues.

### 7.2.4. *The concept of the limit state function and the safety margin*

Before proceeding, we will define some important concepts related to stochastic modeling. The limit state function is defined as the function dividing the event space for the basic random variables between a failure zone and a safe zone. The fatigue failure limit state is defined by:

$$g(z_1, z_2 \ldots \ldots z_n) = 0 \ . \tag{7.1}$$

When the basic variables are treated as stochastic variables, the limit state function becomes a stochastic variable and is defined as the safety margin. This margin is then written $\mathbf{g(z_1, z_2 \ldots \ldots z_n)}$ where the bold letters denote that both the basic variables and the function itself have become stochastic variables. Failure is defined by the event given by $g(z_1, z_2 \ldots \ldots z_n) = 0$, whereas $g(z_1, z_2 \ldots \ldots z_n) > 0$ corresponds to a safe area. The failure of probability then reads:

$$P_f = \int_{g(z_1, z_2 \ldots z_n) < 0} f_z(z_1, z_2 \ldots \ldots z_n) dz_1 dz_2 \ldots dz_n \tag{7.2}$$

where $f_z(z_1, z_2 \ldots \ldots z_n)$ is the multivariate PDF of all the n basic variables. The concept is visualized in Figure 7.2 with two variables only. As can be seen, the two-dimensional PDF $f_z(z_1, z_2)$ has its center in the safe zone, but there is a low probability that a set of the two variables may occur at the failure zone.

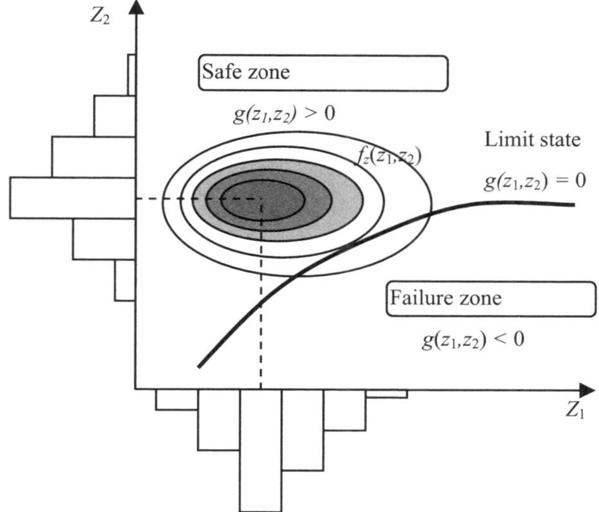


**Figure 7.2.** *Visualization of the two-dimensional probability function with safe space and failure space*

For fatigue life reliability the safety margin may read:

$$\mathbf{g}(\mathbf{z}_1,\mathbf{z}_2.....\mathbf{z}_n) = \mathbf{N}(\mathbf{z}_1,\mathbf{z}_2,....\mathbf{z}_k) - \mathbf{n}(\mathbf{z}_{k+1},\mathbf{z}_{k+2},....\mathbf{z}_n) \qquad (7.3)$$

where $\mathbf{N}$ is the number of cycles to failure and $\mathbf{n}$ is the occurring number of cycles during a given time period with a given stress spectrum. The variables $z_1$ to $z_k$ are the variables that influence the fatigue life, whereas the variables in $z_{k+1}$ to $z_n$ are the variables that have an influence on the number of cycles the welded detail is subjected to during a given time. If a fracture mechanics approach is used, the first set of variables may consist of the initial crack depth, the growth rate parameters and the stress range. The latter set of variables is mainly related to the load frequency. Most of the major uncertainties are found in $\mathbf{N}$, whereas $\mathbf{n}$ is sometimes treated as deterministic. An alternative formulation of the safety margin reads:

$$\mathbf{g}(\mathbf{z}_1,\mathbf{z}_2......\mathbf{z}_n) = \mathbf{a}_c(\mathbf{z}_1,\mathbf{z}_2,....\mathbf{z}_k) - \mathbf{a}(\mathbf{z}_{k+1},\mathbf{z}_{k+2},....\mathbf{z}_n) \qquad (7.4)$$

where $\mathbf{a}_c$ is the critical crack depth and $\mathbf{a}$ is the current crack depth at a given time and stress spectrum. Regardless of the choice between the formulation given by equations (7.3) or (7.4), the probability of failure has to be found through equation (7.2). However, equation (7.2) is almost impossible to integrate directly in cases of practical interest. The integration is often carried out numerically by a Monte Carlo simulation. When carrying out a Monte Carlo simulation, the fraction of the realizations that gives $g(z_1,z_2........z_n)$ zero or negative value will, by definition, be an estimate for $P_f$. An analytical approximation based on the first- and second-order reliability methods (FORM/SORM) is an alternative to the simulation approach. We shall briefly present this approach in the next section before going more into the details of a Monte Carlo simulation.

### 7.2.5. *The first and second order reliability methods (FORM/SORM)*

The FORM/SORM gives approximate analytical solutions of the probability of failure given by the integral in equation (7.2). These solutions are exact if the failure function is linear and the basic random variables are independent standard normally-distributed variables. If the basic variables $(z_1,z_2........z_n)$ are dependent non-normal variables, they will have to be transformed from the physical **z**-space to the standard normal **u**-space described by the Gaussian vector **u**. The failure function $g(\mathbf{u})$ will then, in general, be non-linear and must be approximated by a tangent hyperplane (FORM) or a hyper-paraboloid (SORM) at the point $\mathbf{u}^*$ on the surface $g(\mathbf{u}) = 0$ closest to the origin. The basic principles are shown in Figure 7.3; elaborated details are found in Ref [4].

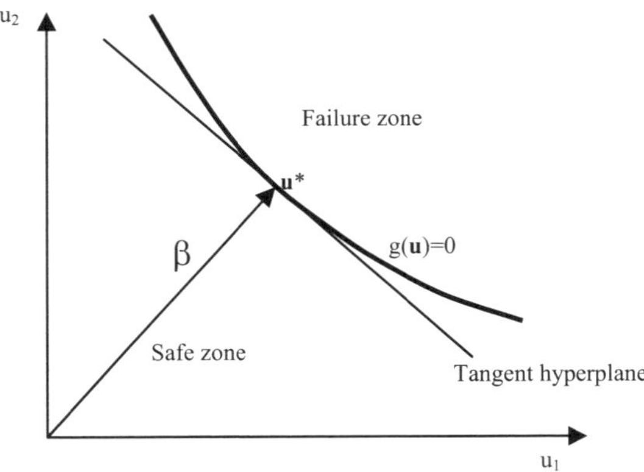


**Figure 7.3.** *Sketch showing the principals of the FORM*

The limit state function based on fracture mechanics is often written in the form (see Ref [4]):

$$g(z_1,z_2....z_n) = \Psi\left(a_0;a_C\right) - \Psi\left(a_0;a(t)\right) = \int_{a_0}^{a_C} \frac{da}{\left(Y(a)\,M_k(a)\sqrt{\pi a}\right)^m} - \mathrm{Cn}(t)\,\Delta\sigma^m \quad (7.5)$$

As can be seen, this is a modification of equation (7.3). The shortest distance from the origin to the design point **u*** defines the probability of failure; see Figure 7.3. The corresponding probability of failure reads:

$$P_f \approx \Phi(-\beta) \quad \Rightarrow \quad \beta \approx -\Phi^{-1}(P_f). \tag{7.6}$$

There is often a discussion of which method to favor, Monte Carlo simulation or FORM/SORM. The former approach will, for remote probabilities, demand a large number of simulations, while the latter method may in some cases give erroneous results. In this book we will emphasize the use of the Monte Carlo simulation as it is easier to illustrate how this technique works in conjunction with our physical fatigue models. The mathematical formalism of the FORM/SORM conceals to some extent the physics involved in the analysis. We will however apply FORM/SORM in Chapter 12.

## 7.3. Elementary reliability models

### 7.3.1. *General considerations*

Before proceeding with stochastic modeling as outlined above, we will undertake a short review of the most common distribution functions used to model the variables involved in the fatigue problem. The most common distributions are the lognormal and the Weibull distribution functions. Experimental fatigue life data often fit directly these distributions. The obtained PDFs are then simple fatigue life models from which the probability of failure can be determined. This is the simplest approach, and it will only reflect the scatter found under laboratory conditions. In other cases these distributions can be used to model one of the involved basic variables given in Figure 7.1. In Chapter 5 we used the lognormal distribution to model the fatigue life at a given stress range, whereas we used the Weibull distribution to model the stochastic loading. Both models have two important features in common with regard to the PDF:

– the PDF is set to zero for negative values of the variable in question;

– the PDF can model positively skewed data.

Most physical variables do not have negative realizations and this fact excludes the normal distribution. Furthermore, fatigue life data are positively skewed, i.e. there is a clear limit of how short a fatigue life can get, but there is almost no limit to how long some joints can last. A general trend is that the Weibull distribution is most frequently used to model stress spectra, whereas the lognormal distribution is used to model the fatigue life. However, there is a debate as to whether the Weibull model can replace the lognormal distribution in the latter case. Although the reader may already have knowledge of the two distributions, a short repetition of important features will be given.

### 7.3.2. *The Lognormal distribution*

In what follows we have designated the random variable as **t** which can be thought of as the time (or cycles) to failure of a welded detail under given loading conditions. When **t** is lognormal distributed it means that $\mathbf{x} = \ln \mathbf{t}$ is normal distributed. We then get:

$$f(t) = \frac{1}{\sqrt{2\pi}s_{\ln t}\cdot t}e^{-\frac{1}{2}\left(\frac{\ln t - \mu_{\ln t}}{s_{\ln t}}\right)^2} \quad t \geq 0$$

$$f(t) = 0 \quad \text{for} \quad t < 0$$

$$F(t) = \Phi(u) \quad , \quad u = \frac{\ln t - \mu_{\ln t}}{s_{\ln t}} = \frac{\ln \frac{t}{t_0}}{s_{\ln t}} \tag{7.7}$$

$$R(t) = 1 - \Phi(u)$$

$$\lambda(t) = \frac{f(t)}{1 - \Phi(u)}$$

where $\mu_{\ln t}$ and $s_{\ln t}$ are the mean and standard deviation for lnt. Where f(t) is the PDF, F(T) is the cumulative distribution function (CDF) and R(t) is the reliability function. The function $\lambda(t)$ is denoted by the failure-rate function. The expression $\lambda(t)\Delta t$ can be interpreted as the probability of failure during the interval t+$\Delta$t given that the detail has survived up to the time t. The parameter $t_0$ is the median value for the distribution defined by $F(t_0) = 0.5$. This value can be determined by $\Phi(u) = 0.5$:

$$t_0 = e^{\cdot \mu_{\ln t}} \ . \tag{7.8}$$

Hence, $\ln t_0 = \mu_{\ln t}$. If the mean and the standard deviation $\mu_t$ and $s_t$ for $t$ are known we can calculate the mean and standard deviations for lnt:

$$\mu_{\ln t} = \ln\left(\frac{\mu_t}{\sqrt{V_t^2 + 1}}\right) \tag{7.9}$$

$$s_{\ln t}^2 = \ln(V_t^2 + 1)$$

For small values of $V_t$ ($V_t$ = COV, less than 0.3) we have approximately $\sigma_{\ln t} = V_t$. Figure 7.4 shows the main characteristic functions pertaining to the lognormal model.

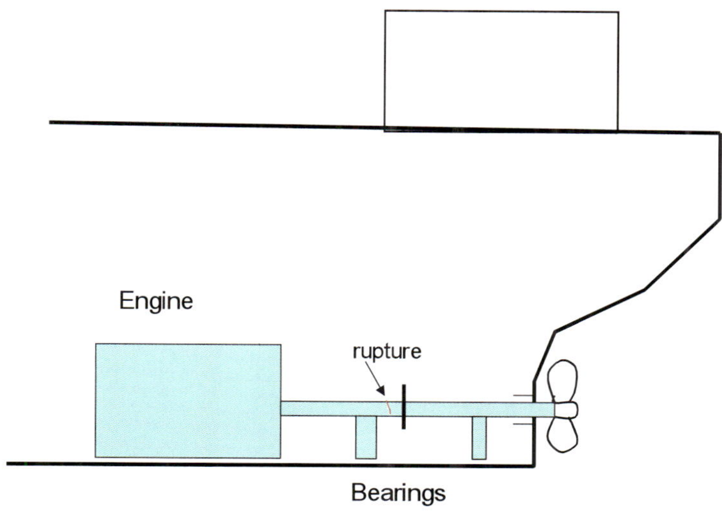

**Figure 2.1.** *Fatigue failure in a propeller shaft in a shuttle tanker. Crack has initiated from a weld arch strike at the surface*

**Figure 3.2.** *Tubular joint with an X geometry subjected to in-plane bending of the braches. Upper part: setup in testing rig. Lower part: detail of branch-chord intersection at the crown point with strain gauges*

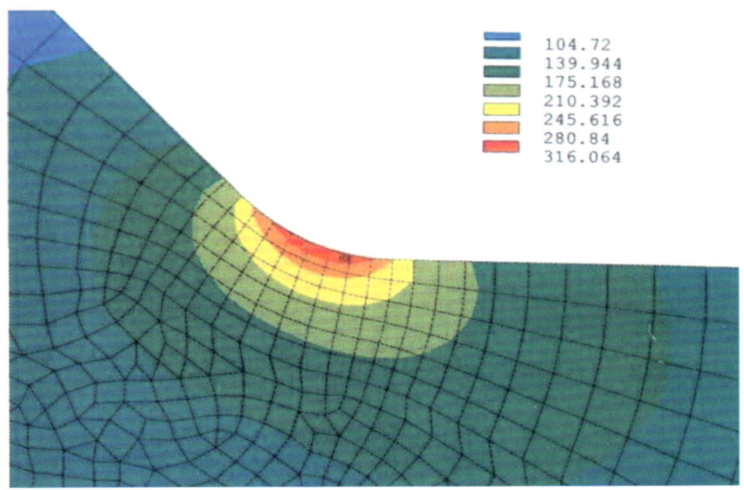


**Figure 3.4.** *Results from a local FEA model for a cruciform fillet-welded joint*

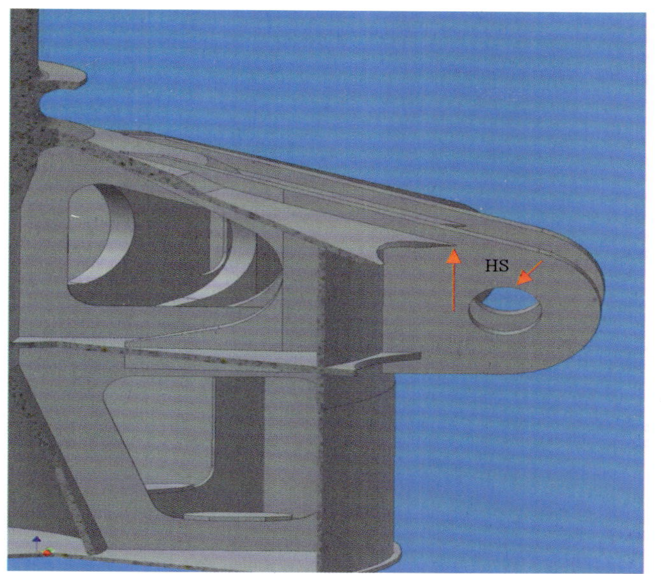

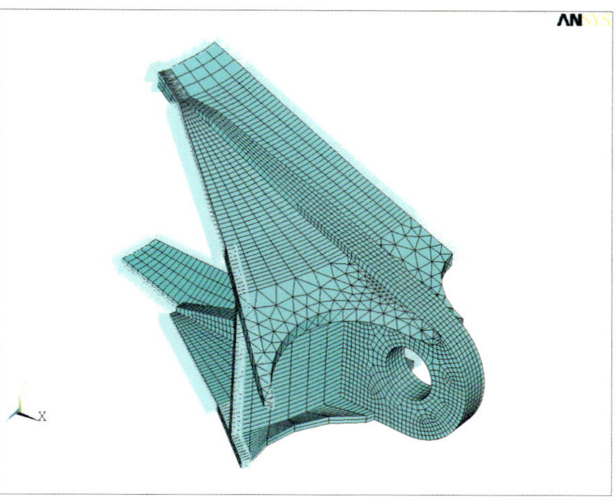



**Figure 5.27.** *Upper: lug in central turret for mooring line connection in an offshore loading buoy. Hot spot at nose of welded plate. Lower: finite element model to determine geometrical stresses at hot spots. Courtesy of Advanced Production and Loading AS, Arendal, Norway*

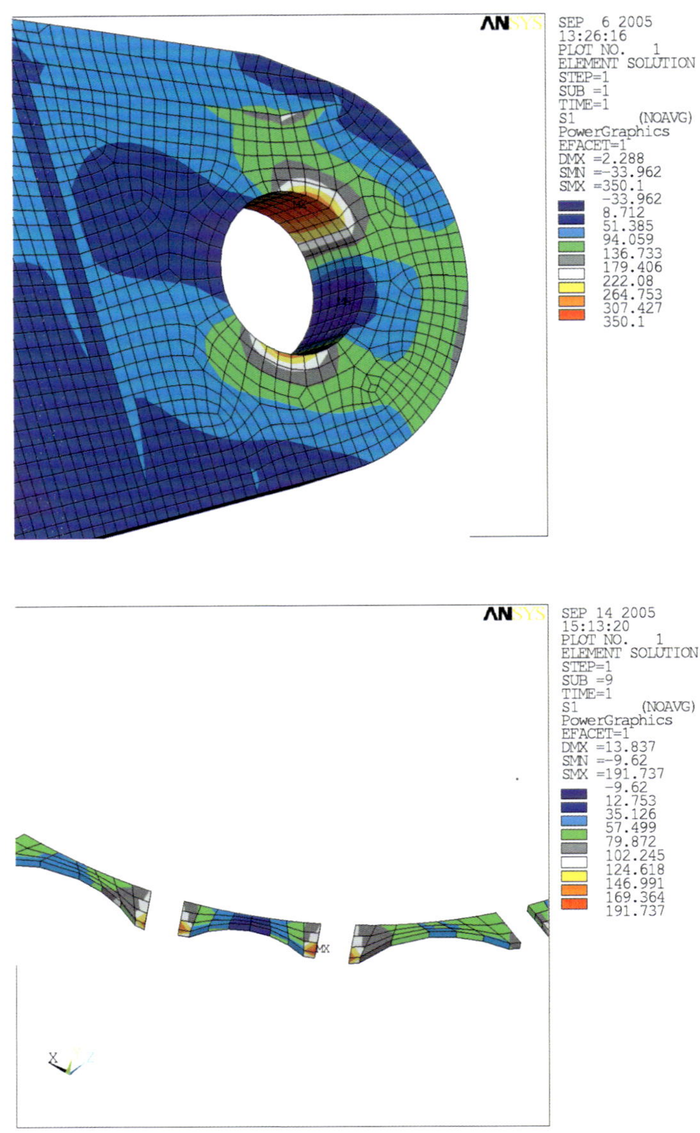

**Figure 5.28.** *Result for stress analysis tension 7374 KN. Upper: lug. Lower: bracket plate*

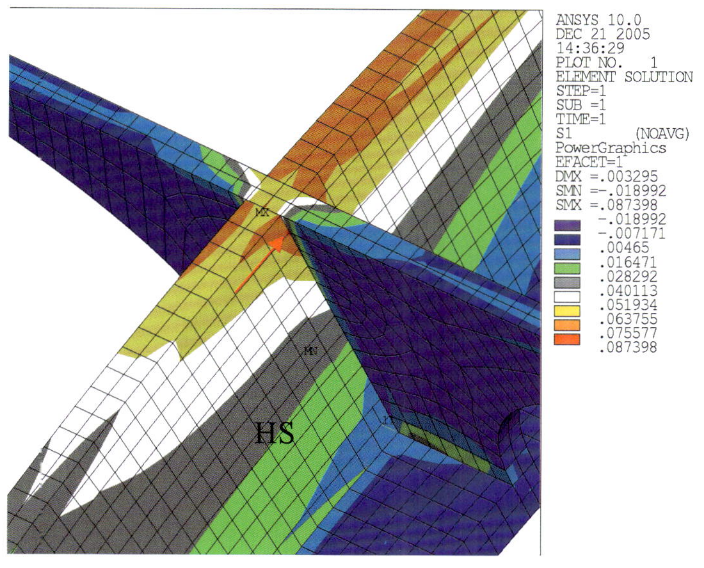


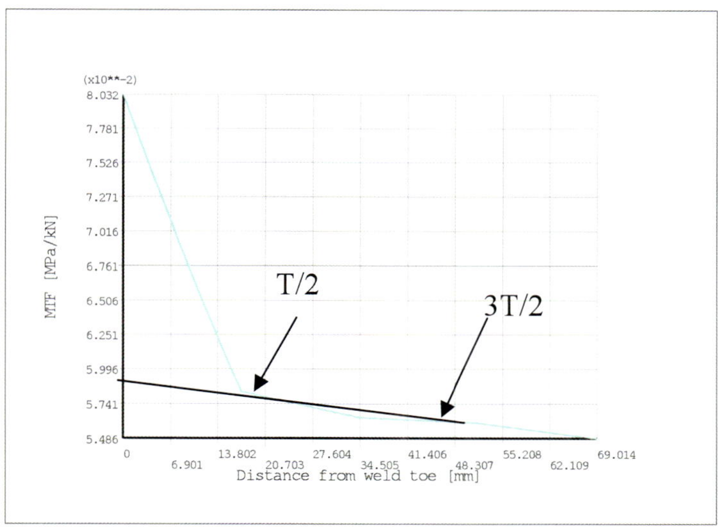


**Figure 5.29.** *Refined sub-model of a welded intersection between two plates within the turret. Upper: hot spot and interpolation line. Lower: interpolated stress values. Courtesy of APL*

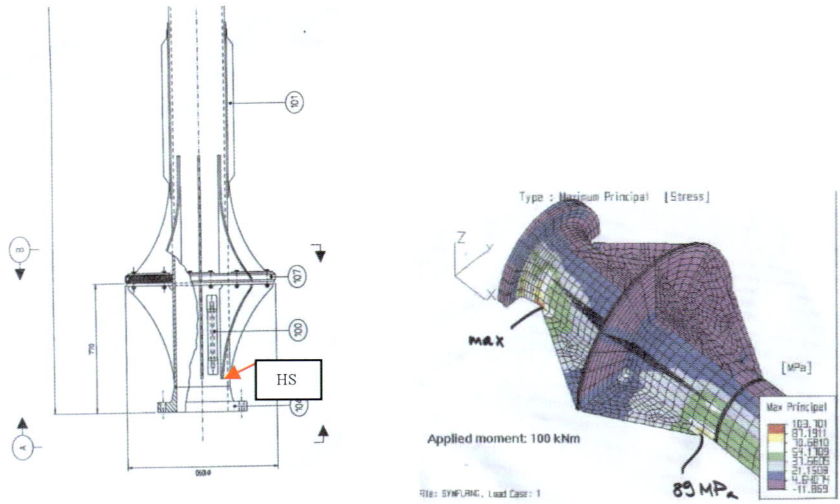



**Figure 6.28.** *Drawing of flow-line (left) and stress plots from FEA (right). Courtesy of APL*

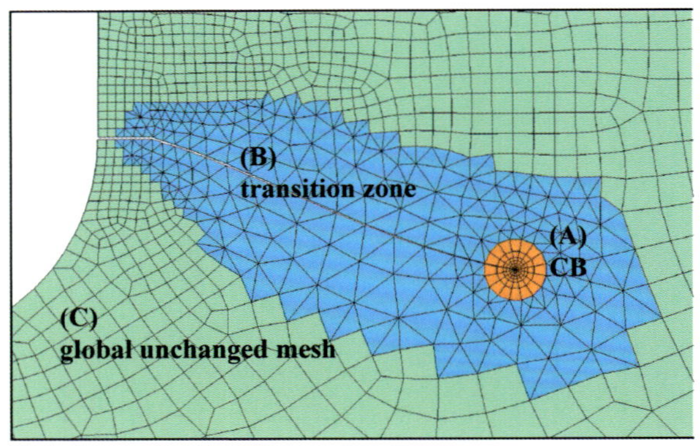



**Figure 11.1.** *Crack box in a structure (regions A, B and C)*

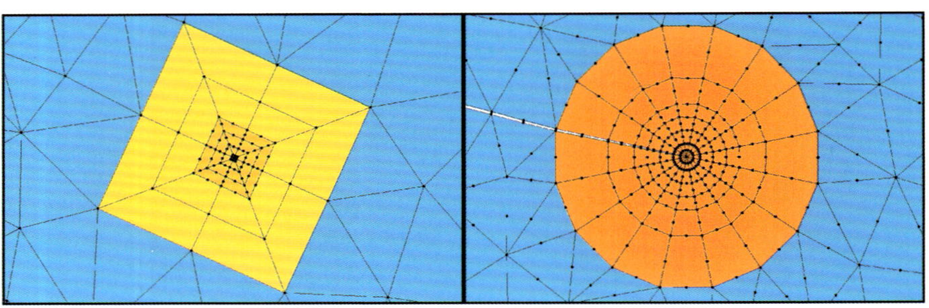

**Figure 11.2.** *Refined crack box*        **Figure 11.3.** *Coarse crack box*

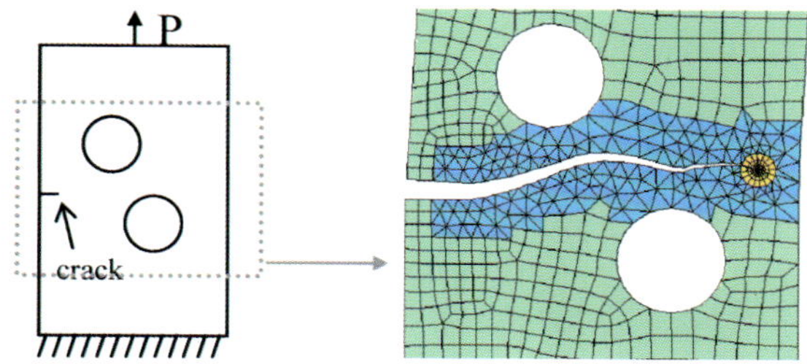

**Figure 11.4.** *Crack growth numerical results using crack box technique (CBT): two-hole specimen*

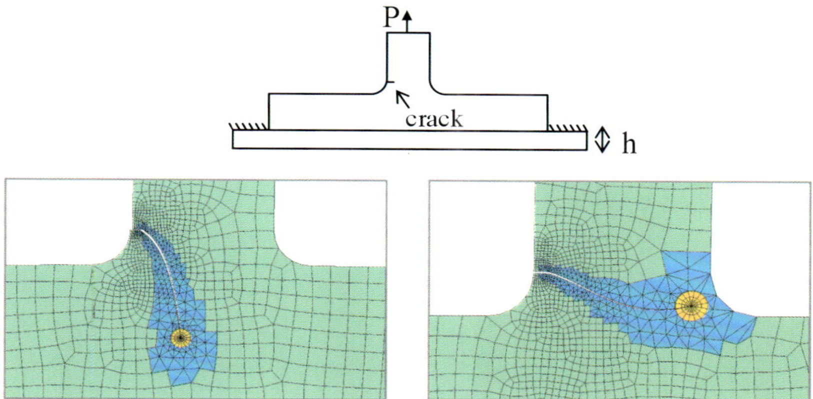

**Figure 11.5.** *Crack growth numerical results using CBT: fillet specimen (left: h low values; right: h high values)*

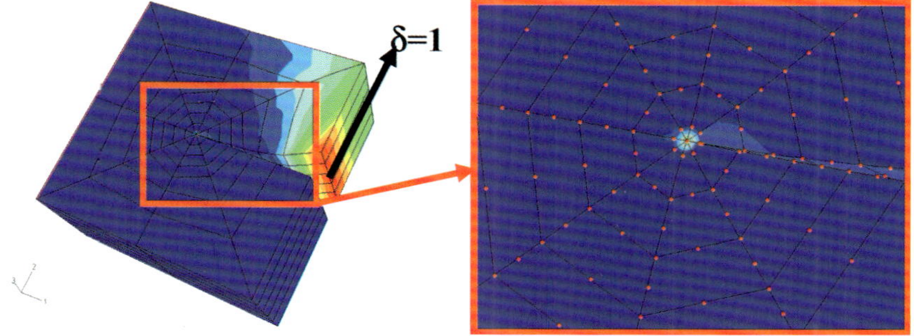


**Figure 11.25.** *Sub-model of the structure: 3-D crack box (left); zoom on crack tip (right)*

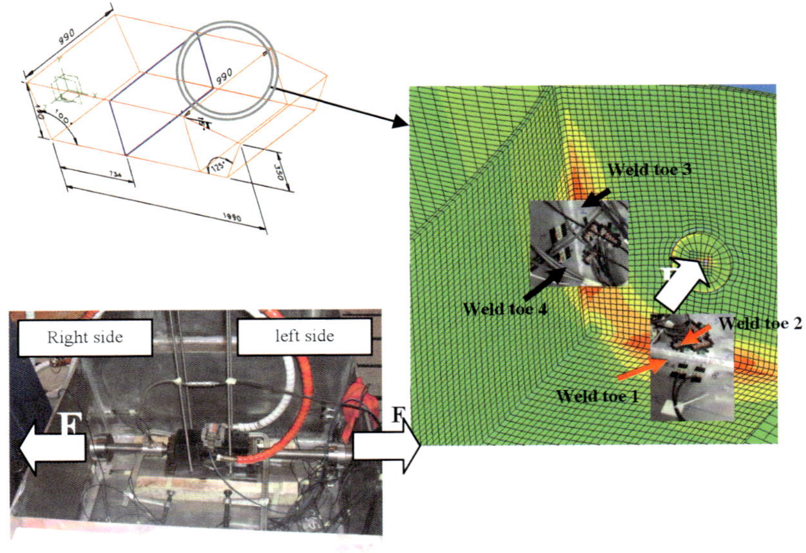


**Figure 11.26.** *Structure dimensions (mm), crack gauge position and picture of the experimental assembly*

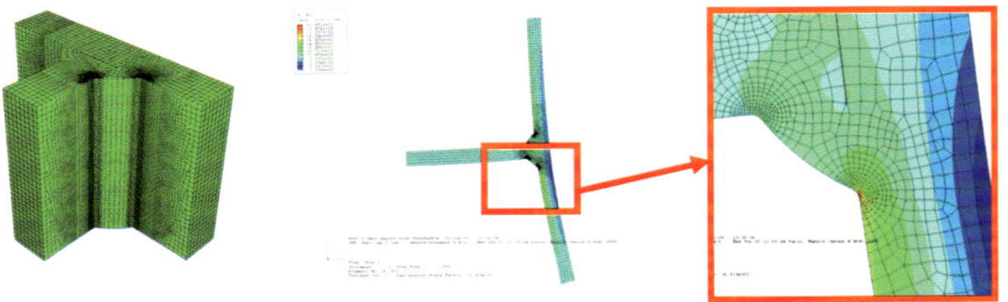

**Figure 11.29.** *Zoom of the weld toe in 3-D and in 2-D modeling*

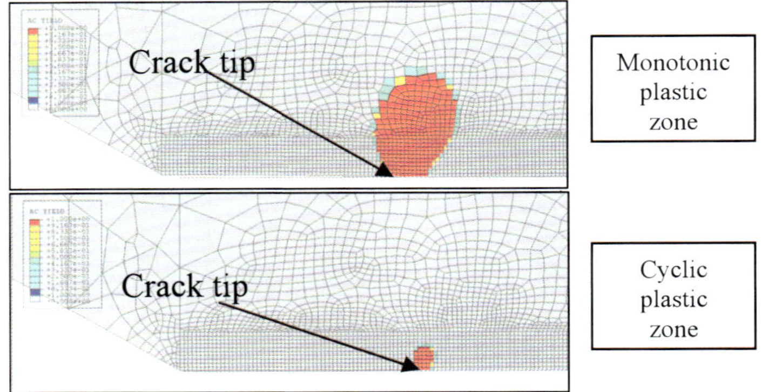


**Figure 12.5.** *Monotonic and cyclic plastic zones*

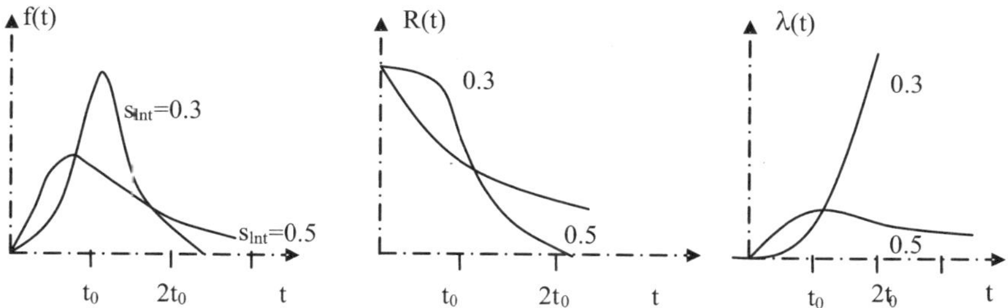

**Figure 7.4.** *Sketches of the characteristic curves for the lognormal model*

It is the curves for $S_{Int} = 0.5$ that are particularly interesting for welded joints. As we remember, $V_t$ is close to 0.5 for most welded details; see Table 5.2, Chapter 5. We see the positively skewed distribution of $f(t)$. Furthermore, the failure rate function $\lambda(t)$ has a peculiar shape. As can be seen to the right in Figure 7.4, this curve is increasing up to time $t_0$. It then levels off and starts to decrease. This peculiar behavior has been used as an argument against the lognormal distribution ability to model fatigue life. One would expect the failure rate function to increase throughout the time the damage accumulated. However, the explanation for the evolution of the failure rate function is that the longer the detail sustains the loading process, the more it proves its fatigue quality.

### 7.3.3. *The Weibull distribution*

We have already used the Weibull model for distribution of stress ranges when describing a stochastic loading process. In this book, we will think of the distribution as a stochastic model for time to failure. The frequency function (PDF) reads:

$$f(t) = \frac{h}{q}\left(\frac{t}{q}\right)^{h-1} \cdot e^{-\left(\frac{t}{q}\right)^h} \qquad , t \geq 0$$

$$f(t) = 0 \qquad , t < 0 \tag{7.10}$$

The two parameters defining the function are denoted as:

– $h$: shape parameter;

– $q$: scale parameter.

By integration of equation (7.10) we get:

$$F(t) = \int_0^t f(t')dt' = \left[ -e^{-\left(\frac{t}{q}\right)^h} \right]_0^t = 1 - e^{-\left(\frac{t}{q}\right)^h}$$

$$R(t) = 1 - F(t) = e^{-\left(\frac{t}{q}\right)^h} \tag{7.11}$$

$$\lambda(t) = \frac{f(t)}{R(t)} = \frac{\left(\frac{h}{q}\right)\left(\frac{t}{q}\right)^{h-1} e^{-\left(\frac{t}{q}\right)^h}}{e^{-\left(\frac{t}{q}\right)^h}} = \frac{h}{q}\left(\frac{t}{q}\right)^{h-1}$$

As can be seen, the reliability function R(t) can be found analytically. The failure rate function $\lambda$(t) will have various evolutions depending on the value of the shape parameter h:

– $h = 1$, $\lambda(t)$ is constant;

– $h > 1$, $\lambda(t)$ is increasing;

– $h < 1$, $\lambda(t)$ is decreasing.

For $h = 1$ we get a special case, which is called the exponential distribution. In this case we will have:

$$\lambda = \frac{1}{q}. \tag{7.12}$$

We used that distribution for the long-term stress model in Chapter 4. The two parameters $h$ and $q$ can be related to the mean and the variance of $t$ by the equations:

$$E(\mathbf{t}) = \mu_t = q\Gamma(1+\tfrac{1}{h})$$

$$Var(\mathbf{t}) = s_t^2 = q^2\left\{\Gamma(1+\tfrac{2}{h}) - \left[\Gamma(1+\tfrac{1}{h})\right]^2\right\} \tag{7.13}$$

where $\Gamma(x)$ is the Gamma function of x. This function is found in standard tables. As can be seen from equation (7.13), the mean value is proportional to the scale parameter $q$. The second equation in (7.13) is rather involved and a useful approximation reads:

$$h \approx V_t^{-1.08} = \frac{1}{V_t^{1.08}} \, . \tag{7.14}$$

From this equation it is clearly seen that when $V_t$ is decreased, the shape parameter $h$ will increase, In other words, a high value for the shape parameter $h$ means low standard deviation if the mean value is kept constant. The relationships are visualized in Figure 7.5.

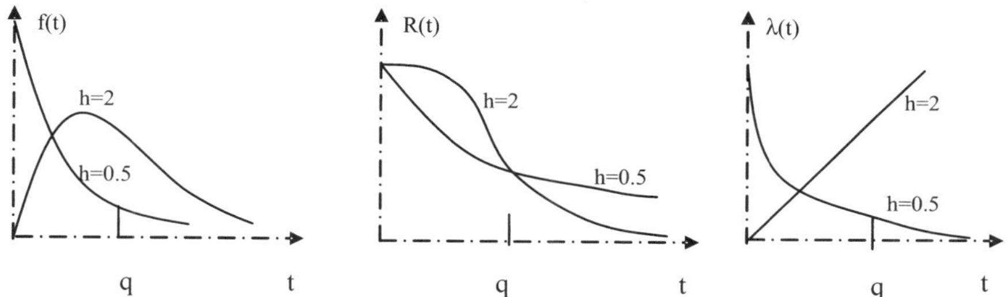


**Figure 7.5.** *Principal sketches for the Weibull model for two values of shape parameter*

Again, knowing that the $V_t$ is close to 0.5 for most welded joints, it is the model with $h \cong 2$ that is the most interesting one. Compared to the lognormal model, it can be seen that the failure rate function is increasing linearly. This is an important difference compared to the lognormal model when it comes to describing the fatigue life behavior. Another important difference is that the Weibull distribution has a stronger tail on the left part of the frequency function f(t). As a consequence, it will predict a higher probability of failure for a relatively low number of cycles. In this low probability area there are not any experimental data to corroborate or exclude one of the models in favor of the other. This is a dilemma when choosing between the lognormal model and the Weibull model. In Figure 7.6 the characteristic functions are shown for a butt joint subjected to a stress range of 150 MPa. Data for this D-class joint are found in Table 5.2, Chapter 5. The median life is approximately $1.2 \ 10^6$ cycles. As can be seen, the left tail of the Weibull model is stronger than that for the lognormal model and as a consequence the lognormal model predicts higher reliability in the beginning. Furthermore, it can be seen that the failure rate function is higher for the Weibull model in the very beginning, before the lognormal failure rate increases and passes it. As can be seen, the failure rate function for the lognormal model levels off when the median value of $1.2 \ 10^6$ cycles has been passed.

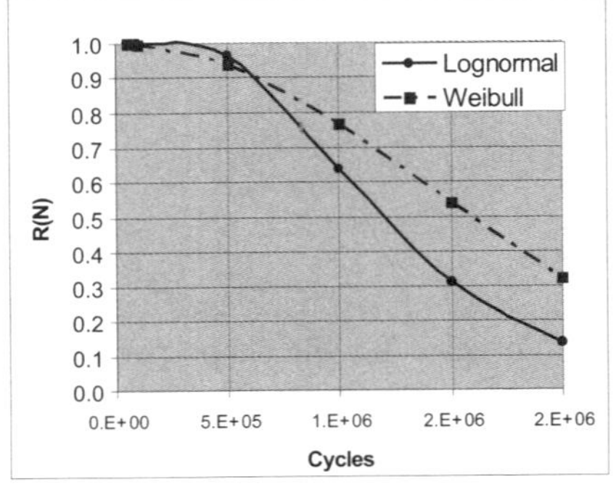
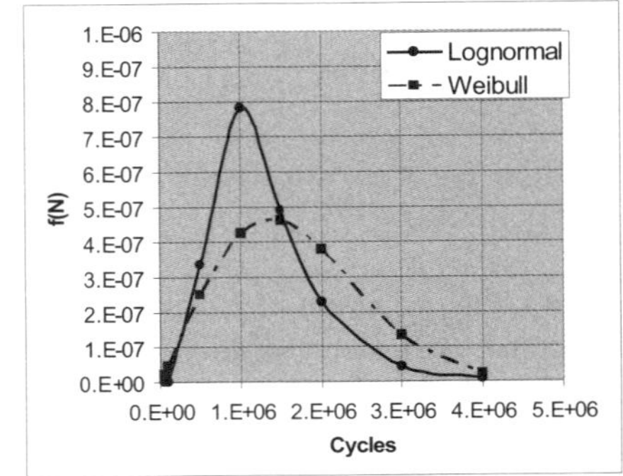
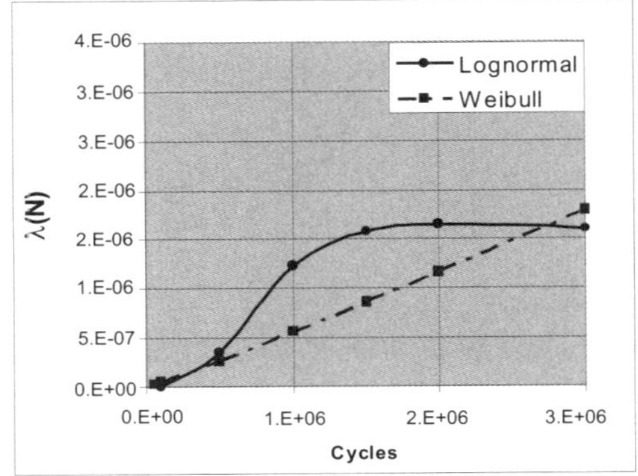

**Figure 7.6.** *Frequency function, reliability function, and failure rate function for the fatigue life of a butt joint subjected to CA stress range of 150 MPa*

## 7.4. The random variable model using simulation methods

### 7.4.1. *General considerations*

As we have discussed, the result variable y in Figure 7.1 can be calculated by an S-N approach or a fracture mechanics approach as outlined in Chapters 5 and 6. The random aspect of the problem can be dealt with by a simulations technique. This is done by repeating the calculation for various sets of the variables on the left-hand side of Figure 7.1. Each set of these variables will contain values that are in accordance with the frequency functions for the variables. This means that most of the sets will have values of the variables near the peak of each frequency function; see Figure 7.2. Values of the variables found out on the tails of the frequency functions will appear less frequent, but these events can be even more important as they may lead to accelerated fatigue damage and reduced fatigue life. The calculation can practically be carried out by the method of Monte Carlo simulation.

The two basic variables $z_1$ and $z_2$ in Figure 7.1 may, for example, be the initial crack depth $a_0$ and the growth parameter C. The calculation scheme in this case can then be based on fracture mechanics. The result variable y may be the crack depth at a chosen time or the time to reach a given crack depth. If the latter is chosen as the critical crack depth, then y will be fatigue life. If many sets of the source variables are generated according to their frequency function $f_1(z_1)$ and $f_2(z_2)$, we will get a corresponding number of fatigue lives y. These generated lives can be ordered in a histogram and a frequency function $f_y(y)$ is obtained.

### 7.4.2. *The realization of a random variable by the Monte Carlo method*

If we know the $f(z)$ (PDF) and F(z) (CDF) for a random variable z, we can simulate realizations of this variable by starting to generate a random number r having a rectangular distribution with possible outcomes in the range [0-1.0]. It can then be shown that a number z defined by:

$$F(z) = r$$
$$\text{or} \tag{7.15}$$
$$z = F^{-1}(r)$$

is a random realization of the variable z according to its frequency function $f(z)$.

The principle is illustrated in Figure 7.7. As can be seen from the sketch, the rectangular distributed r is generated on the vertical axis between 0 and 1.0 and shot into the back of the CDF curve. The realization of z is then found if we proceed vertically down the z-axis. If several realizations are carried out, the reader can be convinced that most of the values will be close to the mean value of the z variable. The realizations will in fact form a histogram that approaches the $f(z)$ function shown in Figure 7.7.

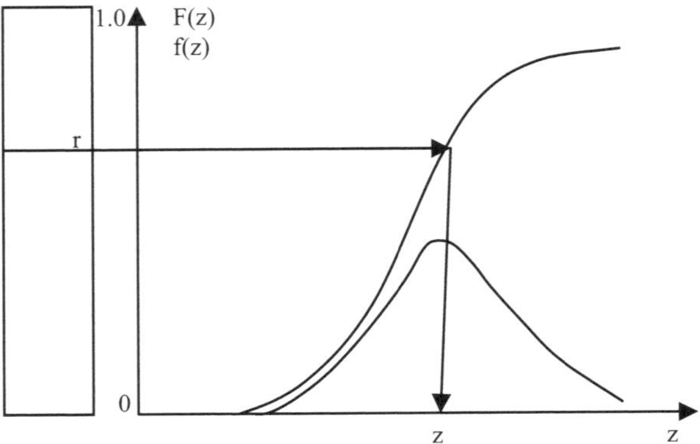

**Figure 7.7.** *Random number realization for a given basic variable*
*according to the Monte Carlo method*

For achieving numerical solutions we will have to invert the CDF in question, equation (7.15). This cannot always be done analytically. For a normal distribution model with mean value $\mu$ and standard deviation s we first make two realizations of a standard normal distributed variable u:

$$u_1 = \sqrt{-2\ln(r_1)}\,\cos(2\pi r_2)$$
$$u_2 = \sqrt{-2\ln(r_1)}\,\sin(2\pi r_2) \tag{7.16}$$

Two realizations for the sought variable z can then be found by the transformation:

$$z_1 = \mu + u_1 \cdot s$$
$$z_2 = \mu + u_2 \cdot s \tag{7.17}$$

In the case where we have two variables that are bi-normal distributed with correlation factor $\rho$, the results will read:

$$z_1 = \mu_1 + u_1 \cdot s_1$$
$$z_2 = \mu_2 + s_2\left[\sqrt{1-\rho^2}\cdot u_1 + \rho\cdot u_2\right] \tag{7.18}$$

Hence, we are able to take the correlation between variables into account. The method can also be used to realize lognormal variables. In this case we first generate the logarithmic values according to equation (7.17), before transforming it to a nonlogarithmic value. For a Weibull distributed variable, equation (7.15) can be inverted directly:

$$1 - \exp\left[-\left(\frac{z}{q}\right)^{h}\right] = r$$

o r:

$$z = \left[q\left(-\ln(1 - r)\right)^{1/h}\right]$$

(7.19)

The purpose of these realizations is, of course, not to reproduce the histogram for each of the random variables, as we already have the PDFs for these random variables. The aim is, as we have discussed, to apply the realizations in the calculations to obtain the histograms of the result variables, such as the crack depth or the fatigue life. Hence, the known simulated variables will enter into a calculation scheme, a main result of which is to give the calculated realization for this unknown result variable. The principle is shown in Figure 7.8.

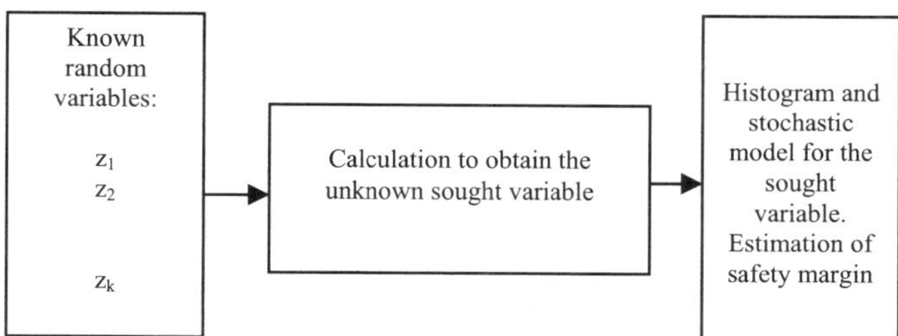


**Figure 7.8.** *General procedure for determining the result variable*

The basic variables can, for example, be fracture mechanics crack growth rate parameters. The calculations can be carried out by the Paris law to obtain the number of cycles to failure. As this results variable will appear as a random variable, the probability of failure can readily be obtained as a function of time.

## 7.5. Random variable models based on the S-N approach

### 7.5.1. *The lognormal format for the S-N fatigue life*

In the previous section the lognormal and Weibull models were used to model fatigue life taking into account the scatter found during S-N testing only. The two models can be extended to take into account additional scatter introduced under in-service conditions. The presentation for the lognormal model is based on the work of Wirsching (see Ref [5]). It is a random variable model in the way that each source variable is modeled by its mean value and scatter. This method originates in the fact that the fatigue life under controlled laboratory condition is lognormal distributed with a mean value and scatter given by $\mu_{\ln N}$, $s_{\ln N}$. An important measure of scatter is the coefficient of variation given by:

$$\mathrm{COV_N} = \frac{s_N}{\mu_N} = V_N .$$ (7.20)

$\mathrm{COV_N}$ (or just $V_N$) is, as we have seen, close to 0.5 for most welded details (see Table 5.2, Chapter 5). The following relation is given:

$$s_{\ln N} = \sqrt{1 + \mathrm{COV_N^2}} .$$ (7.21)

This equation will show that the $s_{\ln N}$ is very close to $\mathrm{COV_N}$, as we have seen. In this context we have chosen the natural logarithm for the number of cycles to failure instead of the logarithm with base number 10. The latter is used in rules and regulations. We must be conscious of which format is used and that the following transformation applies:

$$s_{\log N} = 0.434 \cdot s_{\ln N} .$$ (7.22)

When we go from controlled constant amplitude (CA) laboratory conditions to in-service variable amplitude (VA) conditions, Wirsching suggests that we should take into account the following addition uncertainties:

– bias and scatter in Miner's rule given by $\Delta$;

– stress modeling error given by B.

Each of the two variables is characterized by their median value and COV. Based on the lognormal assumption, it can be proven that the total scatter in fatigue life can be found by law of multiplication:

$$s_{\ln N} = \sqrt{\ln\{(1 + V_N^2)(1 + V_\Delta^2)(1 + V_B^2)^{m^2}\}}$$ (7.23)

and that the median value is the product of the median values:

$$\tilde{N} = \frac{\tilde{\Delta} \cdot \tilde{A}_0}{\tilde{B}^m E(S^m)}$$ (7.24)

where median values given for the various parameters. $E(S^m)$ is the equivalent stress range calculated numerically or analytically based on a Weibull distribution. Typical data for the various variables are shown in Table 7.1 (Ref [5]).

| Parameter | Median | COV |
|---|---|---|
| A | Table 5.2, Chapter 5 | Table 5.2, Chapter 5 |
| M | 3.0 | Deterministic |
| Δ | 1.0 | 0.3 |
| B | 0.9 | 0.25 |

**Table 7.1.** *Typical scatter in an S-N calculation (Ref [5])*

It is only the first variable A that is extensively tested and proven. It seems optimistic that the median value of the Miner's sum at fracture is equal to 1.0; see our discussion in section 5.3.2. From recent research it seems more likely that the median value is close to 0.5 and that the scatter is greater than 0.5. The values for B must be interpreted as if most stress analysis overestimated the stresses by 10% and that typical scatter is 25%. Of course all these figures must be estimates from case to case depending on how much effort is made to determine the figures accurately. The probability of failure at a given service time corresponding to N cycles is given by:

$$P(N < N) = \Phi(\frac{\ln N - \mu_{\ln N}}{s_{\ln N}}) = \Phi(\frac{\ln(N / \tilde{N})}{s_{\ln N}}). \tag{7.25}$$

Based on this expression, the reliability curves ($P(N > N)$) can also be determined.

*7.5.1.1. Example: full-penetration butt joint in an offshore structure*

A butt joint in an offshore structure in the North Sea will, during a target service life (TSL) of 20 years, be subjected to approximately $10^8$ cycles. According to the D-class S-N curve, the joint will have a permissible equivalent stress range of 24.7 MPa to obtain a fatigue design factor of FDF = 1. If we had only accounted for the uncertainty related to the inherent S-N scatter, the lognormal model would give a reliability curve as shown on the left-hand side of Figure 7.9. The reliability curve pertaining to a Weibull model has been drawn for comparison. As can be seen, this curve is substantially lower. If the additional uncertainties for in-service condition, as given in Table 7.1, are added to the model, we will get the reliability curve shown in right-hand side of Figure 7.9. We have used equations (7.23, 7.24 and 7.25) to obtain the results. As can be seen, the reliability at the end of service life has dropped from 0.977 to 0.940 for the lognormal case. It can also be seen that the probability passes through 0.99 at half the service life, i.e. $5 \cdot 10^7$ cycles (10 years) with this additional uncertainty. This means that for FDF = 2, the probability of failure will be $10^{-2}$. As we shall see in what follows, the FDF is often used as a key to the achieved probability level. Hence, if this probability should be obtained at the end of the service life, the predicted fatigue life should have been twice the service life, i.e. $N = 2 \cdot 10^8$ cycles, FDF = 2. If the predicted fatigue life had been set to $10^9$ (i.e. FDF = 10, permissible equivalent stress range 11.5 MPa), then the reliability at the end of service life would have been 0.99999, i.e. $P_f = 10^{-5}$. Based on such figures we can make decisions on which FDF to require. This is shown in Figure 7.10 where the accumulated probability of failure is given as a function of the chosen FDF. A given target reliability will give the required FDF.

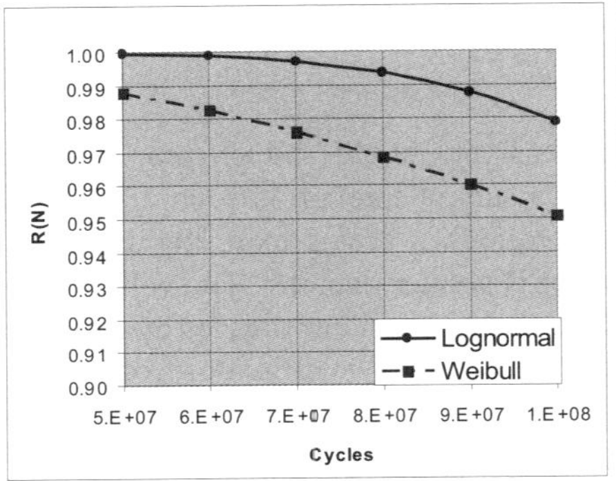
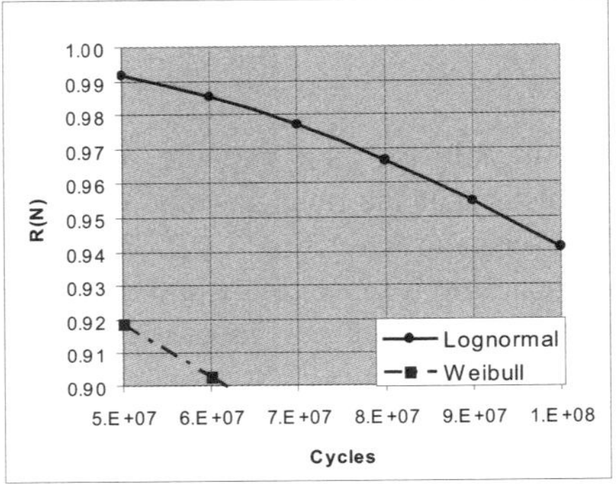

**Figure 7.9.** *Reliability curves for a butt joint with FDF = 1.0 at $10^8$ cycles according to the lognormal model. Left: inherent S-N scatter only; right: with additional scatter for stresses and Miner summation rule*

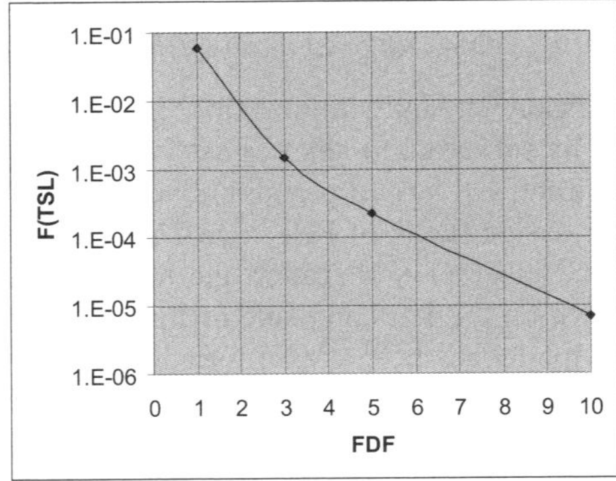

**Figure 7.10.** *Accumulated probability of failure at the end of service life as a function of FDF; based on the lognormal model with all uncertainties in Table 7.1 included*

### 7.5.2. *Monte Carlo Simulation of the S-N fatigue life*

The lognormal model, as proposed by Wirsching, is an approximation because it supposes that all involved variables are lognormal distributed and statistically uncorrelated. The benefit of such a simple model is that the solution is analytical. We could have solved the same problem using the Monte Carlo simulation technique described in section 7.4.2. In this case the involved parameters can have any frequency function and we can also simulate the correlation between them. The calculations scheme is as follows:

– calculate the equivalent $E(S^m)$ stress as accurately as possible;

– generate one value of **A** according to S-N data;

– generate a set of the variables **B** and $\Delta$ according to Table 7.1;

– calculate the number of cycles $N$ to failure from equation (7.24) using the generated random number instead of the median values.

The parameter m is treated as deterministic in order to scale fatigue life correctly according to applied stress levels. After having done the calculation for examples with M = 10,000 sets of random numbers we can establish a histogram of the M-generated fatigue lives. This is shown in Figure 7.11. Let $d_i$ be the number of lives in interval number i, whereas $\Delta N$ is the width of the intervals. The ratio $d_i/M\Delta N$ is then the numerical approximation for the frequency function. At a chosen time limit $N_L$, the probability of failure can be found by counting the number of simulations with shorter fatigue lives than $N_L$. If this number is designated $M_f$, the probability of failure before that time limit will read:

$$g(A,B,\Delta,N_L) = \Delta - \frac{N_L B^m E(S^m)}{A} \quad P_f = \frac{M_f}{M} \quad (7.26)$$

This probability corresponds to the relative number of times the limit state function in equation (7.26) has been less than or equal to zero. If the number of $n_i$ that survives each interval as a fraction of M is drawn, we will get a curve as shown in Figure 7.12. This is a numerical approximation for $R(t)$ comparable with the one obtained analytically in Figure 7.9. The difference is that M must be very large to get the early failures in correct relative number, i.e. get the early, small decrease in $R(N)$.

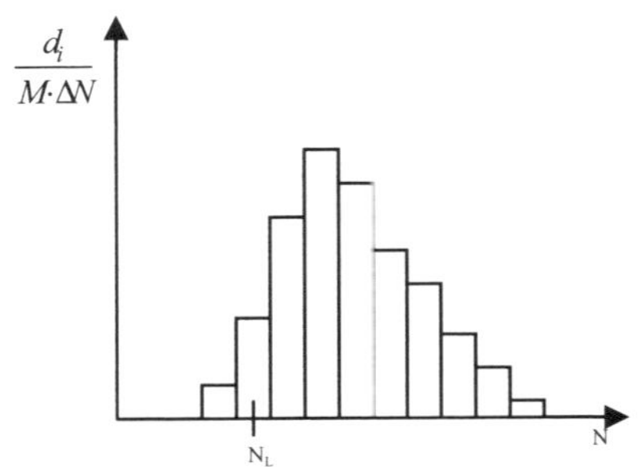

Figure 7.11. *Histogram of simulated fatigue lives*

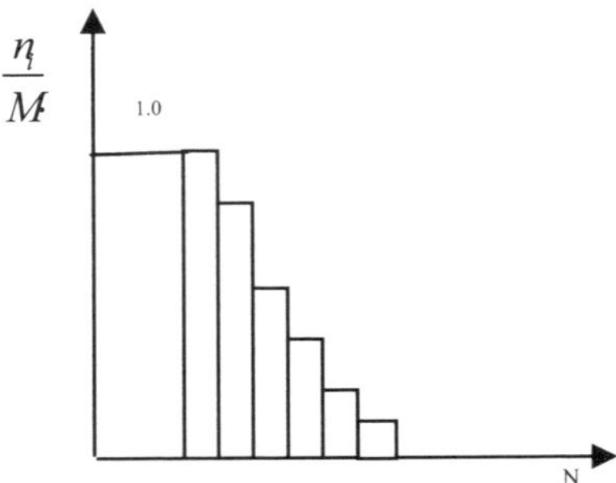

Figure 7.12. *Numerical approximation of the reliability function*

## 7.6. Random variable models based on fracture mechanics

### 7.6.1. *General considerations*

Thus far we have just dealt with the uncertainty in the fatigue life based on the S-N predictions. We have to carry out the stochastic analysis based on fracture mechanics to account for the effect of in-service inspection. We can then obtain the most likely frequency function for the crack depth at various time stages when inspections are planned. The failure criterion based on the fracture mechanics analysis can, at a given time, be written by the safety margin:

$$g(z_1, z_2, ....z_n) = N(z_1, z_2, ....z_k) - n(z_{k+1}, z_{k+2}, ....z_n) \qquad (7.27)$$

where N is the number if cycles for which the critical crack size $a_C$ is reached, whereas *n* is the occurring number of load cycles. N will be a function of all the variables entering into the fracture mechanics calculations. We can illustrate the problem by choosing the three variables $z_1$, $z_2$, and $z_3$ as the variables $a_0$, $a_c$, and C. The simulation scheme is shown in Figure 7.13. As the main result variable, we have chosen the number of cycles N to reach critical crack size. For several realizations we will get the corresponding number of the sought variable N. If we seek the probability of failure at a given service time t, we simply take the fraction of all the realizations where N is inferior to n. If this number of realizations is $M_f$ and the total number of realizations is M, then a point estimate for the probability of failure is again given by equation (7.26).

If we seek low probabilities we will have to carry out a large number of realizations M to have any confidence in the result. As a rule of thumb, M has to be 100 times greater than the inverse on the probability. Hence, if we are operating in the area $P_f = 10^{-4}$, the number of realizations M has to be $10^6$. A step further would be to fit a frequency function to the histogram in Figure 7.13. Then we can calculate the probability of failure for any given time N simply as:

$$P_f = F(N) \tag{7.28}$$

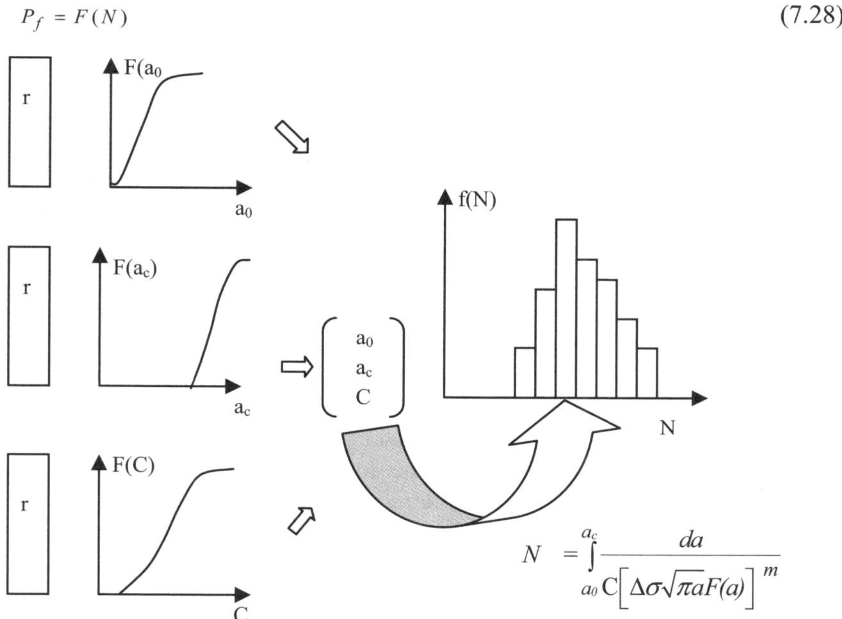


$$N = \int_{a_0}^{a_c} \frac{da}{C\left[\Delta\sigma\sqrt{\pi a}F(a)\right]^m}$$

**Figure 7.13.** *Monte Carlo simulation of fatigue crack growth*

### 7.6.2. *Taking account for future inspections and inspection results*

Simulations with the S-N model and with the fracture mechanics model may have the same goal: simulation of fatigue life. We may choose freely between them for calculating the probability of failure if no inspection strategy is involved during the service life. We must, however, use the fracture mechanics model to simulate the effect of a scheduled in-service inspection program. The scheduled inspection program is characterized by an inspection technique and a time interval between the inspections. The objective is of course to detect cracks and repair them before they become critical. We say that it is a stochastic analysis to corroborate the *damage tolerance* concept. The damage tolerance concept allows a structure to contain cracks if it can be proven that it is highly likely that they will be detected before they become critical. This means that the crack sizes, for which the structure still can sustain the most severe loading, must have a high probability of detection at the time of planned inspections during service.

To carry out the necessary analysis, we must first determine the performance of the inspection technique. Then we must combine it with the crack growth simulation shown in the previous section. Before proceeding let us be quite clear about two different cases related to inspection and repair:

– reliability predictions taking into account future planned inspections;

– reliability updating based on results from inspections that have been carried out.

The first case predicts the most likely influence on the fatigue reliability of future events. The second case gives more information due to the fact that inspection results are at hand. The inspection results lead to updating of the probabilities related to the crack size distribution. The principal difference of the reliability results is shown in Figure 7.14 and Figure 7.15 for future inspections and updating, respectively. In the first case, the reliability function will decrease all the time, but changes to a more favorable slope at each planned inspection. In the latter case, we have more information. We know that the structure has kept its integrity and the result of some inspections. In the case shown in Figure 7.15 it is assumed that we know the results from the first inspection only. If cracks are detected and not repaired, the reliability will follow the lowest curve C in the time after the inspection. If no cracks are found, the reliability curve will follow curve B in the figure. If cracks are detected and repaired to an as-good-as-new condition, it will follow curve A in the figure. There is much confusion between the two cases shown in Figures 7.14 and 7.15. The reader should always be very clear of what case he or she is analyzing. If the aim is to analyze the effect of future inspection at the design stage, the designer is standing at the beginning of Figure 7.14 and is predicting the future. When the service engineer is present at the first inspection and has additional information at hand, we have the updating case in Figure 7.15.

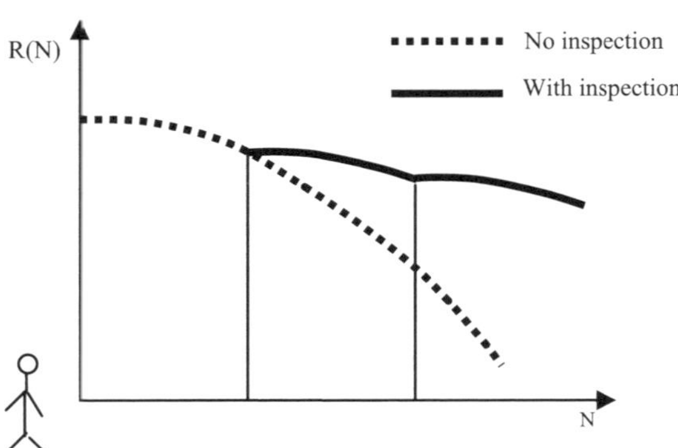


**Figure 7.14.** *Reliability predictions with the influence of future inspections*

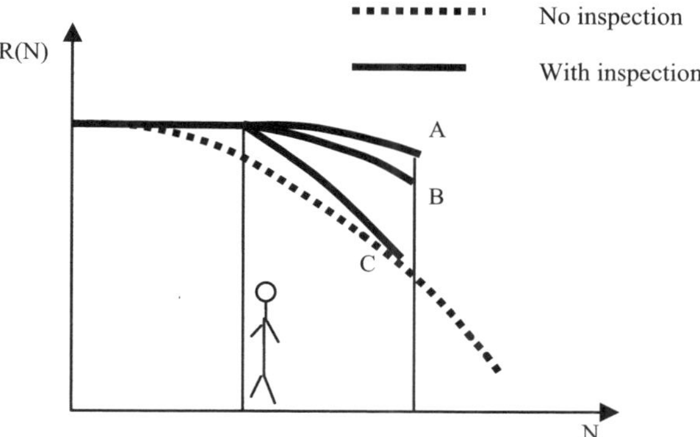


**Figure 7.15.** *Reliability prediction with the influence of updating based on inspection results*

### 7.6.3. Characterization of the performance of the non-destructive inspection technique

In order to prevent fatigue failure and to ensure the safety of a welded structure during service life, non-destructive inspection (NDI) may have to be carried out. If cracks are found, necessary actions regarding repair must be taken. Inspection planning regarding the choice of NDI methods and inspection intervals can be formulated as an optimization problem with the target reliability level as a constraint.

One of the main problems when trying to incorporate the influence of NDI into a stochastic model is to quantify the performance (accuracy) of each inspection technique for a given detail under realistic environmental conditions. This quantification often goes through statistical trials and determination of a POD curve. This curve gives the probability of detection as a function of crack size. When trying to reproduce the in-service inspection condition on trial series for welded details, it is important that: (a) the welded details are representative of the structure in question; (b) the number of cracks is sufficient; (c) the crack population is realistic regarding crack geometry, orientation, and size; and (d) the same environment condition is simulated, e.g. under water. The inspection tests are carried out as blind tests. Several different operators inspect the cracked items without knowing the condition beforehand. The point-estimate for the POD may simply be estimated as the proportion of the cracks that are detected within a given crack size interval. Alternatively, the POD can be estimated based upon a correlation analysis of the crack signal given by the applied technique against real crack size. In the first case where only two outcomes are possible (detection/no detection), a statistical analysis of the test results is based upon binomial distribution model in each crack interval.

The POD curves with corresponding confidence intervals are obtained. The final POD curve is often fitted by the equation (see Ref [6]):

$$POD(2c) = 1 - \frac{1}{1 + \left(\dfrac{2c}{x_0}\right)^{v}}. \tag{7.29}$$

The curve is valid for semi-elliptical cracks with length $2c$ at the surface. The reason for this format is that the performance of most NDI techniques is primarily sensitive to crack surface length, not the crack depth. Typical examples are the magnetic particle inspection (MPI) and the Eddy current method. Typical values for the parameters $x_0$ and $v$ are given in Table 7.2.

| Method | $x_0$ | $v$ |
|---|---|---|
| MPI under water | 2.95 | 0.905 |
| Eddy current | 12.28 | 1.79 |

**Table 7.2.** *Parameters for the performance of MPI and Eddy current for crack inspection in tubular joints, Ref [6]*

In the analysis we will have to relate the crack length $2c$ to the crack depth a, which is our main parameter characterizing the growth rate and the criticality of the crack. It is often assumed that the aspect ratio a/2c is in the area 0.1 - 0.25 (see equation (6.35), Chapter 6. Hence, equation (7.29) can be transformed to a POD curve given as a function of crack depth a, instead of crack length $2c$. A common format is:

$$POD(a) = P_0 (1 - e^{-g(a - a_B)}) \tag{7.30}$$

where $P_0$ is the asymptotic probability of detecting larger cracks, $a_B$ is the smallest crack which can be detected, and g is a shape parameter for the POD curve. The shape of the curve is shown in Figure 7.16. For practical purposes, the crack depth that has a probability of detection of 90% at a confidence level of 95% is regarded as a "sure" value. This crack size characterizes the NDI technique, but in the following sections we will take into account the shape of the entire POD curve.

If the crack is sized, it can be assumed that the measured depth is normally distributed with mean value equal to the true depth $a_M$ and with a standard deviation (SD) designated $s_M$. The SD is assumed constant for all values of $a_M$. If the coefficient of variation for $a_M$ is large, the normal distribution has to be truncated between a = 0 and a = T for a part-through crack. Regarding the fatigue quality after repair, it is assumed that joints repaired by grinding or repair welding have the same fatigue behavior as a new joint. In fact, repair grinding alone may give a significant increase in fatigue life compared to as-welded joints, whereas, the fatigue strength of joints repaired by welding is marginally lower than not-repaired joints.

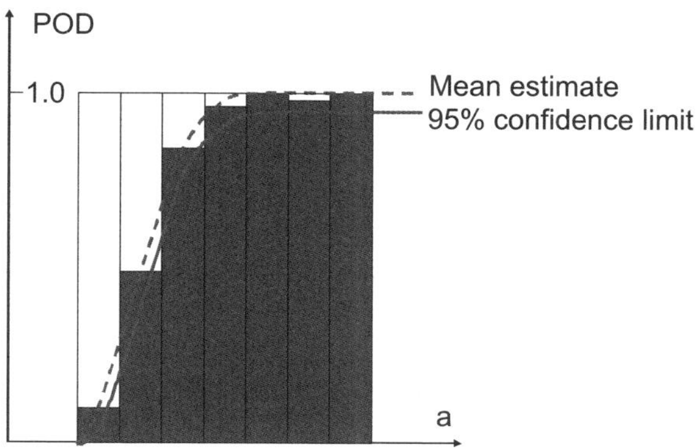


**Figure 7.16.** *POD curve for a given inspection technique*

### 7.6.4. *Simulation with account for future planned inspections*

#### 7.6.4.1. *A first approximation to the inspection problem*

We are now ready to combine our knowledge of crack growth simulations and the stochastic characterization of the inspection technique. In other words, we shall combine the Paris crack growth law with the POD curve. As a first approximation we shall use just one crack growth history to get an estimate of the reliability when future inspections are taken into account. The method is illustrated in Figure 7.17. As can be seen, for one crack history, three scheduled inspections are planned. The time interval between the inspections is designated I. At each inspection the POD value is estimated from the crack depth pertaining to the crack history used. These crack depths are designated $a_1$, $a_2$, and $a_3$. No account is taken of the possible scatter in crack depth at the given inspection times.

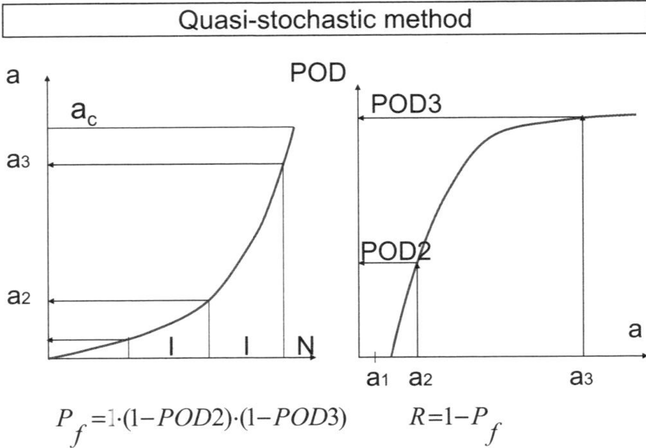


**Figure 7.17.** *Illustration of the quasi-stochastic method*

The probability of missing the crack at all the scheduled inspections is given by:

$$P_f = \prod_{i=1}^{k} (1 - \text{POD}(a_i)) \tag{7.31}$$

where $i$ is the number of the current inspection, $a_i$ is the crack size at the inspection and $k$ is the total number of inspections. In the case shown in Figure 7.17, $k$ is equal to 3. As can be seen, the crack size $a_1$ at first inspection is so small that the crack is impossible to detect, whereas at the last inspection it has a high probability of detection. The method is denoted quasi-stochastic because it ignores the scatter in fatigue-crack growth. If the method should have any credibility, the $a$-$N$ curve used must be an unfavorable one. The curve can be chosen as the one pertaining to a crack growth parameter C corresponding to the mean plus two standard deviations, as discussed in Chapter 6. This curve will then have a 97.7 confidence level, i.e. it is 97.7% probable that the crack will grow more slowly. If the POD curve is based on a 95% confidence level, the obtained reliability against fatigue failure is based on high confidence levels. However, the obtained reliability level is a rough approximation and one should establish a model where the crack growth is random, as is the case in reality. This will be addressed in the next section.

### 7.6.4.2. *Full stochastic simulation*

The calculation scheme for a full stochastic simulation is shown in Figure 7.18 (see Ref [7]). The crack growth is now random according to experimental measurements and fracture mechanics modeling. As can be seen, it is based on the simulation technique that was shown in Figure 7.13, but the simulations are now carried out in time steps corresponding to one inspection interval. The effect of the planned inspections at the end of each interval is included through the POD concept. The flowchart is based on the following assumptions and inspection strategy:

- the target service life is given as $N_S$;
- inspection is planned at regular times $N_i$ at interval $\Delta N$;
- C, $a_0$ and $\Delta\sigma$ are random variables;
- the POD curve is known;
- all cracks detected are to be repaired.

As can be seen from Figure 7.18, each simulation runs through all planned inspections. At each inspection the crack depth is determined based on the depth at the previous inspection and the stress level. Hence, the crack depth is increased from one inspection to another according to the increment in the number of cycles $\Delta N$. Detection is defined to occur if $POD(a_i) > r$. As can be seen, a simulation is ended either when the number of cycles passes the target service life (survival) or the crack size passes the critical crack size (failure). All possible crack sizes at the various inspections have to be simulated through a large number of simulations. To obtain a better understanding of the physics in the problem, typical outcomes are illustrated a growth histories in Figure 7.19. For simplicity, only two inspections are scheduled before end of service life. The outcomes are:

- realization 1: slow crack growth without any crack detection. No failure;

- realization 2: relatively slow crack growth with crack detection and repair at second inspection. No failure;

- realization 3: rapid crack growth without detection at first inspection and failure before the second inspection is reached. Failure.

In a large simulation with, say, M = 100,000, there will of course be most realizations of the first type for a well-designed joint where a reasonable requirement for the FDF is met. The total number of simulations of the third type, relative to the total number of simulations M, will give us the probability of failure. If the FDF is set to 10 there will be very few realizations of the second or third types and inspection can be avoided. If we let $N_S$ vary we will get reliability curves, as has already been illustrated in Figure 7.14.

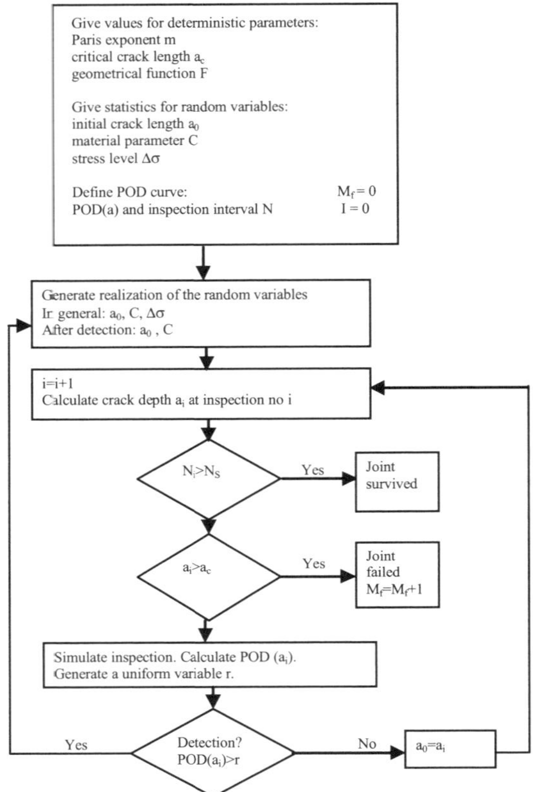



**Figure 7.18.** *Flowchart for simulation model with planned inspections*

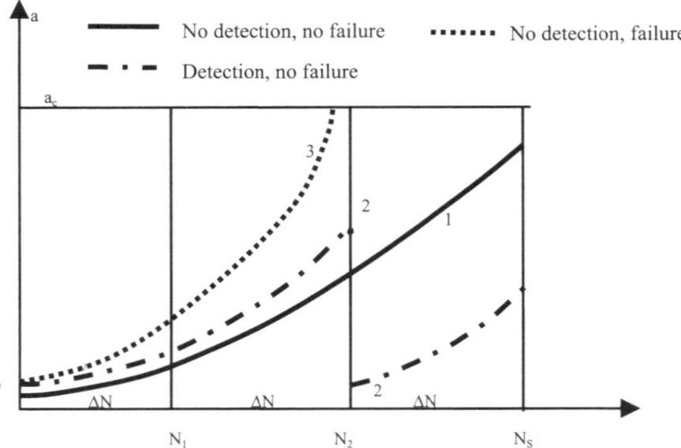



**Figure 7.19.** *Illustration of possible outcomes of the crack growth simulations
with planned inspections*

### 7.6.5. Simulation of planned inspections for a fillet welded joint

As an example we shall consider a typical detail in an offshore structure subjected to wave-induced loading. The number of cycles encountered during the target service life of 20 years is approximately $10^8$ cycles. The detail is a fillet welded joint (F-class) where the crack will emanate for the weld toe. We shall simulate the crack growth with the influence of planned future inspections. The joint is designed with low design factor, alternatively FDF = 1.0 and FDF = 3. With a requirement of a FDF = 1.0 we will have a permissible equivalent stress level of $\Delta\sigma$ = 18.5 MPa according to equation (5.9) in Chapter 5. If the required FDF is set to 3, then the permissible equivalent stress has to be lowered to $\Delta\sigma$ = 12.8 Mpa.

*Crack growth*

The parameters and variable in Tables 7.3 and 7.4 are based on the experimental statistics for database 1 given in Chapter 6. The crack growth parameters and variables reflect the crack behavior of database 1; see Chapter 6 and Figure 6.16. Based on experimental results, the number of cycles to reach a crack depth of 0.1 mm is treated as a stochastic variable. As the experimental data pertain to laboratory conditions, a variable $Z_F$ is introduced to account for additional scatter found in service, such as fabrication quality and workmanship. Both the time to crack initiation and the parameter C in the Paris law are multiplied with this variable. The total scatter in fatigue life will then coincide with the F-class scatter (COV = 0.54). The uncertainty in applied stress is taken into account directly though the scatter in $\Delta\sigma$.

| | | |
|---|---|---|
| m | 3.0 | (deterministic) |
| $a_1$ | 0.1 mm | (deterministic) |
| $a_c$ | 12.5 mm | (deterministic) |
| $N_I$ | random variable | |
| C | random variable | |
| $\Delta\sigma$ | random variable | |
| $Z_F$ | random variable | (random variable according to fabrication quality) |

**Table 7.3.** *Definition of parameters and random variables in the Monte Carlo model*

| Variable | Mean | COV | Distribution | Comment |
|---|---|---|---|---|
| $N_I$ | $145 \cdot 10^3 (150/\Delta\sigma_E)^3$ | 0.34 | Weibull | Database 1, Ch. 6 |
| LnC | -25.98 | 0.22 | Normal | Database 1, Ch. 6 |
| $\Delta\sigma$ | 12.8 | 0.2 | Lognormal | Database 1, Ch. 6 |
| $Z_F$ | 1.14 | 0.5 | Lognormal | Additional scatter for an F-class |

**Table 7.4.** *Definition of random variables in the Monte Carlo model*

As can be seen from Table 7.4, the mean value used for lnC has been slightly modified compared to the statistics for database 1 in Chapter 6. This was done to make the final fatigue life calculated by linear elastic fracture mechanics (LEFM) coincide with the F-class predictions.

The geometry function is approximated by:

$$F = 0.77 \left( \frac{T}{a} \right)^{0.233} \tag{7.32}$$

The present Monte Carlo model is applied using the figures in Table 7.4. The figures correspond to a median $a$-$N$ curve as shown in Figure 7.20 with FDF = 1. As can be seen, for a predicted rule-based life of $10^8$ cycles, the median life is close to $2.7 \times 10^8$ cycles, i.e. close to a factor of 3 compared to the design life. The mean crack growth curve deviated somewhat from the experimental measurements. To make the LEFM-obtained curves coincide with the experimental ones, the geometry function has to be given in a numerical form. This means that the computational efforts will increase substantially.

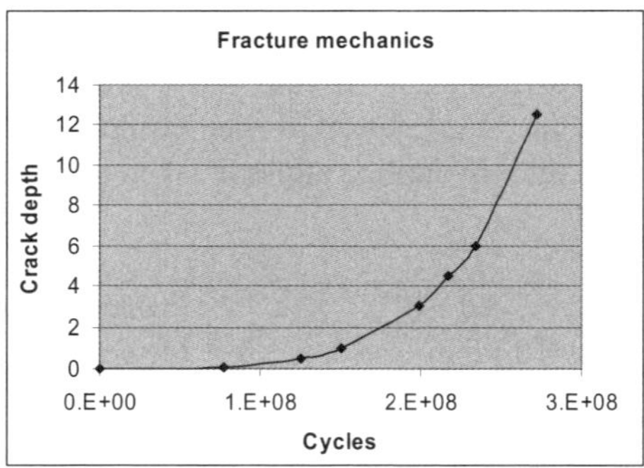


**Figure 7.20.** *Median a-N curve for the Monte Carlo simulation, FDF = 1*

*Performance of the inspection*

The performance of the inspection technique is given by its POD which is a function of the crack depth:

$$\text{POD}(a) = P_0 [1 - e^{-g(a-a_B)}] \quad a > a_B \tag{7.33}$$

For MPI under water, realistic parameters are $a_B = 1$ mm, $g = 1$, and $P_0 = 0.9$. The figures are based on an investigation carried out by ELF Aquitaine; the results are presented in Ref [8].

*Simulation and results*

The Monte Carlo simulation has been carried out according to the flowchart in Figure 7.18. It is assumed that for all detected cracks the weld is repaired to a state that is as good as new. Failure is defined each time the calculated number of cycles N is inferior to the target service life of $N_S = 10^8$ cycles. The main results from the calculations are given in Table 7.5 for FDF = 3. As can be seen, the accumulated probability of failure at the end of service life is $3.3\ 10^{-3}$ if no inspections are scheduled. If inspections are planned every other year in the present case, the probability of failure will drop from $3.3\ 10^{-3}$ to $2.2\ 10^{-5}$. This is a verification of the efficiency of the planned inspections.

| Inspection program | FDF = 3 |
|---|---|
| No inspection | 3.3 E-3 |
| With 9 inspections | 2.2 E-5 |

**Table 7.5.** *Cumulative probability of failure at the end of*
*TSL = 20 years, FDF = 3*

### 7.6.6. *Updating based on inspections results*

The principle reason why this case is different from the simulation shown in the previous section is because we have to introduce updating based on inspection results. All updating must be based on the law of conditional probability:

$$P(E_I | E_J) = \frac{P(E_I \cap E_J)}{P(E_J)} \tag{7.34}$$

where the left-hand side is the posterior probability of event $E_I$ given that event $E_J$ has occurred. The following events $E_j$ are considered:

- $E_0$: failure has not occurred;
- $E_1$: no cracks detected;
- $E_2$: crack detected, but not sized or repaired;
- $E_3$: crack detected and sized, but not repaired;
- $E_4$: repair is carried out.

Equation (7.34) is used to update the crack-size distribution after the inspection has been carried out. To do so we must keep track of the histogram of possible crack sizes at the actual inspection time; see Figure 7.18. Let us use the following notation:

– $H_{iprior}$: probability that the crack has a size confined in histogram column number i before inspection;

– POD($a_i$): probability of detecting a crack size confined in histogram column number i;

– $H_{ipost}$: probability that the crack has a size confined in histogram column number i after inspection/repair has been carried out.

Event $E_0$ is normally the basis for all the other events; it is known that no failure has occurred. The updated damage state probabilities, knowing that failure has not occurred reads:

$$H_{ipost} \mid no\ failure) = \frac{H_{iprior}\delta_i}{R(t)} \quad i = 1,k \tag{7.35}$$

The nominator in the expression represents the joint probability of being in histogram column number i and that no failure has occurred. For histogram columns containing critical crack depths, the parameter $\delta_i$ will be zero whereas it will be 1.0 for other cases. The correction has to be carried out for all the histogram columns. The correction is usually benign and can be omitted.

According to equation (7.34) the updating based on a no-find result will be:

$$H_{ipost} \mid no\ find) = \frac{H_{iprior}\cdot(1-POD(a_i))}{\sum_{j=1}^{k} H_{iprior}\cdot(1-POD(a_j))} \quad I = 1,k \tag{7.36}$$

The nominator is the probability that the crack size is confined in the range of column number i and not detected. The denominator is the probability of not detecting the crack in any of the various columns of the histogram. The equation will change the histogram for the crack size in the way that columns containing larger cracks (easy to detect) will decrease, whereas column representing smaller cracks (easy to miss) will increase. In other words, as no cracks have been detected it is more likely that the sizes are smaller than originally simulated. The modification is visualized to the left-hand side in Figure 7.21.

The other events follow the same logic. In the event of find and not repaired, we can use equation (7.36) again, but (1-POD) will be replaced by POD:

$$H_{ipost} \mid \text{find}) = \frac{H_{iprior} \cdot \text{POD}(a_i)}{\sum_{j=1}^{k} H_{iprior} \cdot \text{POD}(a_j)} \quad I = 1,k \tag{7.37}$$

In this case the crack size histogram will change towards larger crack sizes after the inspection has been carried out. After the updating, the simulation in Figure 7.18 can restart with the updated histogram as initial crack-size distribution; see Figure 7.21.

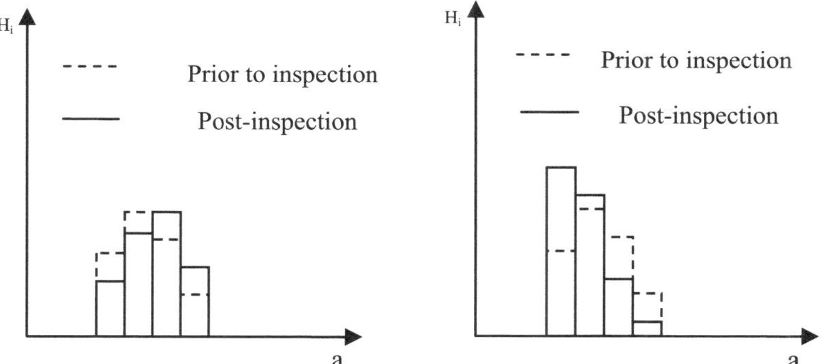


**Figure 7.21.** *Updating of crack histogram: left: no-find event; right: find event*

In the case of repair, it is often assumed that the joint is as good as new and the crack size will again follow the original initial crack-size distribution.

The final event is the detecting and sizing of the crack. In this case we can analyze the situation in two steps. The first step is given in equation (7.37) and is based on the fact that the crack is detected. The sizing of the crack will always be uncertain and this is illustrated in Figure 7.22, on the left-hand side. The inspection may report the most likely size and a possible size range. This can be modeled by a rectangular distribution or a normal distribution as shown in Figure 7.22. It is assumed that the size measurement is unbiased, i.e. the mean value $a_{zm}$ is the real, expected crack size. Based on equation (7.34) we will again have:

$$P(H_{ipostsize} \mid \text{sizing} = \frac{H_{ipost} \cdot P_{size}(a_i)_i}{\sum_{j=1}^{k} H_{jpost} P_{size}(a_j)} \quad i = 1,k \tag{7.38}$$

where $P_{size}(a_i)$ is the probability that the sized crack is in histogram column number i:

$$P_{size}(a_i) = f(a_i)\Delta a .\qquad (7.39)$$

$H_{postsize}$ is shown on the right-hand side in Figure 7.22.

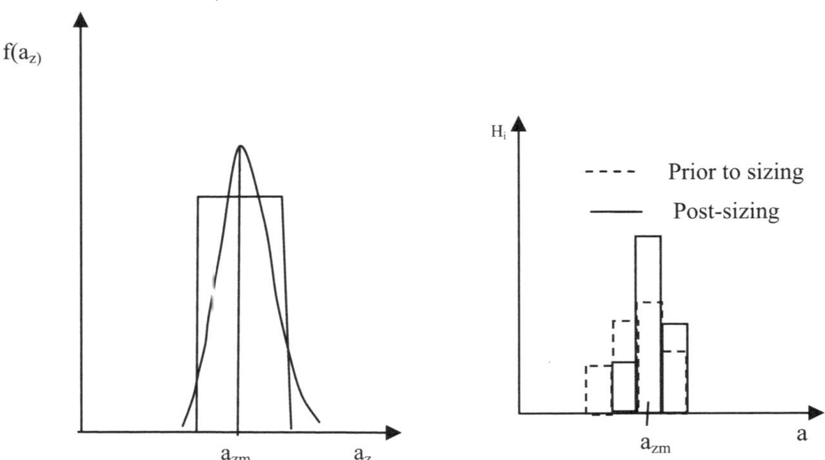


**Figure 7.22.** *Left: model for uncertainty of sizing; right: updating of histogram*

The updating according to equation (7.38) will, as a result, eliminate histogram columns with crack sizes outside the distribution given on the left-hand side in Figure 7.22. Histogram columns close to $a_{zm}$ will be blown up, as shown on the right-hand side on Figure 7.22. If the inspection technique had been very accurate, the post-crack size would of course have been $a_{zm}$. The crack histogram on the right-hand side in Figure 7.22 would then been degenerated to one, single peak value.

In all the cases treated in equations (7.35 to 7.38) we must finally carry out a normalization of the histogram so that the sum of the columns remains 1.0.

## 7.7. The Markov chain model

### 7.7.1. *Basic concepts*

The Markov chain model was first proposed by Bogdanoff and Kozin (1985); see Ref [2]. Instead of carrying out the integration of the Paris law with associated basic variables, the stochastic waiting time between discrete damage states was modeled directly. It is by definition a stochastic process model. The model will be discussed in some detail here and extensions of the model will be formulated to cope with the large scatter in time to crack initiation, the inter-specimen variability, the uncertainty in applied load level, and the updating through inspection events. The model has to be established for a given welded detail, stress spectrum, and environment. The basic assumption for the model is that the crack propagation is without memory. The future random propagation of a crack with a given size is independent of the history of how the crack reached the given crack size. Hence, cracks that initially grow slowly may start to accelerate and vice versa. This assumption is questionable, but the Markov model still fulfils all the four requirements given in section 7.2.3. The only exception is that the crack growth histories are more irregular than measured crack histories, but not significantly different. We will first present the method by a simple example before going into details for more realistic modeling of the fatigue process. At the end we shall formulate the model in a generic manner and apply it to some industrial cases. The generic model is meant as a general damage tolerant supplement to rules and regulations.

### 7.7.2. *Simple illustration on how the model works*

Let us assume for simplicity that the condition of a welded joint can be characterized with only four discrete damage states. The states are related to the depth of a fatigue crack at the weld toe and the states are defined as:

- state 1: as good as new, no cracks are possible to detect;
- state 2: the first small crack has appeared and can be detected;
- state 3: the crack is larger and has reduced a portion of the cross section;
- state 4: the onset of fracture.

The damage states are visualized in Figure 7.23. Damage state 1 cannot be shown as a crack depth.

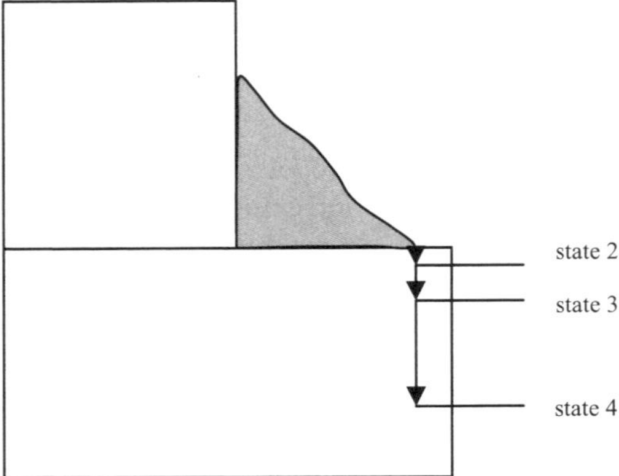


**Figure 7.23.** *Definition of fatigue damage states in a fillet welded joint where cracks emanate from the weld toe*

State 3 is an imminent failure state, whereas state 4 is the failure state. The probability of being at the different states at a given time t are given by the state vector:

$$\boldsymbol{P}(t) = \{p_t(1), p_t(2), p_t(3), p_t(4)\} \tag{7.40}$$

where $p_t(j)$ is the probability of being in state $j$ at time $t$. At all time stages we will have:

$$\sum_{j=1}^{b} p_t(j) = 1 \tag{7.41}$$

where $b$ is the failure state, in our case $b = 4$. This state is denoted the absorbing state; the process cannot leave this state unless repair is carried out. At time $t = 0$ it is usually assumed that there is a good initial quality of the joint and the state vector reads:

$$\boldsymbol{P}(0) = \{1,\ 0\ ,0\ ,0\}. \tag{7.42}$$

As can be seen, the probability of being in state 1 is equal to 1.0. If we doubt the initial quality of the joint, this assumption can be modified. Some of the probability of being in state 1 must then be transferred to the higher neighbor states. To know $\boldsymbol{P}(t)$ at later time stages we must know the probability transition matrix (PTM) $\boldsymbol{Q}$ which reads:

$$Q = \begin{bmatrix} q_{11} & q_{12} & q_{13} & q_{14} \\ 0 & q_{22} & q_{23} & q_{24} \\ 0 & 0 & q_{33} & q_{34} \\ 0 & 0 & 0 & q_{44} \end{bmatrix} \tag{7.43}$$

where $q_{ij}$ is the probability of going from state $i$ to state $j$ during a time interval $\Delta t$. The matrix $Q$ is valid for a given time interval and loading condition. To determine the various figures in the matrix, we have to study crack behavior in the laboratory and observe how the crack propagate. The mean value and scatter in fatigue growth will determine the parameters in equation (7.43). Let us take an example where $\Delta t = 12$ months. Let us suppose that we know the parameters in the transition matrix:

$$Q = \begin{bmatrix} 0.7 & 0.2 & 0.1 & 0.0 \\ 0 & 0.5 & 0.3 & 0.2 \\ 0 & 0 & 0.3 & 0.7 \\ 0 & 0 & 0 & 1.0 \end{bmatrix} \tag{7.44}$$

As can be seen from equation (7.44), there is a probability of 0.7 of remaining in state 1 during a period of 12 months, whereas there is a probability of 0.2 of jumping from state 1 to state 2. The latter event means that the first cracking has taken place. There is even a probability of 0.1 for jumping directly from state 1 to state 3 where the cracking is already severe. It is however impossible to jump from state 1 directly to 4 during the course of 12 months. It is noted that the sum of the probabilities in each row is 1.0, i.e. the crack has to end up somewhere. With the given figures we can now calculate the state vector after 12 months:

$$p_{12}(1) = 1.0 \times 0.7 = 0.7 \tag{7.45}$$

$$p_{12}(2) = 1.0 \times 0.2 = 0.2$$

$$p_{12}(3) = 1.0 \times 0.1 = 0.1$$

$$p_{12}(4) = 1.0 \times 0 = 0.0.$$

As can be seen, the figures are simply obtained by multiplying the probability of being in various states at the beginning of a time interval with the probability of jumping to other states during the same interval. The first probabilities are given from the state vector at the beginning of the interval, whereas the transition matrix gives the second type of probabilities. After 24 months the vector reads:

$$p_{24}(1) = 0.7 \times 0.7 = 0.49$$

$$p_{24}(2) = 0.7 \times 0.2 + 0.2 \times 0.5 = 0.24 \tag{7.46}$$

$$p_{24}(3) = 0.7 \times 0.1 + 0.2 \times 0.3 + 0.1 \times 0.3 = 0.16$$

$$p_{24}(4) = 0.7 \times 0.0 + 0.2 \times 0.2 + 0.1 \times 0.7 + 0.0 \times 1.0 = 0.11.$$

The probability of being in various states as a function of time is shown graphically in Figure 7.24.

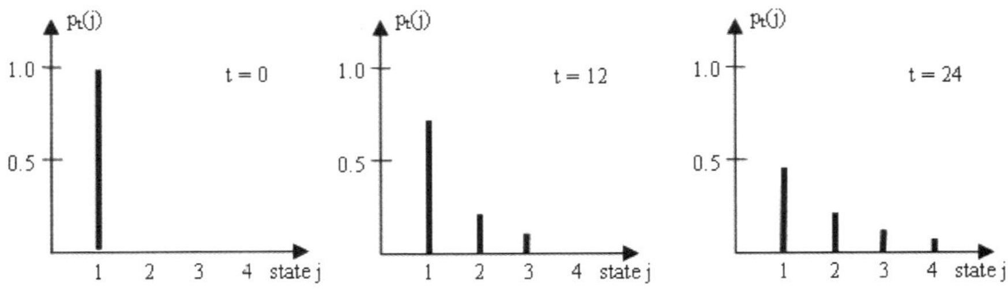


**Figure 7.24.** *Evolution of the state vector as a function of time*

As can be seen from the figure, it becomes more and more likely that the imminent failure state 3 is reached. At $t = 24$ months, there is a probability of 0.11 that the failure state 4 has been reached. The probability multiplication just shown can, in general terms, be expressed by matrix multiplication:

$$\mathbf{P}(t) = \mathbf{P}(0)\mathbf{Q}^{k} \tag{7.47}$$

where $k = t/\Delta t$, i.e. the number of time intervals that have passed from the beginning up to the time we are considering. If $k = 2$, we will get the same results as presented in Figure 7.24. The probability of failure as a function of time is equal to the probability of being in the absorbing failure state 4, $p_t(4)$. This figure is, by definition, the cumulative probability of failure at any time $t$. A typical evolution can be as shown in Figure 7.25.

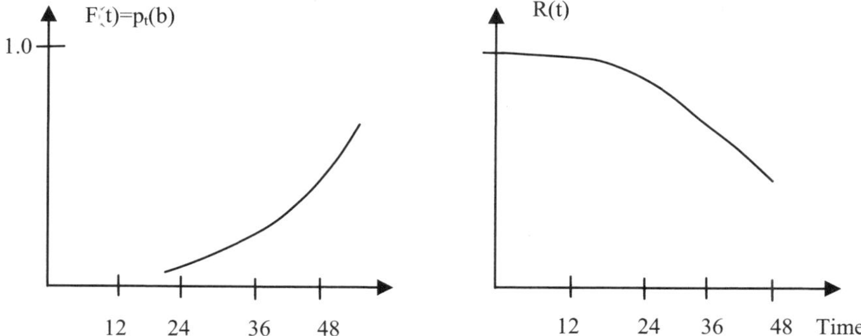


**Figure 7.25.** *Evolution of cumulative probability of failure and the reliability function*

The reliability is by definition given by $1 - p_t(4)$, and the results can be presented as is shown on the right-hand side in Figure 7.25. The advantage of the model is that the influence of in-service inspection can readily be accounted for. Let us, for example, assume that we are planning an inspection at time $t = 24$ months, and that the strategy is to repair all cracks that are found. Let us assume for simplicity that only state 3 is possible to detect. The probability of detection and repair is then:

$$P_R(24) = p_{24}(3)\ \text{POD}(3) \tag{7.48}$$

where POD(3) is the probability of detecting state 3. If we assume that POD(3) = 0.95 we will get:

$$P_R(24) = 0.16 \times 0.95 = 0.152 \tag{7.49}$$

If the repair is carried out by welding, it can often be assumed that the repair weld is as good as new. Hence, the state vector after repair will read:

$$p_{24}'(1) = 0.49 + 0.152 = 0.642 \tag{7.50}$$

$$p_{24}'(2) = 0.24 = 0.24$$

$$p_{24}'(3) = 0.16 - 0.152 = 0.008$$

$$p_{24}'(4) = 0.11 = 0.11.$$

The probabilities are simply obtained by transferring probability from state 3 to state 1. This is the most likely influence of a future planned inspection. This kind of probability transfer will, over time, give less probability of ending up in the failure state 4. This is of course the very goal of inspection and repair. The effect is as shown in Figure 7.26.

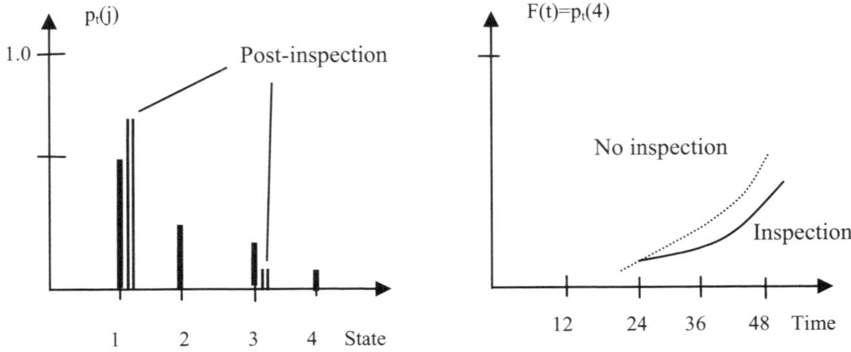



**Figure 7.26.** *Influence of planned inspection at t = 24 months*

The probability of failure as a function of time is shown on the right-hand side in Figure 7.26. This is the complementary curve to the reliability curve that was shown in Figure 7.14.

We can, using these curves, study the influence of various inspections strategies. Both inspection techniques and time intervals can be changed to obtain the required probability. The results from our numerical example are shown in Table 7.6 and Table 7.7.

| Time (months) | $p_t(1)$ | $p_t(2)$ | $p_t(3)$ | $p_t(4)$ | R(t) |
|---|---|---|---|---|---|
| 0 | 1 | 0 | 0 | 0 | 1 |
| 12 | 0.7 | 0.2 | 0.1 | 0 | 1 |
| 24 | 0.49 | 0.24 | 0.16 | 0.11 | 0.89 |
| 36 | 0.343 | 0.218 | 0.169 | 0.27 | 0.73 |
| 48 | 0.2401 | 0.1776 | 0.1504 | 0.4319 | 0.568 |
| 60 | 0.16807 | 0.13682 | 0.12241 | 0.5727 | 0.427 |
| 72 | 0.117649 | 0.102024 | 0.094576 | 0.685751 | 0.414 |

**Table 7.6.** *Reliability function without inspection*

| Time (months) | $p_t(1)$ | $p_t(2)$ | $p_t(3)$ | $p_t(4)$ | R(t) |
|---|---|---|---|---|---|
| 0 | 1 | 0 | 0 | 0 | 1 |
| 12 | 0.7 | 0.2 | 0.1 | 0 | 1 |
| 24a | 0.49 | 0.24 | 0.16 | 0.11 | 0.89 |
| 24b | 0.642 | 0.24 | 0.008 | 0.11 | 0.89 |
| 36 | 0.4494 | 0.2484 | 0.1386 | 0.1636 | 0.837 |
| 48a | 0.31458 | 0.21408 | 0.16104 | 0.3103 | 0.690 |
| 48b | 0.467 | 0.21408 | 0.009 | 0.3103 | 0.690 |
| 60 | 0.3129 | 0.19644 | 0.111624 | 0.359416 | 0.640 |
| 72 | 0.21903 | 0.1608 | 0.1237092 | 0.4768408 | 0.523 |

**Table 7.7.** *Reliability with inspection (a: before inspection, b: after inspection)*

The calculations can be done very rapidly on a spreadsheet. If we compare the reliability functions in the right column of Tables 7.6 and 7.7, we can see that the effect of the inspection is benign in the present case. To improve the situation we can either decrease the time interval to 12 months and/or use a better inspection technique that may detect state 2 and increase the POD for state 3. The reader may try to do the calculation for POD(2) = 0.8 and keep POD(4) at 0.95. The general expression for repair at time $t = 24$ months is given by:

$$P_R(24) = \sum_{j=2}^{3} p_{24}(j)POD(j) \tag{7.51}$$

where POD(j) is the probability of detecting state j. We have not included state b because this is repair of a failure. Even if such repair is possible, we have excluded it because our aim is to study the effects of preventive inspection and repair only. Instead of the probability of failure, we can draw the reliability curves as shown in Figure 7.27.

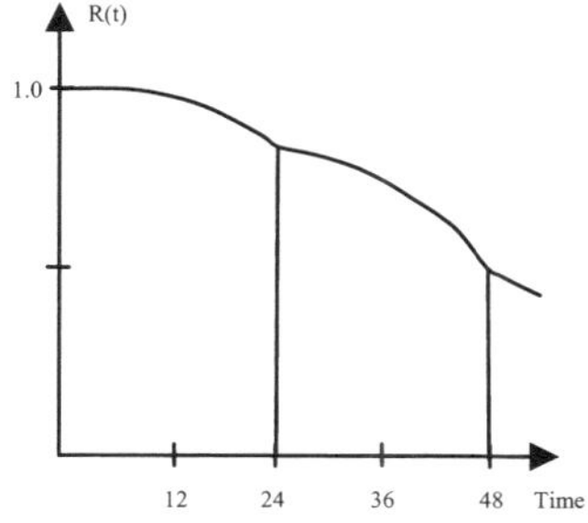


**Figure 7.27.** *Reliability as a function of time with impact of future inspection and repair*

### 7.7.3. *Elaboration of the model*

The example just shown was a fictitious case tailored for pedagogical demonstration. In the present section we will improve the model so that it will be able to describe realistic crack growth in a welded joint. We will also show how the parameters in the PTM are determined in real cases. The first modification is to split the model into crack initiation and crack propagation. The accumulation of the fatigue damage is modeled by a Weibull distribution for the time to crack initiation and a Markov chain model for the subsequent propagation. The damage states are labeled by j = 0, 1, 2....., b where state 0 is defined as the initial damage that cannot be related to a measurable crack size. The higher model states (j = 1, 2,..., b) correspond to defined crack depths in the material. It is only the transition between state 0 and 1 that is modeled by a Weibull distribution; the subsequent states belong to a pure Markov chain as shown in the previous section. The time n in number of cycles is scaled into t that is measured in number of so-called duty cycles (DC). A DC may represent a repetitive period of operation during the service of the joint. In the present work, one DC corresponds to a certain number of loading cycles, $\Delta N$, with constant amplitudes. Hence, the total number of load cycles at time t will be t multiplied with $\Delta N$. The time $t_b$ is the number of DCs to fatigue failure, i.e. to reach the absorbing state b. The state of the damage is considered only at the end of each DC. The damage accumulation during a DC can only be zero or one unit. This means that for a given damage state at the start of one DC, the cumulative damage can either remain in the same state or jump to the next higher state. This is referred to as a one-jump model. The time $N_I$ to crack initiation is assumed to be governed by a two-parameter Weibull distribution:

$$P(\mathbf{N_I} \le N) = F_{N_I}(N) = 1 - \exp[-(\frac{N}{q})^h] \tag{7.52}$$

The reason for using a Weibull distribution for the time to reach the first state is that this distribution is more suited to modeling the high COV found in this crack regime. The Markov Chain (MC) is not able to model COVs that are higher than 1.0 for the waiting time between two neighboring states, as will be discussed further below. The probability that the damage state is transferred from the initial state 0 to the damage state 1 within a time interval $\Delta N$ corresponding to the last DC at time t is then given by:

$$\Delta F_{N_I}(t) = \exp[-(\frac{(t-1)\Delta N}{q})^h] - \exp[-(\frac{t\Delta N}{q})^h] \tag{7.53}$$

This probability will be transferred into the first state of the MC according to equation (7.53). If crack initiation is considered, equation (7.47) will be modified:

$$\mathbf{P}_t = \mathbf{I}_t + \mathbf{P}_{t-1}\mathbf{Q} \tag{7.54}$$

where the term $\mathbf{I}_t$ takes into account the contribution from crack initiation during the last DC. According to equation (7.53), $\mathbf{I}_t$ reads:

$$\mathbf{I}_t \;=\; [p_0(0)\Delta F_n(t),0,0,.....,0] \tag{7.55}$$

As the number of DCs increases, state 0 will act as a feeder to the probability of being in state 1 through the first term in this expression. This incubation process will fade away when the Weibull distribution $F_{NI}(t)$ approaches unity, but will reappear in the case of an inspection with repair. Given the parameters in the Weibull distribution, the PTM and the initial crack population, a description of the evolution of the cumulative damage by simple matrix multiplication, is possible; see equation (7.54). Using the calculation procedure described above, the PDF for crack depth at a specified lifetime and the PDF for lifetimes for specified crack depths are derived.

State 1 corresponds to a measurable crack size $a_1$. Further damage accumulation is characterized by crack propagation and is approximated by a Markov chain model. If a given damage state is denoted by j, the probability of remaining in the state during a DC is $p_j$ and the probability of jumping to state j + 1 is $q_j$. Since these two events are defined to be the only two possible outcomes, we will have $p_j + q_j = 1$. The definition of the damage states and the flow diagram are illustrated in Figure 7.28. This figure also shows the physical deterministic equations that are related to the process. These equations will be used to modify the stochastic model for various loading modes and stress levels. The parameters $p_j$ and $q_j$ are given by the PTM which at a given stress level reads:

$$\mathbf{Q} = \begin{bmatrix} p_1 & q_1 & 0 & 0 & ....... & 0 & 0 \\ 0 & p_2 & q_2 & 0 & ....... & 0 & 0 \\ .. & .. & .. & .. & .. & .. & .. \\ .. & .. & .. & .. & .. & .. & .. \\ 0 & 0 & 0 & 0 & .. & p_{b-1} & q_{b-1} \\ 0 & 0 & 0 & 0 & ....... & 0 & 1 \end{bmatrix} \tag{7.56}$$

$$p_j + q_j = 1 \;,\; p_j > 0 \;\; q_j > 0$$

If it is assumed that no initial cracks exist, $p_0(0)$ is set to 1. If one takes into account the possibility of pre-existing cracks due to poor production quality or accidental damage, $p_0(0)$ is reduced and the first terms in $\mathbf{P}_0$ (j $\geq$ 1) are increased correspondingly.

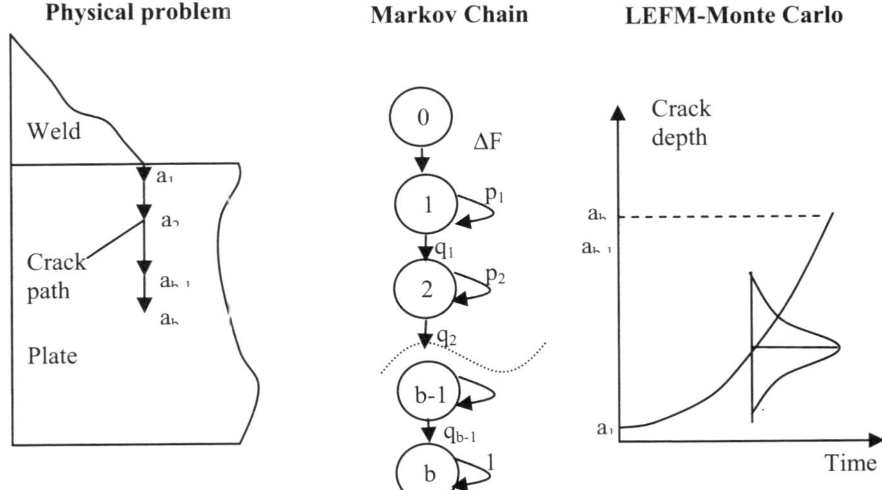



**Figure 7.28.** *Illustration of the stochastic modeling of crack growth. Cracks starting at the weld toe propagating through the plate thickness (left drawing) are related to discrete damage states (middle drawing) or LEFM simulation (right drawing). Note that $a_b = T/2$*

### 7.7.4. *Influence of scheduled inspection and repair*

One of the main advantages of the Markov model is that the crack depth distribution at a given time is given by the state vector $P_t(j)$ is available at any time t measured in number of DCs. This makes it easy to predict the influence of a given inspection strategy and make decisions when planning a scheduled inspection program. The performance of the inspection technique is given by its POD which is a function of crack size. The probability of repair for a given planned inspection at time t reads:

$$P_R = \sum_{j=L}^{j=b-1} P_t(j)POD(a_j) \tag{7.57}$$

where damage state j correspond to crack depth $a_j$. As can be seen, the strategy is to repair every detected crack larger than $a_L$, whereas the failure state b is not assumed to be repairable. We have excluded state b due to the desire to study the effect of preventive inspections only. The most likely damage distribution after a future inspection will read:

$$p_t(j) = p_t(j) + P_R\, p_0(j), \quad j = 0, L-1 \tag{7.58}$$

$$p_t{}'(j) = p_t(j)(1 - \text{POD}(a_j)) + P_R\, p_0(j), \quad j = L, b-1 \tag{7.59}$$

where $L$ is the damage state corresponding to $a_L$. The most common case is to set $a_L$ equal to $a_1$, i.e. all detected cracks are to be repaired. The influence of future inspections, as given by equations (7.57)–(7.59), is shown in Figure 7.29. Each damage state is symbolized by a bubble and the probability of being in a state corresponds to the degree the bubble is filled.

The updating based on inspection results follows the same procedure as outlined for the histogram in the case with Monte Carlo simulation, equations (7.35)–(7.38). In fact, a histogram column for crack size in the Monte Carlo simulation model is analogous to a damage state in the Markov model. Hence, equations (7.35) and (7.38) can be used directly for updating of the damage state vector. The results are visualized in Figure 7.30 where only the three first damage states are included. This figure is analogous to Figure 7.21.

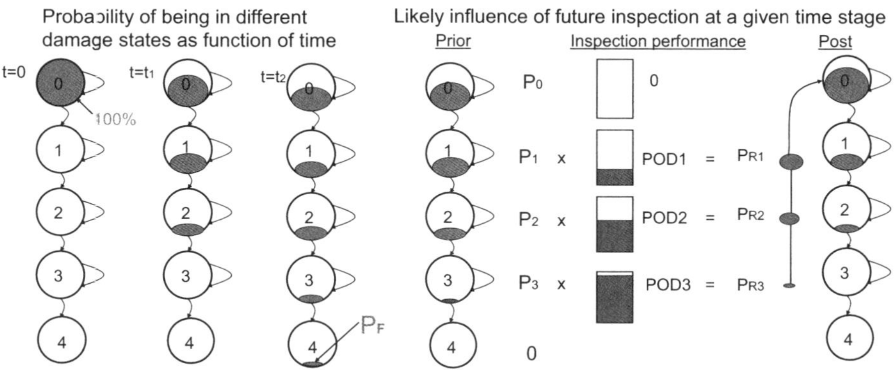


**Figure 7.29.** *The random evolution of damage states with the influence of future inspections*

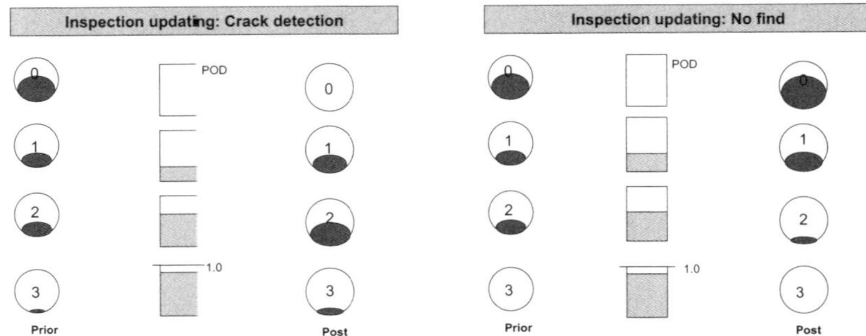



**Figure 7.30.** *Updating of state probabilities based on inspection results*

### 7.7.5. *Parameter estimation*

To construct a model for a given joint, experimental statistics for the time to reach given crack depths have to be known for a given load spectrum. The parameters in the PTM are determined from the first- and second-moments of time to reach given crack depths. It follows from the assumption of a one-jump model that the time spent in each damage state j ≥ 1 has a geometric distribution:

$$P(\mathbf{t}_j = t) = q_j r_j^{t-1}. \tag{7.60}$$

Hence:

$$E(\mathbf{t}_j) = \frac{1}{q_j} = (1 + r_j) \tag{7.61}$$

$$Var(\mathbf{t}_j) = \frac{p_j}{q_j^2} = r_j(1 + r_j) \tag{7.62}$$

where $r_j = p_j/q_j$. From these equations and the Markov assumption, it follows that the mean and variance for the time from state 1 to b reads:

$$E(\mathbf{t}_{1,b}) = \sum_{j=1}^{b-1} (1 - r_j) \tag{7.63}$$

$$Var(\mathbf{t}_{1,b}) = \sum_{1}^{b-1} r_j (1 + r_j) . \tag{7.64}$$

These equations are based on the assumption that $p_0(1) = 1$. It is seen that the time statistics are obtained by simply adding the mean and variance of the time spent in each damage state. Hence, the correlations between times spent in different intervals are ignored as a consequence of the Markov assumption.

The experimental statistics must be put into the left-hand side of equations (7.63) and (7.64) to find b and the $r_j$-values. To describe the entire mean a-N curve, the equations must be solved for the time between state 1 and several other damage states arriving before the final absorbing state b. These damage states must be associated with selected crack depths for which the experimental time statistics are known. To improve the flexibility of the model, Bogdanoff and Kozin (1985) proposed to allow for some additional damage states between the states that correspond to the *a priori* selected crack depths. Equations (7.63) and (7.64) are then replaced by:

$$E(\mathbf{t}_{1,b}) = \sum_{k=2}^{M} (d_k - d_{k-1})(1 + r_{k-1}) \tag{7.65}$$

$$Var(\mathbf{t}_{1,b}) = \sum_{k=2}^{M} (d_k - d_{k-1}) r_{k-1}(1 + r_{k-1}) \tag{7.66}$$

where $d_k$ are damage states which correspond to chosen crack depths $a_k$. The equations are solved successively for k = 2 to M, M being the total number of selected crack depths. For each k, a pair of equations is given from which $d_k$, $p_k$, and $q_k$ are found. The auxiliary damage states between $d_k$ and $d_{k-1}$ are related to the crack depths by interpolation. It follows that $d_1$ equals damage state 1 and $d_M$ equals state b, which often is set to half the plate thickness. Equations (7.65) and (7.66) are based on the assumption that the parameter $r_k = p_k/q_k$ is constant between the damage states associated with the selected crack depths. Other assumptions are possible and will result in other model parameters, but not in significantly different derived reliability parameters. A more thorough discussion of this uniqueness problem is given by Bogdanoff and Kozin (1985). The statistical moments on the left-hand side of equations (7.65) and (7.66) may be based on point estimates or on a given confidence levels. If experimental data are not available, the first and second moment statistics, could be obtained by Monte Carlo simulation using a fracture mechanics model for the crack propagation.

It is *a priori* necessary to make some reasonable assumption when selecting the number of crack depths and number of cycles $\Delta N$ in one DC. As already stated, the number of crack depths should be large enough to describe the mean $a$-$N$ curve. This is particularly important when the influence of inspection is to be taken into account. On the other hand, if one uses too many values of $a_k$, the model will end up with more parameters than justified by the data. Bogdanoff and Kozin (1985) chose six to eight crack depths for test series where the number of specimens ranged from 23 to 68. Short intervals between these depths are chosen at the start of the propagation phase and longer intervals as the growth acceleration occurs. Low values of $\Delta N$ will give high $r_j$-values, which implies that there will be a high probability that the crack will remain in the already occupied state during one DC. This is a natural choice for a one-jump model.

*Discussion of model scatter*

If there exists a positive correlation between the times spent in each damage state, then the Markov approximation will overestimate the variance for each intermediate waiting time for given values on the left-hand side of equations (7.65) and (7.66). This means that the total scatter in fatigue life is obtained by $a$-$N$ curves which are too irregular compared to the experimental ones. It follows from equations (7.61) and (7.62) that the COV for the time spent in one, single state equals the square root of $(r/1 + r)$. Hence, this COV can never exceed 1.0. Furthermore, for the total time spent in the MC, that is the sum of the times spent in each state, equations (7.63) and (7.64), the COV will be even lower. In fact, if the number of model states exceeds 10, it will be difficult to model COVs higher than about 0.5. This fact may create a problem when models are to be constructed from data from welded joints. Typical COV values for the fatigue life of the welded classes in the building codes are, as we have seen, in the range 0.5-0.6. A sufficiently large number of damage states must be chosen to describe the mean $a$-$N$ curve from the experimental data. Hence, a problem may arise in cases where the COV is large, as in in-service conditions. The MC may not be able to model both the mean $a$-$N$ curve and the final high scatter. This problem was not discussed by Bogdanoff and Kozin (1985) due to the fact that they analyzed non-welded metals with COVs in the range 0.1-0.2. The subject will be further discussed in the next section.

### 7.7.6. *Hybrid model to account for additional scatter*

The model thus far described is well suited for crack growth in cases where scatter is mainly dominated by weld toe variability and material inhomogenities, i.e. typical laboratory conditions. To account for uncertainty in global geometry and uncertainty in applied stresses, an external variable $Z_E$ is introduced. The model developed above is treated as a conditional one for a given value of $Z_E$. The mean waiting times and the corresponding parameters in the PTM equation (7.56) are scaled linearly with $Z_E$. Hence applying the total law of probability we get:

$$p_t(j) = \int_0^\infty p_{t,Z_E}(j) f_Z(Z_E) dZ_E \tag{7.67}$$

where $p_{t,Z_E}$ is the damage state probability obtained from the Markov model for a given value of $Z_E$, and $f_z(Z_E)$ is the frequency function for the external variable $Z_E$. The variable $Z_E$ may in turn be treated as a product of several sources of uncertainty, such as global geometry and tolerances given by $Z_F$ and possible errors in stress analysis given by $Z_S$. This will give a total COV for the external variable that reads (see equation (7.23)):

$$COV_E = \sqrt{V_F^2 + m^2 V_S^2}\ . \tag{7.68}$$

This scatter comes in addition to the scatter inherent in the Markov model discussed in the previous section. A lognormal distribution is assumed for $Z_E$. If the applied stresses are unbiased, the external variable has a median value of 1.0.

*Obtained reliability parameters*

Once the model parameters in the Markov chain and the external variable have been determined, the reliability function is readily obtained. The cumulative probability of failure as a function of time is obtained directly from the probability of being in damage state b:

$$P(\mathbf{t}_b \leq t) = F\ (\mathbf{t}_b)\ =\ p_t(b)\ . \tag{7.69}$$

The failure rate function (FRF) is defined by:

$$\lambda(t.b) = \frac{F\ (t_b) - F\ (t_b - 1)}{1 - F\ (t_b - 1)}\ . \tag{7.70}$$

This is a conditional probability of failure, i.e. the probability of failure during one DC given the condition that the welded joint has survived up to the start of that DC. This is, as we have seen, a useful measure for age-reliability under in-service inspection: see Figure 7.6.

### 7.7.7. Analysis of a fillet welded joint

Let us return to the example with the fillet welded joint in an offshore structure that was treated by Monte Carlo simulation in section 7.6.5. In the present section we shall carry our reliability predictions of the joint by applying the Markov Chain model. From the data given in database 1 (Figure 6.16, Chapter 6), the model parameters can be determined, but we have to have more information than we have used before.

### 7.7.7.1. *Short review and elaboration of database 1*

The fillet welded joints were subjected to a constant stress range of 150 MPa. The whole test series (34 specimens) of database 1 had a mean value of 468,000 cycles and a coefficient of variation of 0.22 for the total fatigue life $N_T$. This is defined as the total time from test start to the stage where the specimen is subjected to ductile fracture at a crack depth corresponding to about half the plate thickness. The mean value of $N_T$ is close to the statistics given for the S-N curves in the building codes for this type of joint (F-class; see Chapter 5). This median value is equal to 514,000 cycles i.e. 10% longer than our mean value. For the F-class the COV is as high as 0.54. The high COV is assumed to be due to a greater variation of both local and global geometry of the specimens for the large sample size pertaining to the F-class. Thus, the present test series can be regarded as representative of this type of welded steel joint, although the standard deviation is less. Based on experimental evidence, the initial flaws or intrusions created by the welding process were found to be in the sub 0.1 mm range. The number of cycles to reach $a_1 = 0.1$ mm is defined as cycles to crack initiation $N_I$. $N_I$ had a mean value of 145,000 cycles and a COV of 0.34. The mean value corresponds to 31% of the total fatigue life. The number of cycles elapsed after the crack initiation is defined as cycles spent in crack propagation $N_P$. $N_P$ had a sample mean of 323,000 cycles and a COV equal to 0.22. There was a correlation coefficient $\rho = 0.48$ between $N_I$ and $N_P$, i.e. cracks that initiate slowly have a tendency to propagate slowly as well. The statistics to reach given crack depths are given in Table 7.8. Further details are found in Ref [9].

| Crack depth (mm) | Sample mean | Standard deviation | COV |
|---|---|---|---|
| 0.1 | 145 | 50 | 0.34 |
| 0.5 | 244 | 69 | 0.28 |
| 1.0 | 302 | 80 | 0.26 |
| 3.0 | 404 | 99.5 | 0.25 |
| 6.0 | 446 | 104.4 | 0.234 |
| Failure (12.5 mm) | 468 | 105.5 | 0.23 |

**Table 7.8.** *Time (1,000 cycles) statistics to reach chosen crack depths, $\Delta\sigma = 150$ MPa*

*7.7.7.2. Determination of parameters in the Markov model*

The parameters $p_j$ and $q_j$ can now be estimated from the time statistics given in Table 7.8. The crack depths in the left column correspond to the chosen crack depths $d_k$ in equations (7.65) and (7.66).The model will be able to simulate the fatigue crack initiation and growth in fillet-welded joints under CA loading in laboratory conditions where the observed scatter is dominated by variability in the local toe geometry, surface condition at the fusion line, and heterogeneities of the HAZ. The simulated crack depths as a function of time have a similar global shape to the experimental ones, although the paths are somewhat more irregular; see Figure 6.16, Chapter 6. This is because the experimental crack growth does not exhibit fully Markovian behavior; for details see Refs [7, 9]. Crack behavior exhibits a certain memory, e.g. if a crack initiates slowly it will have a tendency to propagate slowly as well.

Before determining the model parameters we will scale the time statistics from accelerated laboratory loading condition ($\Delta\sigma$ = 150 MPa) to in-service condition ($\Delta\sigma$ = 18.5 MPa, FDF = 1). The mean time on the left-hand side of equation (7.65) is derived directly from experimental data by scaling the mean time in each damage state by a factor of $(150/\Delta\sigma)^3$ according to the F-class S-N curve. Furthermore, all mean waiting times are increased by 14% to calibrate the mean fatigue life predicted by an F-class design curve. The increase of 14% for the propagation part will give an increase of 10% for the mean total fatigue life. This was the discrepancy between the present test series and the F-class curve, as discussed in the previous section. The total mean fatigue life is 2.73 $10^8$ cycles, whereas the design life is $10^8$. The scaled waiting times are given on the left-hand side in Table 7.9 where the time to crack initiation is subtracted. The time is measured in 100,000 cycles that correspond to a DC in the Markov model. The initiation time will be modeled separately by a Weibull model, as explained. The parameters of the Markov model are given on the right-hand side of Table 7.9. They are found from equation (7.65) and (7.66). If we take the time statistics to reach a crack depth of 0.5 mm (k = 2) we get:

$$602 = (d_2 - 1)(1 + r_1)$$
$$289^2 = (d_2 - 1)r_1(1 + r_1) \tag{7.71}$$

This pair of equations can be solved approximately only, due to the fact that $d_2$ has to be an integer. The approximate solution sought should, first, give the correct expected waiting time and, secondly, model the scatter. The deviation in scatter can then be adjusted for in the next pair of equations, i.e. for k = 3. The solution to equation (7.71) is then:

$$d_2 = 5$$
$$r_1 = 150$$

The damage states appearing between the chosen crack depths can be related to crack depths by a linear interpolation. This is shown in Table 7.10.

| Time statistics related to models states | | | | | Parameters in transfer matrix | |
|---|---|---|---|---|---|---|
| Crack depth (mm) | Time ($10^5$ cycles) SM  SD | | k | $d_k$ | States | $r_k = p_k/q_k$ |
| 0.1 | 0 | 0 | 1 | 1 | 1–4 | 150 |
| 0.5 | 602 | 289 | 2 | 5 | 5–6 | 173 |
| 1 | 954 | 379 | 3 | 7 | 7–9 | 208 |
| 3 | 1,574 | 523 | 4 | 10 | 10–11 | 128 |
| 6 | 1,829 | 557 | 5 | 12 | 12–13 | 66 |
| 12.5 | 1,963 | 564 | 6 | 14 | | |

**Table 7.9.** *Determination of model parameters for $\Delta\sigma = 18.5$ MPa, DC = 100,000 cycles*

| Damage state | 1 | 2 | 3 | 4 | 5 | 6 | 7 | 8 | 9 | 10 | 11 | 12 | 13 | 14 |
|---|---|---|---|---|---|---|---|---|---|---|---|---|---|---|
| Crack depth (mm) | 0.1 | 0.2 | 0.3 | 0.4 | 0.5 | 0.75 | 1.0 | 1.67 | 2.33 | 3.0 | 4.5 | 6.0 | 9.25 | 12.5 |
| Waiting times | 151 | 151 | 151 | 151 | 174 | 174 | 209 | 209 | 209 | 129 | 129 | 67 | 67 | |

**Table 7.10.** *Mean waiting time (number of DC = 100,000 cycles) in various damage states; $\Delta\sigma = 18.5$ MPa, FD 1*

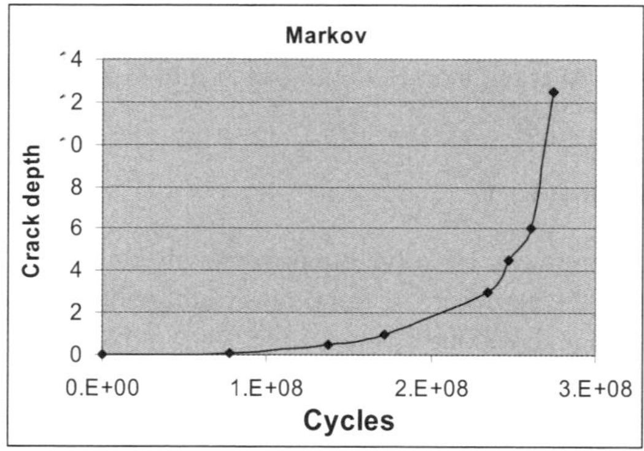


**Figure 7.31.** *Mean a-N curve for the Markov model*

The final mean life coincides with that of an F-class for $\Delta\sigma = 18.5$ MPa, i.e. FDF = 1. So far we have modeled the crack propagation by using the Markov chain. The time to crack initiation is given in the first row of Table 7.11. The corresponding Weibull parameters can be found from equations (7.13) and (7.14). As we shall use the constructed model to analyze fatigue reliability under in-service conditions, uncertainties related to in-service condition must be introduced. The variable $Z_F$ is an external variable accounting for additional scatter in overall geometry, fabrication tolerances, and workmanship. The uncertainty in applied stresses must be assessed from the uncertainties in load and in the stress calculations. The variable $Z_S$ accounts for the uncertainty in the applied stress. The model is quite similar to the Monte Carlo model for this example.

| Variable | Mean | COV | Distribution | Comment |
|---|---|---|---|---|
| $N_I$ | $145\ 10^3(150/\Delta\sigma_S)^3$ | 0.34 | Weibull | |
| $Z_F$ | 1.14 | 0.5 | lognormal | Median = 1.0 |
| $Z_S$ | 1.03 | 0.2 | lognormal | Median = 1.0 |

**Table 7.11.** *Other variables in the Markov model*

The analysis is carried out using equations (7.53), (7.54) and (7.55). Each time an inspection is planned, the damage state vector is modified according to equations (7.57), (7.58) and (7.59). The POD curve used is given by equation (7.33). It is assumed that every crack found is repaired. The final results are given by equations (7.67), (7.69) and (7.70).

### 7.7.7.3. *Reliability results and discussion*

The Markov chain model was used to obtain the reliability as a function of time, as shown on Figure 7.32. The first curve was determined for FDF = 1 without any inspection. The results are compared with the figures obtained by the lognormal and Weibull models presented in earlier sections; see Figure 7.9. As these models were based on the S-N approach they have a different basis, but the comparison is interesting in any event. As can be seen from Figure 7.32, the Markov model gives results very close the lognormal model. The reliability at the end of target service life is 0.91 according to the Markov model, whereas it is 0.94 according to the lognormal model. Hence, the Markov model is somewhat more pessimistic in the prediction of the reliability at the end of the target service life. The results are quite different from the prediction given by the Weibull model. Both the design cases, FDF = 1 and FDF = 3, are treated with various inspection strategies. The results are given as accumulated probability of failure, as shown in Figure 7.33. These types of probability graphs will reveal the influence on the accumulated probability of failure from chosen FDF and inspection strategies at any time during service life. As an example, FDF = 1 without any inspection will correspond to a final cumulative probability of failure at the end of target service life close to $10^{-1}$, as can be seen from Figure 7.33. In most cases this probability level is too high. To achieve a probability level of $10^{-2}$, one could recommend inspections at an interval of 2 years with an FDF = 1. Alternatively, one may increase dimensions to obtain FDF = 3. In this case, a probability level of $2 \cdot 10^{-3}$ is obtained without an inspection program, see Figure 7.33, right-hand side.

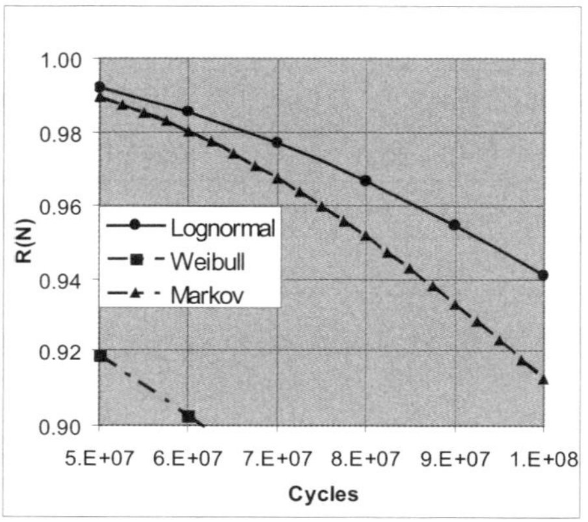

**Figure 7.32.** *Markov-obtained results compared with lognormal and Weibull S-N format, FDF = 1*

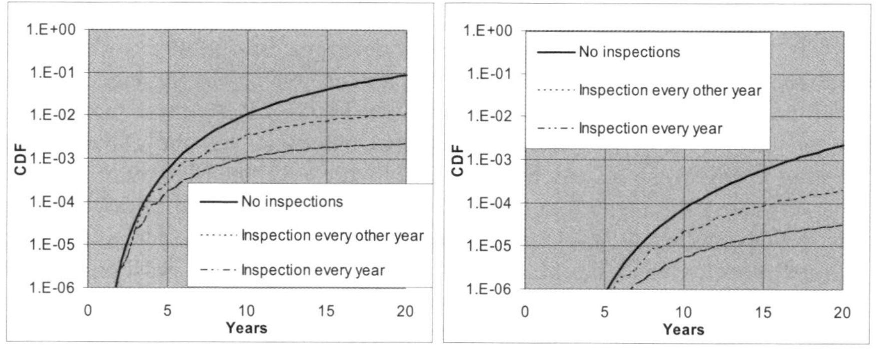

**Figure 7.33.** *Cumulative probability of failure derived by Markov chain model for various FDFs and inspection strategies; left: FDF = 1, right FDF = 3*

The accumulated probabilities of failure at the end of target service life given in Figure 7.33 are listed in Table 7.12. As can be seen, the figures are higher than the figures obtained from the Monte Carlo simulation for the same example; see Table 7.5. This is mainly due to the fact that the two models have different mean $a$-$N$ curves. This can be seen by comparing Figure 7.20 with Figure 7.31. The fracture mechanics model used in the simulation predicts a curve that is easier to detect, whereas the Markov curve has a more hidden crack path on the way to failure. In addition, the Markov generated $a$-$N$ curves are generally more irregular and may more easily escape detection. It is important to be conscious about these matters when choosing the growth model. Several different models may predict the same fatigue life, but as they may have different crack paths to reach the failure, they may predict very different effects of inspections.

| FDF = 1 | | | FDF = 3 | | |
|---|---|---|---|---|---|
| No inspection | 9 inspections | 19 inspections | No inspection | 9 inspections | 19 inspections |
| $8.7 \ 10^{-2}$ | $1.1 \ 10^{-2}$ | $2.2 \ 10^{-3}$ | $2.2 \ 10^{-3}$ | $2.0 \ 10^{-4}$ | $3.1 \ 10^{-5}$ |

**Table 7.12.** *Accumulated probability of failure at the end of target life*

## 7.8. A damage tolerance supplement to rules and regulation

### 7.8.1. *Introduction*

The Markov chain model presented in the previous section will now be elaborated to serve as a general damage-tolerance supplement to the rule-based S-N calculations presented in Chapter 5. The model is valid to begin with for fillet-welded joints with plate thicknesses of 25 mm under axial-loading mode with cracks emanating from the weld toe. This is the test series for database 1 and we already have presented this example at the end of the previous section. However, the model can be used as a generic model for any welded joint with toe cracking and load condition if following assumptions are made:

1) *Calibration to an F-class detail:* the model is based on statistics from experimental crack growth histories. The mean times to reach given crack depths are increased by 14% to match the final life of an F-class detail. For the same reason, the scatter is increased from COV = 0.23 to COV = 0.54. These topics were discussed in the previous section.

2) *Normalization of the crack growth data to other weld classes:* the experimental statistics for the time spent in various crack stages pertain to a fillet welded cruciform joint, i.e. an F-class detail in the original classification system. If the hot-spot concept pertaining to the P-curve is adopted, the mean waiting time $E(t_j)$

in any damage state j can be scaled from laboratory condition $E(t_{jL})$ to in-service condition $E(t_{jS})$ by the equation:

$$E(t_{jS}) = \left( \frac{1.34 \cdot 150}{SCF \cdot \Delta\sigma} \right)^{m} E(t_{jL}) \,. \qquad (7.72)$$

Here 150 MPa is the nominal stress range applied under laboratory conditions, whereas $\Delta\sigma$ is the in-service nominal stress range. SCF = 1.34 in the nominator is taken from Table 5.6, Chapter 5, and the stress concentration factor (SCF) for the joint in question can be taken from the same table or found by FEA. The waiting times and the corresponding Markov model can now be derived for any plated joints using the actual SCF. Hence, the P curve approach for the entire fatigue life has now been extended to the time spent in each crack stage through equation (7.72).

3) *Other thicknesses:* regarding the plate thickness, the present model is valid for 25 mm thicknesses only. For greater thicknesses the higher damage states in the model will correspond to larger crack depths than the ones used in Table 7.10. This will give a higher probability of detection when used in conjunction with a POD curve. However, the time spent in this crack regime (close to half of the thickness of the plate) is very short considering the fact that the $a$-$N$ curves are so steep. It was found that the effect could be ignored for thicknesses in the range 16 to 32 mm. For greater thicknesses the last damage state (state 14 in Table 7.10) is set to half the plate thickness and the depth of damage state 13 is found by interpolation between 6 mm and half of the plate thickness. The increased propagation rate due to the thickness effect is addressed under 5) below.

4) *Critical crack size:* as already discussed, the critical crack size is kept at half of the plate thickness, as in laboratory conditions. For very high-strength steel with low fracture toughness, critical crack size may be less, but this is very rare. On the other hand, for joints in statistically undetermined structures, the joints may be stress relieved as the crack propagates due to alternative load paths. This will give greater critical crack depths and longer fatigue lives than observed in the present test series. This will increase the reliability levels, especially in the case of scheduled inspections. This is due to the fact that the cracks will not have the steep acceleration as observed for deep crack depths in the test specimens. This means that the cracks will spend more time in the deep crack regime where they are easily detected. This positive effect is not accounted for. One may conclude that our model is based on the same assumption as the model for the S-N curve. This seems a wise choice as they will be used in conjunction with these curves.

5) *FDF as the key parameter:* various joints that have the same FDF must have a permissible nominal stress range in-service that is inversely proportional to the SCF for the joint. Hence, the product of SCF$*\Delta\sigma$ will be constant and so will the mean waiting time in each damage state according to equation (7.72). It follows that joints

with the same FDF according to the P-curve will have the same Markov model parameters. One may, in fact, give the following relationship:

$$E_{FDF}(t_{js}) = FDF \cdot E_1(t_{js})$$  (7.73)

where $E_1(t_{js})$ is the waiting time for FDF = 1, whereas $E_{FDF}(t_{js})$ is the waiting time for the actual structural detail with a specific FDF value. $E_1(t_{js})$ can be found from equation (7.72) by choosing SCF·$\Delta\sigma$ so that the FDF equals 1.0 for the total life according to the S-N curve. By using equation (7.65) and (7.66), the Markov model parameters can be found. Reliability calculations can then be carried out. The obtained reliability curves can be presented as a function of dimensionless TSL. This is true for the design S-N curve where the final probability of failure is 2.3% with FDF = 1, regardless of the absolute value of TSL and applied stress level. The same holds true for times given as fraction of the final TSL. This means that the reliability curves can be presented in a dimensionless form and as a function of the derived FDF. This is a great advantage because the FDF is always available because a fatigue analysis starts with a design life verification based on an S-N curve. For offshore structures in the North Sea, it is required that a design, fabrication, and installation (DFI) résumé is issued after the installation phase. This document identifies all major joints in the structure that are critical to fatigue and the joints FDFs are listed. All effects that influence the FDF through the S-N curves will automatically have the correct impact on the DTS model through equation (7.73). When applying the FDF, with equation (5.9) in Chapter 5 as a basis, the thickness effect will be accounted for. The benefit from a lesser slope at a lower stress will likewise be accounted for.

Summing up points 1) to 5) above, the only limitation for the Markov DTS model is that it is obtained from fatigue cracks emanating from the weld toe under axial loading mode. However it was found that experimental curves for a bending loading mode did not behave very differently. In fact, the model can be used in conjunction with any design curve that is based on mean value minus two standard deviations and a scatter close to COV = 0.54 (i.e. 0.5 – 0.6.). Hence, they are not limited to steel joints only.

The results of the present reliability analysis are presented as the CDF as a function of time in Figure 7.34 This is the dimensionless presentation as an alternative to Figure 7.33.

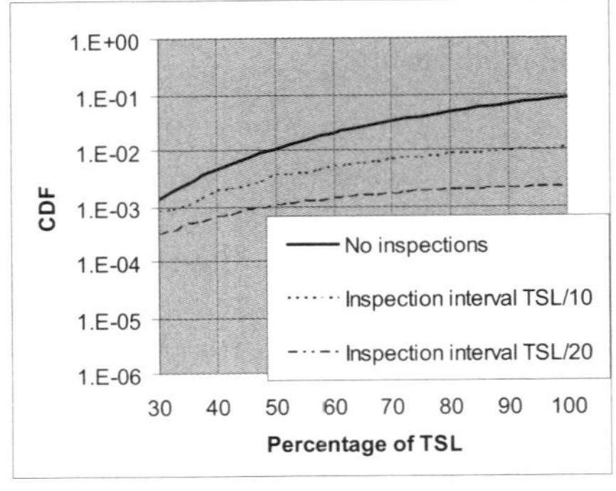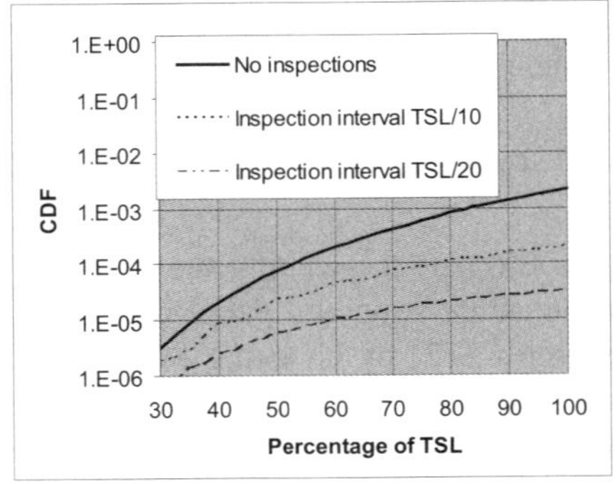

**Figure 7.34.** *Generic Dimensionless Reliability curves for welded joints, Markov Chain model. Left FDF=1, Right FDF=3. Based on $COV_S=0.2$ and Magnetic Particle Inspection (MPI) underwater*

The reliability curves in Figure 7.34 can be related to annual acceptable probabilities given the DNV Guidelines; see Table 7.13, Ref [1]. These types of recommendations must be used in conjunction with some type of stochastic model as presented herein. In the present model, the annual probability is simply the difference between CDF values with a one-year time span in Figure 7.34.

| Type of structure | Consequence of failure | |
|---|---|---|
| | Less serious | Serious |
| Redundant structure | $10^{-3}$ | $10^{-4}$ |
| Non-redundant with significant warning | $10^{-4}$ | $10^{-5}$ |
| Non-redundant; no significant warning | $10^{-5}$ | $10^{-6}$ |

**Table 7.13.** *Acceptable annual probabilities of failure; see Ref [1]*

The results given in Figure 7.34 may also be presented as the cumulative probability at the end of the TSL, as shown in Table 7.14 and Table 7.15. The results in Table 7.14 are without any uncertainty in the calculated stresses, i.e. $COV_S = 0.0$, whereas Table 7.15 is applicable for an uncertainty of 20%, i.e. $COV_S = 0.2$.

| FDF | No inspection | Inspection I = TSL/10 | Inspection I = TSL/20 |
|---|---|---|---|
| 1 | **3.0 10$^{-2}$** | 2 10$^{-3}$ | 2 10$^{-4}$ |
| 3 | 4.9 10$^{-5}$ | VR | VR |
| 6 | VR | VR | VR |
| 10 | VR | VR | VR |

**Table 7.14.** *Cumulative probability of failure at the end of TSL, COV$_S$ = 0.0, MPI underwater (VR = very remote i.e. less than 10$^{-6}$)*

As can be seen from Table 7.14, at FDF = 1 without inspection the CDF is 3 10$^{-2}$. The corresponding probability level inherent in the S-N design curve is 2.3 10$^{-2}$. The discrepancy is due to the difference in the lognormal model (S-N curves) and the Markov model (present model). See also the difference in the reliability curves given in Figure 7.32.

| FDF | No inspection | Inspection I = TSL/10 | Inspection I = TSL/20 |
|---|---|---|---|
| 1 | 8.7 10$^{-2}$ (1 10$^{-2}$) | 1.1 10$^{-2}$ (6 10$^{-4}$) | 2.2 10$^{-3}$ (10$^{-4}$) |
| 3 | 2.2 10$^{-3}$ (5 10$^{-4}$) | 2.0 10$^{-4}$ (2 10$^{-5}$) | 3.1 10$^{-5}$ (2 10$^{-6}$) |
| 6 | 7.5 10$^{-5}$ (2 10$^{-5}$) | 6 10$^{-6}$ (1.5 10$^{-6}$) | VR |
| 10 | 3.2 10$^{-6}$ (1 10$^{-6}$) | VR | VR |

**Table 7.15.** *Cumulative probability of failure at the end of TSL (greatest probability in 1/20 fraction of TSL in parentheses), COV$_S$ = 0.2, MPI underwater*

One may regard Tables 7.14 and 7.15 as decision matrices with columns and rows. If one moves vertically downwards through the no inspection column to obtain high reliability, it is the safe life (SL) philosophy that is the governing principle, i.e. the probability of cracking during TSL gets very low as the FDF increases. Scheduled detailed inspection can then be avoided. If this strategy gives too large dimensions and too heavy associated steel weight, one may choose to reduce dimensions and increase the inspection efforts. This means moving horizontally at the right of the first row in Table 7.15 to achieve high reliability. Then it is the damage tolerance (DT) philosophy that is the overriding principle, i.e. cracks are accepted as long as the probability of detection and repair is high. In most practical cases the designer has to make a tradeoff between these two lines of thought when making the final decision. The designer must consider the feasible geometries of the joint, allowable dimensions, and accessibility for inspection and repair. All these topics are of course related to the life cycle costs (LCC). One may also give guidance on how a specified target reliability may be achieved. Vårdal *et al.* (see Ref [11]) suggested that the target reliability level should be derived by transforming the required FDF given in the regulations into a corresponding reliability level. The transformation is valid for a given reliability methodology. Assuming substantial consequences of fatigue failure and no access for inspection, the required FDF is 10 (see Ref [1]). As can be seen from Table 7.15, this correspond to a accumulated probability of failure of $3.2 \ 10^{-6}$. As can be seen from Table 7.16, there are several strategies to achieve this target value.

| FDF | Inspection interval I |
|-----|----------------------|
| 10 | no inspection |
| 6 | I = TSL/10 |
| 4 | I = TSL/20 |

**Table 7.16.** *Possible design and inspection strategies to achieve $P_F = 3.2 \ 10^{-6}$ with $COV_S = 0.2$, MPI underwater*

If the TSL is set to 20 years it can be seen that the cumulative probabilities given for FDF = 10 in column 2 (no inspection) and for FDF = 6 in column 3 (inspection every other year) of Table 7.15 correspond approximately to the target probability in Table 7.16. It is also seen that annual probabilities given in the parentheses in Table 7.15 for the same cases very closely agree with the acceptable annual probability of $10^{-6}$ in Table 7.13. This figure is valid for non-redundant structures with no significant warning and serious failure consequences, as is our case. Hence, for the applied Markov model the acceptable annual values recommended by DNV in Table 7.13, and the target cumulative probabilities of failure in Table 7.16 obtained by the approach suggested by Vårdal *et al.*, seem to be consistent (see Ref [11]). No fracture mechanics modeling is necessary to verify these reliability levels. Only in cases with geometries, boundary conditions, and loading modes that are not covered by the database and hence for which the validity of the present Markov chain model may be questioned, a full Monte Carlo simulation or FORM/SORM analysis based on LEFM needs to be carried out to substantiate the results.

### 7.8.2. *An industrial case study: single anchor loading system*

The single anchor loading (SAL) system consists of a single pile anchor-single mooring line configuration that provides temporary mooring for a shuttle tanker; see Figure 7.35. The SAL system functions both as a mooring point and as an offloading pipeline termination. A thorough description of the SAL system is found in the technical specification: see Ref [12]. The system has a clump weight with connections for the upper and lower mooring rope segments and loading hoses. A principal sketch of the clump weight is shown in Figure 7.36. The clump weight is a welded steelwork structure subjected to fatigue loading, both from the mooring line (at upper and lower steel thimbles) and the loading hoses (at upper and lower pipeline support). The TSL for the installation is 20 years. A general survey by remote operated vehicle (ROV) is scheduled every year. The fatigue life predictions at the design stage are carried out according to the DNV rules in Chapter 5. The aim of this example is to show how the presented damage tolerance supplement can be used for further assessments, given the results from the S-N predictions. It is assumed that the uncertainty of the calculated stress ranges is 20%, i.e. somewhat less than was suggested by Wirsching (see Ref [5]) and that the actual inspection technique is MPI; hence Table 7.15 applies.

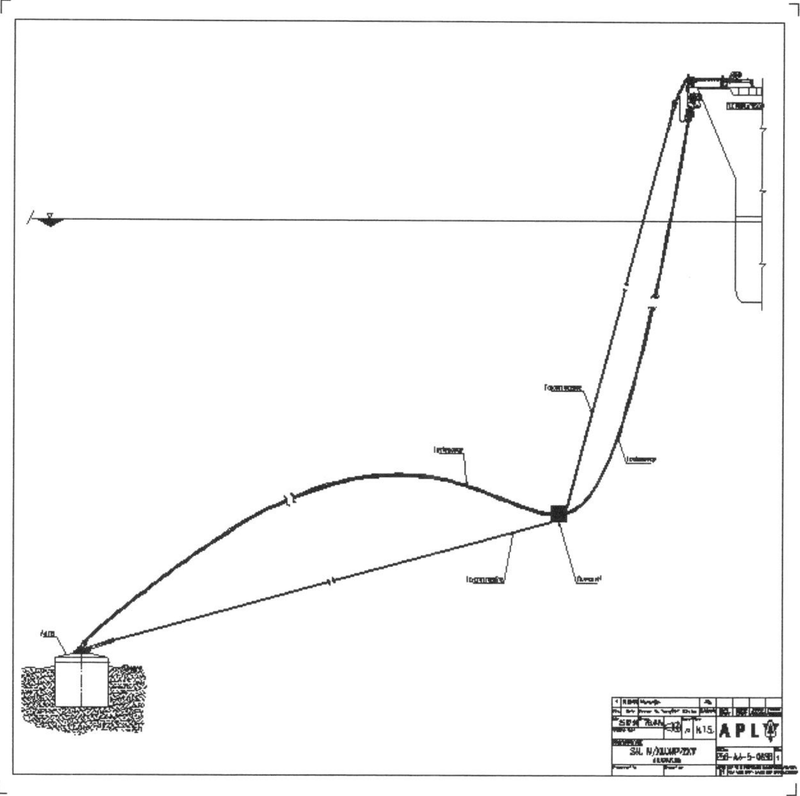

**Figure 7.35.** *SAL system*
*(Courtesy of Advanced Production and Loading AS, Arendal, Norway)*

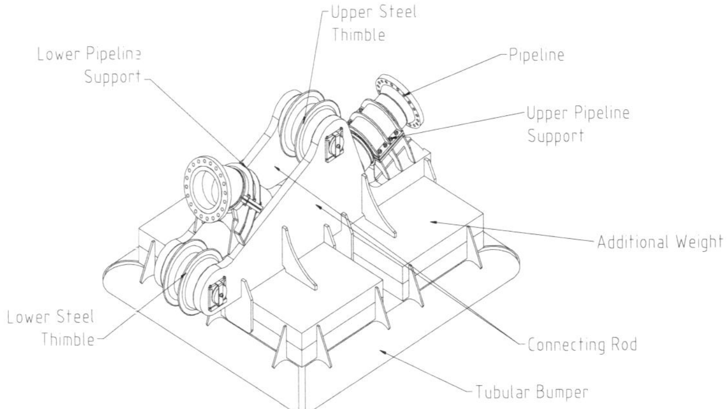


**Figure 7.36.** *Clump weight in a SAL system*

### 7.8.2.1. *Example 1: butt weld in upper pipeline*

This weld is subjected to the fatigue load spectrum from the loading hose. It is a traditional D-class weld with SCF = 1.0. The joint is non-redundant, but there will be a warning (e.g. oil slick on the sea) before the final fracture. The consequence of both the leakage and the fracture is that the system becomes non-operational. Furthermore, the clump weight must be brought up on to the deck of a service vessel to be repaired. This may be regarded as a less serious consequence as there is no threat to safety or severe pollution. According to Table 7.13, the admissible annual probability is then $10^{-4}$. On the other hand, from an operational point of view, it is a serious consequence. Hence, according to Table 7.13, the highest annual probability of failure should be limited to $10^{-5}$ (non-redundant serious consequence). From engineering judgment and the criterion of As Low As Reasonably Practical (ALARP, see section 7.9) one wishes to achieve an annual probability of failure close to $10^{-5}$. The question now becomes how to achieve this goal. It is found that for a wall thickness of 16 mm in the tube, the PFL will be 60 years (i.e. FDF = 3), whereas an increased thickness of about 20 mm will increase the PFL to 120 years (i.e. FDF = 6) according the S-N curves. According to Table 7.15, there are two alternative ways of achieving the prescribed reliability level with these cases:

    – alternative 1: FDF = 3, inspection every other year;

    – alternative 2: FDF = 6, no inspection.

Both alternatives will result in an annual probability of failure of 2 $10^{-5}$, i.e. somewhat higher than the target value of $10^{-5}$, but regarded as in the acceptable range. When looking at the cost of these two alternatives, alternative 2 is by far the most favorable. The first alternative will include nine sub-sea diver (or ROV) interventions to carry out the MPI.

### 7.8.2.2. *Example 2: welded brackets on the main plates*

The welds around the bracket toes are classified as an F2-class, i.e. SCF = 1.52. One may regard these secondary details as a redundant structural system in the way that it is still possible for the structure to transfer load even if one bracket has large visible cracks. Furthermore, these cracks will probably be seen during the scheduled annual general survey by ROV and they can then be repaired by welding at the seabed. Hence, the structure is regarded as redundant for this failure mode, and the consequences are considered less serious. Again, by using Table 7.13 as guidance, the accepted annual probability is $10^{-3}$. Again applying the ALARP concept, one seeks a design where the goal is for a probability level in the range $10^{-3}$-$10^{-4}$. Inspection should be avoided for the same reason as discussed in example 1. The probability levels are not found in Table 7.15, but an extended table will give:

– FDF = 2.2 gives an annual probability level of failure $10^{-3}$;

– FDF = 4.5 gives an annual probability level of failure $10^{-4}$.

Based on cost considerations, the FDF = 4.5 is chosen. The increased dimensions for this alternative gives very little extra weight compared to the FDF = 2.2 alternative. Similar evaluations have been made for other details on the structure.

### 7.8.3. *Conclusions for the damage tolerance supplement*

In the present section, reliability matrices for welded plate joints have been presented. The reliability levels are given in dimensionless form and can be used for any plate welded joints regardless of joint type and TSL. In cases with no inspection and no uncertainty in the applied loading, the model will yield the same reliability as the ones inherent in the S-N curves. The inspection interval is given as a fraction of TSL. The key parameter for entering the matrices is the FDF obtained from rule-based S-N fatigue life prediction. Hence, thickness corrections and corrosive environment are accounted for right from the start. The DTS matrices can be used in conjunction with any design curve that is based on mean value minus two standard deviations and a scatter close to COV = 0.54 (i.e. 0.5 – 0.6). Hence, they are not limited to steel joints only. As has been seen, the reliability levels for various FDFs and inspection scenarios are easily achieved. The information is crucial for decision-making at the design stage. The designer must be able to give estimates for the stress scatter ($COV_S$) and the parameters for the POD curve. It should be emphasized that the $COV_S$ has a strong impact on the probability levels in Table 7.15 and on the curves in Figure 7.34. The probability of failure will increase significantly if the $COV_S$ increases. Various common inspection methods can be dealt with. These types of decision matrices provide useful information to supplement the traditional S-N fatigue life approach. The latter approach does not capture the influence on reliability from chosen FDFs or various inspection strategies.

Only in cases with geometries, boundary conditions, and loading modes that are not covered by the database and, hence, for which the validity of the present Markov chain model may be questioned, a full Monte Carlo simulation based on LEFM needs to be carried out to substantiate the results. For primary structural details for which failure has a direct bearing on structural integrity and could even endanger human life, it would be cost effective to gather the necessary information and carry out the numerical efforts for such analyses. Other joints will not merit such attention and the present DTS approach, based on the use of tables such as Table 7.15 and Table 7.16, is recommended. This could be done based directly on the S-N predictions without any additional calculations.

## 7.9. Risk assessments and cost benefit analysis

So far we have based our decisions on the reliability recommendations given in Table 7.13. The ALARP philosophy, mentioned in the previous section, is based on risk assessments. Risk-based decisions can be made using a risk matrix, as shown in Figure 7.37. Such types of matrices are often used for assessment of risk-based inspection of offshore structures. The probability and consequence classes in the diagram are defined in Tables 7.17 and 7.18. As can be seen, the probability and consequence classes are divided into safety (loss of lives), environmental, and operational classes. In many cases revealed risks will be found in the ALARP region of the risk matrix in Figure 7.37. This means that the effort and cost of the risk reductions measures should not be out of proportions compared to what is obtained by risk reduction. A formal cost-benefit analysis can be carried out to see which efforts (such as inspection plans) should be set into action. The LCC can be optimized based on the following equations when applying the Markov chain model:

$$C = C_0 + \sum_{i=1}^{n}[C_I + C_R\, p_{ri}] + \sum_{i=1}^{n+1} C_F\big(p_{t_i}(b) - p_{t_{i-1}}(b)\big) \qquad (7.74)$$

where:

$C_0$ = cost related to design

$C_I$ = cost related to inspection

$C_R$ = cost related to repair work

$C_F$ = cost related to failure.

The probabilities $p_{t-1}$ and $p_t$ are found from the Markov chain analysis. This cost function will typically have a minimum for a given inspection plan as shown on the sketch in Figure 7.38. The number of planned inspections during service life is varied for chosen inspection techniques. The cost curves are shown for two different inspection techniques. As can be seen, it will be beneficial to choose technique 1 with K inspections during service life. This kind of optimization is straightforward if the consequences of failure are given in terms of reduced operability and economic loss. If the consequences involve possible loss of life, the optimization has to be done under constraints of the permissible probability of failure. Ethical considerations have to be taken into account; see Ref [13].

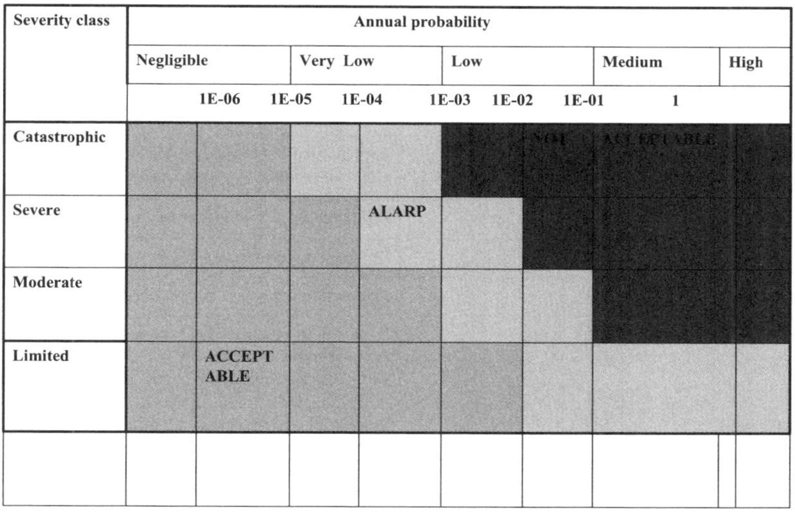

| Severity class | Annual probability | | | | | | | |
|---|---|---|---|---|---|---|---|---|
| | Negligible | | Very Low | | Low | | Medium | High |
| | 1E-06 | 1E-05 | 1E-04 | 1E-03 | 1E-02 | 1E-01 | 1 | |
| Catastrophic | | | | | NOT ACCEPTABLE | | | |
| Severe | | | ALARP | | | | | |
| Moderate | | | | | | | | |
| Limited | ACCEPTABLE | | | | | | | |

**Figure 7.37.** *Risk matrix*

| Classification | Notation | Frequency range per year |
|---|---|---|
| High | H | P>0.5 |
| Medium | M | 0.5>P>10-1 |
| Low | L | 10-1>P>10-3 |
| Very low | VL | 10-3>P>10-5 |
| Negligible | N | P<10-5 |

**Table 7.17.** *Definition of probability classes*

| Classification | Notation | Criteria |
|---|---|---|
| Catastrophic | Catastrophic | There will be deaths or serious injuries to people |
| Severe Safety | SS | May endanger human lives |
| Moderate Safety | MS | Possible injuries to people |
| Severe Environment | SE | Serious oil pollution, typically more than 100 km of coastline affected. Spills with major effect on wildlife. Environmental effect not restored in 5 years |
| Moderate Environmental | ME | Effect on more than 10 km of coastline. Moderate effect on wildlife. Average recovery period 2 years |
| Limited Environmental | LE | Typically 1 km of coastline affected. Spills with limited effect on wildlife. Average recovery period 1/2 years |
| Severe Operational | SO | Production delay longer than 1 months. |
| Moderate Operational | MO | Production delay between 2 weeks and 1 months |
| Limited Operational | LO | Production delay less than 2 week |
| Economic | EC | Repair deferred to earliest convenience<br>No unscheduled shutdown<br>Total cost between 50 000 and 1 mill NOK |
| Non-Classified | NC | Fewer consequences than all other classifications |

**Table 7.18.** *Consequence classes for offshore installations*

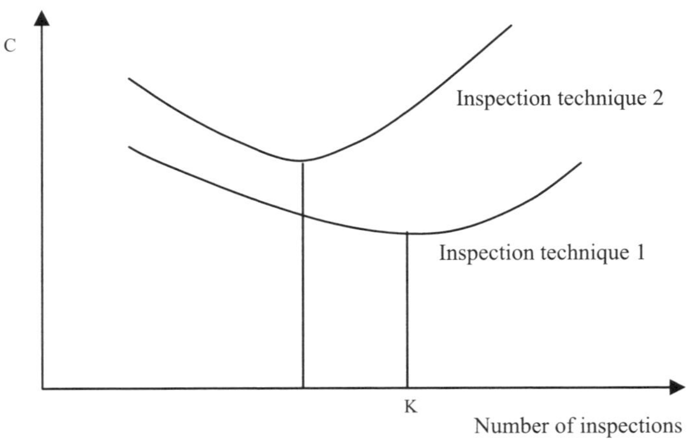


**Figure 7.38.** *Principal sketch for optimization of LCC with respect to the number of planned inspections during service life for two different inspection techniques*

## 7.10. Reliability and risk assessment for the riser steel pipe

Before ending this chapter let us return to the analysis of the riser steel pipe in the loading buoy presented at the end of Chapter 6. The pipe was analyzed to verify a service life extension of 10 years for a second installation (at Field 2) after 5 years in a first installation (at Field 1). In that example we used the S-N curve and a fracture mechanics model in an almost deterministic manner. Inherent uncertainty was accounted for by using the design S-N curve corresponding to a 2.3% probability of failure. Analogously, the mean value plus two standard deviations was used for the growth parameter C. Crack sizes that can be missed during an inspection were discussed. Additional safety margins are obtained by requirements on the FDF. A more comprehensive stochastic analysis can be based on a Monte Carlo simulation or a Markov chain model, as we have discussed in the present chapter. Important additional uncertainty can be introduced by the loading, assuming a coefficient of variation of 0.2 for the applied stresses. Furthermore, the entire history of the POD curve is used without defining a "sure" detectable size. The key parameter for the reliability levels is the achieved FDF, as we just have discussed. For the steel pipe case, the total FDF is equal to 6 for the use at Fields 1 and 2, i.e. for a total service life of 15 years. The results in terms of accumulated probability of failure are shown in Figure 7.39. The highest curve is obtained by one updating of the crack-size distribution after the inspection at five years. The updating is carried out according to Figure 7.30, left-hand side. With only one inspection in shop environment during the replacement after phase 1 (five years), the cumulative probability of failure will be close to $810^{-5}$ at the end of phase 2, as can be seen from Figure 7.39. The corresponding maximum annual probability is close to $3 \ 10^{-5}$. This is somewhat higher than usually recommended; see Table 7.13. The reliability curve with one inspection only is, in fact, almost identical to the curve with no inspection at all. This demonstrates that an inspection carried out at a time stage where the FDF still is superior to 10 (end of phase 1) has very little effect, except that it can reveal gross errors regarding the weld quality that are not included in the stochastic model. The curve with FDF = 10 is drawn for comparison. This curve has a cumulative probability of failure close to $3 \ 10^{-6}$ and with a maximum annual probability close to $10^{-6}$. These reliability levels are fully acceptable for most of the consequence classes given in Table 7.18. The effect of planned sub-sea inspection based on magnetic particle technique is also shown for comparison. Although this method performs less well than the inspection in shop environment, it has a favorable effect when repeated every other year during phase 2. This type of inspection was not possible to carry out for the steel pipe in question, but for other items it is usually feasible. The likely influence of future inspections is estimated through the logic shown in Figure 7.29, right-hand side. The results are shown in Figure 7.39. As can be seen the accumulated probability of failure at the end of service life in this case will be close to $10^{-5}$. The corresponding annual probability will be quite close to $2 \ 10^{-6}$. However, this inspection schedule is not feasible for the steel pipe in question. The conclusion is that if an annual probability of failure greater than $10^{-6}$ is not regarded as acceptable, the pipe must be retrieved and inspected on the deck of the tanker after 10 years of service (5 years at Field 2). At that stage one may update the reliability level based on equation (7.34), based both on the fact that the pipe has survived up to that time and the inspection results – if inspection is carried out. One may then seek out arguments for a life prolongation by a risk assessment based on Figure 7.37 and Table 7.18. These types of assessments form the basis for risk-based inspection.

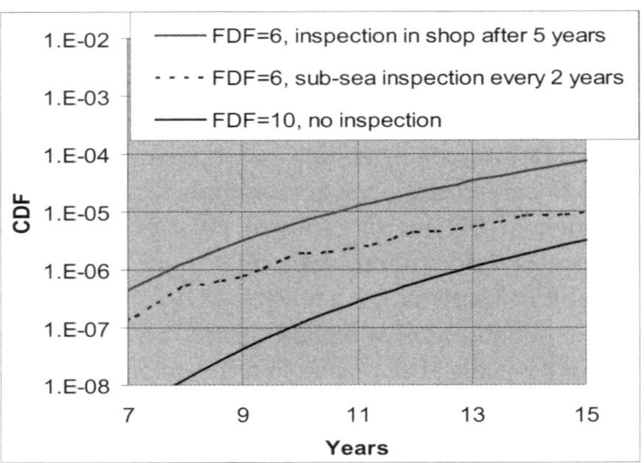


**Figure 7.39.** *Reliability curves for various FDF and inspection strategies*

## 7.11. References


1    DNV, Classification note 30.6: "Structural Reliability Analysis of Marine Structures", Det Norske Veritas, 1322 Høvik, Norway, 1992

2    J.L. Bogdanoff F. and Kozin, *Probabilistic Models of Cumulative Damage*, John Wiley & Sons, 1985

3    T. Lassen, "Experimental Investigation and Stochastic Modelling of the Fatigue Behaviour of Welded Steel Joints", PhD Thesis, AUC, Denmark, 1997

4    H.O. Madsen, S Krenk and NC Lind, *Methods of Structural Safety*, Prentice Hall, 1986

5    P. Wirsching and Y.N. Chen, "Consideration of Probability-Based Fatigue Design for Marine Structures" *Journal of Marine Structures*, 1988, pp 23–45

6    DNV: Guidance for offshore structural reliability analysis, Det Norske Veritas, 1322 Høvik, Norway, 1996

7    P.H. Wirsching, *et al.*, "Fatigue Reliability and maintainability of Marine Structures" Journal of Marine Structures, vol. 3, 1990

8   T. Lassen, J.D. Sørensen, "A Probabilistic Damage Tolerance Concept for welded joints: Part 1: database and stochastic modeling", Journal of Marine Structures, 15, pp 599-613, 2002

9   T. Lassen, "The Effect of the Welding Process on the Fatigue Crack growth in Welded Joints" *Welding Journal* 96, pp 75–85, 1990

10  T. Lassen, J.D. Sørensen, "A Probabilistic Damage Tolerance Concept for welded joints: Part 2: A supplement to the rule based S-N approach", Journal of Marine Structures, 15, pp 615-626, 2002

11  O.T. Vårdal, T. Moan and L.G. Bjørheim, "Application of Probabilistic Fracture Mechanics Analysis for Reassessment of Fatigue Life of a Floating Production Unit-Philosophy and Target Levels" OMAE/MAT-2078, pp 1–9, 2000

12  APL, General Technical Specification, Design Basis, Advanced Production and Loading Doc. No 273-990267SP, Arendal, Norway, 1999

13  G. Ersdal and T. Aven, "Risk management and its ethical basis" submitted to Reliability Engineering and System Safety, 2006

# PART III

# Recent Advances

# Chapter 8

# Proposal for a New Type of S-N Curve

## 8.1. Introduction and objectives

The S-N model is based on the assumption that the number of cycles N to failure is directly correlated to the applied nominal stress range $\Delta S$. All other parameters are assumed to play a minor role. The background for these curves is a simple linear regression analysis in the finite life region. The data for the curve are obtained by accelerated constant amplitude tests at relatively high stress levels. Furthermore, the model assumes the existence of a fatigue limit, given as a stress range below which no failure will take place. This fatigue limit is subsequently added as a horizontal cut-off line to the regression line. The final curve is bilinear for a log-log scale between N and $\Delta S$. These curves are presented as the conventional S-N curves given in rules and regulations; see Chapter 5. At the end of Chapter 3 we showed how S-N data were analyzed to establish the median and the design S-N curve for the fatigue life in the finite life region. The procedure and calculations are quite straightforward; apart from test specimens that are stopped at a given number of cycles before fatigue fracture has occurred. The slope of the mean curve was determined along with the coefficient of variation in fatigue life. We did not go into the problem of determining the fatigue limit. In this chapter, emphasis is laid on the modeling of the fatigue life close to this limit. This stress regime is of major importance because many structures will be subjected to service stresses that correspond to this stress region. Experimental data in this stress regime are sparse and do not fit the knee point of the conventional bilinear S-N curve found in rules and regulations. Consequently, an alternative model where both the fatigue life and the fatigue limit are simultaneously treated as random variables is investigated. The model parameters for this random fatigue-limit model (RFLM) are determined by the maximum likelihood method, and confidence intervals are obtained by the profile likelihood method. The advantages of the RFLM are that it takes into consideration the variations in fatigue limits found from specimen to specimen and that run-out results are easily included.

The objective of this chapter is to investigate the ability of the bilinear S-N curves (BS 5400, Eurocode 3, AISC, Refs [1], [2] and [3]) to describe the fatigue life behavior in general and, in particular, the behavior near the fatigue limit regime. The validity of the upper part of these curves in the finite life area is not questioned as these curves have been corroborated by a large amount of data. The dilemma is that the conventional S-N curves are widely used in a stress regime where the curve changes slope to a horizontal line, as described above, and where very few data exist to corroborate this abrupt shift in slope. Small alterations of the position of the knee point of the curve in the fatigue limit area will have a strong bearing on fatigue life predictions and, as a consequence, fatigue design and final dimensions of welded details. If the ability of the bilinear curve to describe fatigue life behavior is dubious; other statistical models should be applied to obtain more reliable predictions. The RFLM was first suggested by Pascual and Meeker; see Ref [4]. In this model the scatter in finite fatigue life and in the fatigue limit are integrated into one joint model and treated simultaneously. This results in a smooth non-linear S-N curve for a log-log scale. The model has so far been applied with success to describe the fatigue life behavior of epoxy laminate panels and smooth specimens with metal base material (Refs [4] and [5]). The model has been applied to welded joints in Ref [6] and this work will be elaborated in this chapter. To study this problem, experimental data are collected in the actual stress regime for fillet welded joints; see Figure 8.1. For this type of joint the fatigue cracks emanate from the weld toe and through the plate thickness, as we have shown in several examples. The purpose of the study is to compare the conventional S-N curves found in rules and regulations with the alternative approach based on the RFLM for given series of test data. The analysis and discussion are limited to constant amplitude behavior in a dry air environment in the high-cycle fatigue regime.

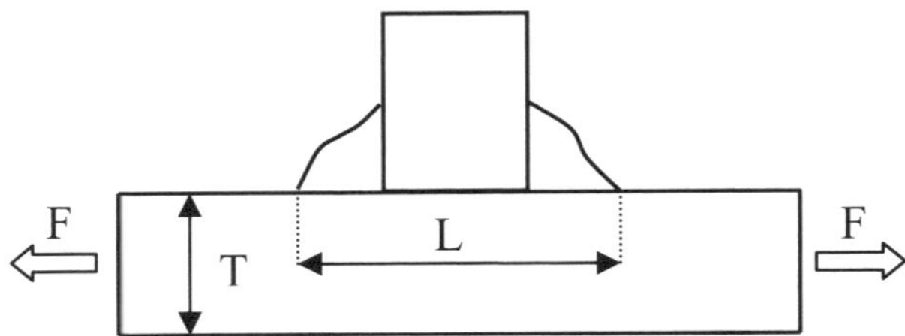


**Figure 8.1.** *Plate with fillet welded attachments*
*(typical F-class in BS5400 and Category 71 in Eurocode 3, Refs [1] and [2])*

## 8.2. General considerations for the conventional S-N approach

### 8.2.1. *Basic assumptions*

Let us start by repeating the basic assumptions for the S-N curves of a given welded detail. These assumptions are:

– there is a strong dependence between the applied stress range $\Delta S$ and fatigue life given as the number of cycles N to failure;

– the standard deviation in fatigue life is not constant and increases as the stress range decreases;

– there is a fatigue limit and it is assumed that below this threshold stress the joints will never fail.

### 8.2.2. *The S-N approach based on BS5400 and Eurocode 3*

In rules and regulations (Chapter 5) the relationship between the number of cycles N and the stress range $\Delta S$ is assumed to be linear for a log-log scale:

$$\log(N) = \log A_0 - m\mathrm{Log}(\Delta S) + \varepsilon \quad \Delta S > \Delta S_0 \tag{8.1}$$

where log denotes the logarithm to base 10. The parameters A and m characterize the fatigue quality for the joint in question, whereas $\varepsilon$ is the error term due to the inherent scatter. $\Delta S_0$ is the fatigue limit and it is assumed that no failure occurs under this threshold value. Hence, the ruled-based S-N curves are bilinear for a log-log scale. The upper curve will have a slope of $-1/m$ whereas the lower line will be horizontal. Examples are given in Figure 8.2, which shows the F-class curve and the Category 71 curve taken from BS5400 and Eurocode 3 respectively. These curves will be discussed below. The standard deviation for the logarithm of fatigue life is assumed constant for all stress levels. As a consequence, the coefficient of variation (COV) for fatigue life will be constant and the standard deviation of the fatigue life will increase towards lower stress levels. The COV in fatigue life is as high as 0.5 and this makes the scatter a major issue. The S-N approach is based entirely on constant amplitude (CA) experimental fatigue life. A linear regression analysis is carried out and the mean curve and standard deviation for the fatigue life are obtained. The design curve is drawn at the median value minus two standard deviations (Ref [1]), alternatively at minus 1.5 standard deviations if the standard deviation has a 75% confidence level (Ref [2]). At a given stress level the fatigue life is assumed to obey a lognormal distribution; see Chapter 7 for details.

This implies that the resulting mean logarithmic curve corresponds to a failure probability of $p = 0.5$, i.e. the median fatigue life. In this chapter we shall work with the median S-N curve unless otherwise stated. The S-N curve taken from BS5400 (Ref [1]) reads:

$$N = \begin{cases} A\Delta S^{-m} & \Delta S > \Delta S_0 \\ \infty & \Delta S \leq \Delta S_0 \end{cases} \tag{8.2}$$

The parameters for the median F-class curve are $\log A = 12.238$ and $m = 3$. The COV is 0.54. The fatigue limit is 56 MPa and the corresponding fatigue life is $10^7$ cycles. The F-class gives almost the same predictions as Category 71 (Ref [2]), except that the latter curve becomes horizontal at $5 \times 10^6$ instead of $10^7$ cycles. This difference pinpoints the uncertainty in the stress region near the knee point of the curves and this is why we have collected test results in this region. The curves are drawn in Figure 8.2, together with collected data points. The data will be presented in detail in the next section.

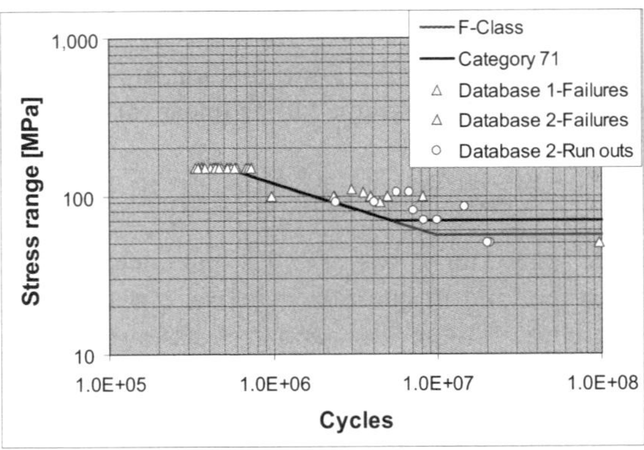


**Figure 8.2.** *Bilinear rule-based median S-N curves together with test data*

As can be seen from the figure, almost all the data are to the right-hand side of the curves near the knee points. The F-class curve is almost identical to the F-class given in AISC; see Ref [3]. A general problem with the S-N curves is that they are based on data compiled without regard to material quality, thickness, and loading ratio. A re-analysis of the data where these aspects are taken into account when defining the test population is given in Ref [7]. More homogeneity classes are chosen and the scatter in fatigue life decreases significantly. The present F-class and Category 71 are comparable with Category 20 (cruciform joint, Ref [6]) and Category 31A (lateral attachment on a flange plate, Ref [6]). The former class predicts somewhat less fatigue life than the F-class, the latter somewhat longer. The difference is not important for the present study, but one should be aware of the fact that the scatter may be reduced by a stricter definition of the test population.

## 8.3. S-N curves based on a random fatigue limit model

Due to the uncertainty and large scatter in fatigue life in the knee point region, an S-N curve based on a random fatigue-limit approach has been proposed; see Refs [4] and [5]. In fact, the sparse data available indicate that there is a variation in fatigue limit from specimen to specimen. Consequently the distribution for the fatigue limit should be sought and incorporated into the statistical model for the fatigue life. It should not be treated separately as done for the bilinear curves. The S-N curve obtained from the RFLM will not have an abrupt change from an inclined straight line to a horizontal line, but a gradually change in slope as stress ranges become very low. Our hypothesis is that this non-linear curve for a log-log scale is more consistent with observed fatigue life data for welded details at low stress ranges. The governing equation is:

$$\ln(N) = \beta_0 - \beta_1 \ln(\Delta S - \gamma) + \varepsilon, \Delta S > \gamma \qquad (8.3)$$

where ln denotes the natural logarithm and $\gamma = \Delta S_0$ is the fatigue limit. $\beta_0$ and $\beta_1$ are fatigue curve coefficients. As can be seen, equation (8.3) is fundamentally different from equation (8.1). Let $V = \ln(\gamma)$ and assume that V has a probability density function (PDF) given by:

$$f_V(v) = \frac{1}{\sigma_\gamma} \phi_V\left(\frac{v - \mu_\gamma}{\sigma_\gamma}\right) \qquad (8.4)$$

with location parameter and scale parameter $\mu_\gamma$ and $\sigma_\gamma$, respectively. $\phi_v(\cdot)$ is the normal PDF. The normal distribution was chosen because it gave the best fit to fatigue data in Ref [4] and due to the fact that it is the usual assumption for fatigue-life distribution in rules and regulations. Let $x = \ln(\Delta S)$ and $W = \ln(N)$. Assuming that, conditional on a fixed value of $V < x$, $W|V$ has a PDF:

$$f_{W|V}(w) = \frac{1}{\sigma} \phi_{W|V}\left(\frac{w - [\beta_0 - \beta_1 \ln(\exp(x) - \exp(v))]}{\sigma}\right) \qquad (8.5)$$

with the location parameter $\beta_0 - \beta_1 \ln(\exp(x) - \exp(v))$ and scale parameter $\sigma$. The marginal PDF of W is given by:

$$f_W(w) = \int_{-\infty}^{x} \frac{1}{\sigma\sigma_\gamma} \phi_{W|V}\left(\frac{w - [\beta_0 - \beta_1 \ln(\exp(x) - \exp(v))]}{s}\right) \phi_V\left(\frac{v - \mu_\gamma}{\sigma_\gamma}\right) dv \quad (8.6)$$

The marginal cumulative distribution function (CDF) of W is given by:

$$F(w) = \int_{-\infty}^{x} \frac{1}{\sigma_\gamma} \Phi_{W|V}\left(\frac{w - [\beta_0 - \beta_1 \ln(\exp(x) - \exp(v))]}{\sigma}\right) \phi_V\left(\frac{v - \mu_\gamma}{\sigma_\gamma}\right) dv \quad (8.7)$$

where $\Phi_{W|V}(\cdot)$ is the CDF of $W|V$. For given sample data $w_i$ and $x_i$ from various test specimens $i = 1, n$, the model parameters can be determined by the maximum likelihood (ML) function:

$$L(Q) = \prod_{i=1}^{n} \left[f_W(w_i; x_i Q)\right]^{\delta_i} \left[1 - F_W(w_i; x_i Q)\right]^{1-\delta_i} \quad (8.8)$$

where:
$\delta_i = 1$ if $w_i$ is a failure
$\delta_i = 0$ if $w_i$ is a censored observation (run out).

The vector $Q$ contains the model parameters:

$$Q = \left(\beta_0, \beta_1, \sigma, \mu_\gamma s_\gamma,\right). \quad (8.9)$$

Once these parameters have been determined from the optimization of equation (8.8), the corresponding confidence intervals can be obtained by a profile likelihood method using the profile ratio of the variables together with chi-square statistics. The integration of equations (8.6) and (8.7) and the optimization of equation (8.8) must be done numerically. Details are found in Refs [4] and [5]. When the parameters are determined we can calculate the fatigue life for a chosen probability p of failure by using equation (8.7). Hence, the median curve and quantile curves for design purposes are obtained.

## 8.4. Experimental data for model calibration

### 8.4.1. *Data for fatigue life at high stress levels (database 1)*

We have presented this database in Chapters 3, 6, and 7, but will give a short review before proceeding. The test series consist of thirty-four non-load-carrying cruciform and T-joints test specimens (see Figure 8.1). All the test specimens were fabricated from C-Mn steel plate 25 mm thick. The nominal yield stress was 345 MPa. The welding procedures were taken from normal offshore fabrication practice. The joints were proven free from cracks and undercuts. The specimens were tested under constant amplitude axial loading at $\Delta S = 150$ MPa with a loading ratio of $R = 0.3$. Experimental details are found in Ref [8]. The total fatigue lives for the 34 specimens have been plotted in Figure 8.2. The median life of the series ($N = 460,000$ cycles) is only 12% less than the prediction of the F-class ($N = 513,000$ cycles). Hence, the test series are of normal quality and are comparable with the population pertaining to the F-class and Category 71. However, due to the homogeneity of the test series, the COV is, as expected, much smaller: $COV = 0.22$. In addition to recording the fatigue life, crack growth measurements were made during the course of each test; see Figure 3.8 in Chapter 3. These data are important information for modeling the fatigue process as will be done in Chapter 9, but not for the present fatigue life statistical analysis.

### 8.4.2. *Data for fatigue lives at low stress levels (database 2)*

Data points pertaining to the F-class S-N curve have the center of gravity at a stress range in the region of 120-150 MPa. This regime is well represented by database 1. In addition, we have in the present work assembled results at lower stress level from fatigue-life test series in several large experimental investigations carried out in Europe; see Ref [9]. The applied stress ranges are between 80 and 105 MPa and the thicknesses of the plates range from 16 to 38 mm. Other thicknesses are excluded to minimize the so-called thickness effect. All the selected specimens are as-welded and the loading ratio $R$ is between 0 and 0.3. These data points are also plotted in Figure 8.2. As can be seen, most of the data points have, with only a few exceptions, substantially longer lives than the predictions of the median F-class curve. Furthermore, some of the results are run outs. The scatter is considerably greater than for database 1, as expected. This is partly due to the fact that database 1 contains only one homogeneity test series in one laboratory, but it is mostly due to the fact that, as already discussed, scatter increases at low stress levels.

## 8.5. Comparison between the F-class curve, the RFLM based curve and the data

By applying equation (8.4) to equation (8.9) for the databases presented above, the parameters for the RFLM are determined and given in Table 8.1. The 90% confidence intervals are also listed.

| Parameter | Point estimate | 90% confidence interval | |
|---|---|---|---|
| $\beta_0$ | 22.4800 | 22.407 | 22.555 |
| $\beta_1$ | 2.100 | 2.084 | 2.118 |
| $\sigma$ | 0.14 | 0.089 | 0.240 |
| $\mu_\gamma$ | 4.100 (60.3 MPa) | 4.044 (57 MPa) | 4.154 (63.7 MPa) |
| $\sigma_\gamma$ | 0.16 | 0.120 | 0.216 |

**Table 8.1.** *Parameters for the RFLM*

The confidence intervals are obtained from plots of the profile ratio as shown for $\mu_\gamma$ and $\sigma_\gamma$ in Figure 8.3. As can be seen, the point estimate for the fatigue limit is near 60 MPa. The BS5400 fatigue limit of 56 MPa is well within the 90% confidence interval and is quite close to the point estimate of 60 MPa. The Eurocode fatigue limit of 70 MPa is far outside our confidence interval. Although the present database is limited, it is a surprise that the Eurocode fatigue limit is so far outside the 90% confidence interval. The median curve (p = 0.5) for the fatigue life pertaining to the point estimates in Table 8.1 is drawn in Figure 8.4 along with the F-class curve and the data points. As can be seen, the F-class curve, the RFLM curve, and the data points at a stress range of 150 Mpa (database 1) are in good agreement. At lower stress ranges the RFLM curve becomes non-linear and predicts substantially longer fatigue lives than does the F-class. At a stress range of 80 MPa the RFLM predicts more than three times longer fatigue life, and at 70 MPa the RFLM predicts close to 10 times longer fatigue life than the F-class. It can also be seen from the figure that the RFLM is obviously more in accordance with the data points. The quadratic sum of the error terms $\varepsilon$ in equation (8.3) will be much less than the corresponding sum pertaining to equation (8.1). As the stress level approaches the F-class fatigue limit of 56 MPa, the RFLM curve becomes so horizontal that it almost coincides with the F-class horizontal line. However, it is only for fatigue lives longer than $10^9$ cycles that the two curves are, for all practical purposes, identical. This illustrates that care must be taken when comparing fatigue limits as we did above. The fatigue limit for the F-class will already appear at $10^7$ cycles, whereas for the RFLM curve will approach the fatigue limit of 60 MPa at 5 $10^9$ cycles. The reason for this major difference is that the RFLM is based on a joint distribution (random fatigue life – random fatigue limit) and this will push the point estimate for the fatigue limit towards a higher number of cycles. When comparing the RFLM curve with the Category 71 curve, it can be seen from Figure 8.4 that the discrepancy is less than was found from comparison with the F-class. In fact, the curves cross each other at 5 $10^7$ cycles at the fatigue limit of 70 MPa given for Category 71. Above this stress range the RFLM curve will predict longer lives than the Category 71, whereas at lower stress ranges the reverse will be true. The quantile curve pertaining to p = 0.025 is shown in Figure 8.5. This curve is intended for design purposes. When comparing the obtained curve with the data points, it can be seen that it is only the two run outs at 2 $10^7$ cycles that are well below the curve, whereas the two failure data points at $10^6$ and $10^8$ cycles are just beneath the curve. All these four points pertain to database 2. The obtained curve cannot be directly compared to the F-class design curve due to the fact that our database is more limited and has less scatter in the finite life region (database 1).

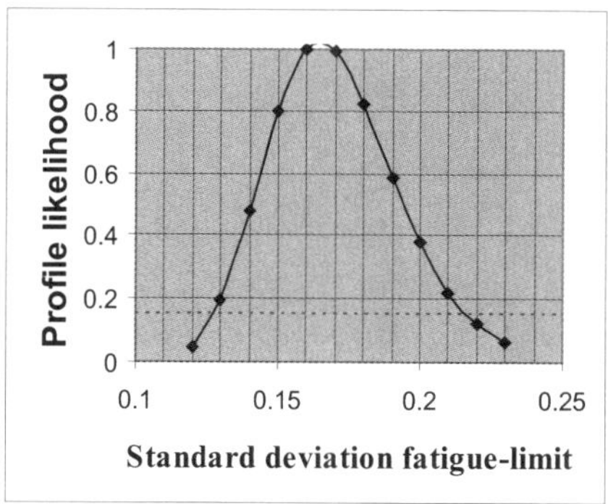


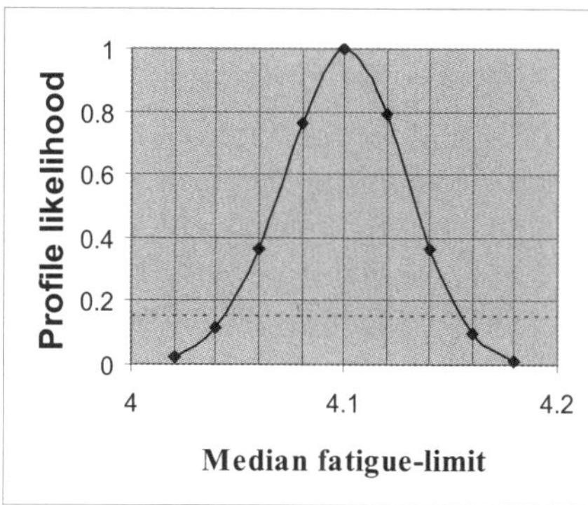


**Figure 8.3.** *Profile likelihood with 90% confidence interval for the median and standard deviation of the fatigue limit*

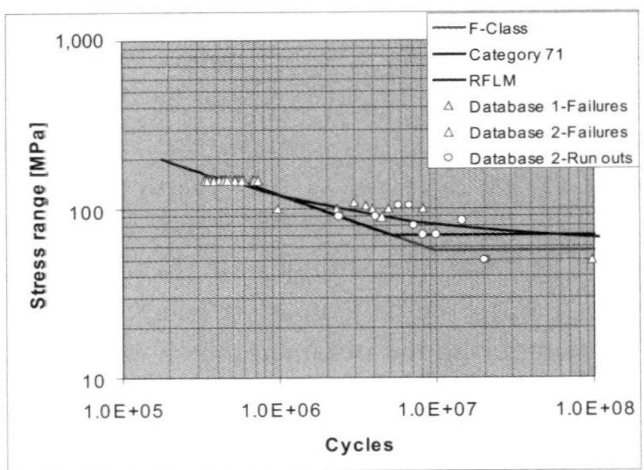

**Figure 8.4.** *Bilinear- and RFLM-based median S-N curves together with test data*

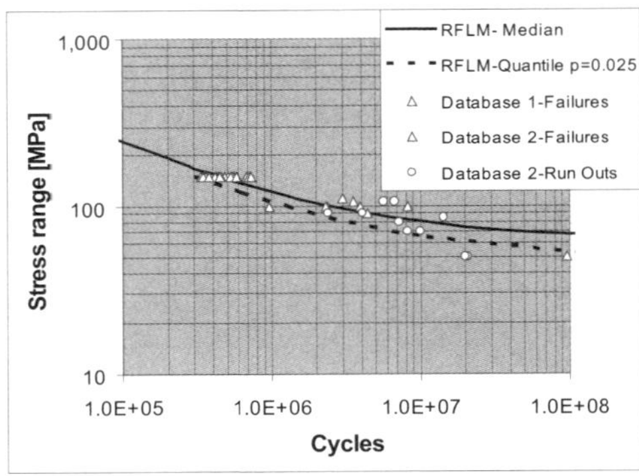

**Figure 8.5.** *Median and quantile S-N curves based on the RFLM together with test data*

To force our model to be valid for the huge amount of data pertaining to the F-class in the finite life region we adjust the parameters given in Table 8.1 so that the mean life and scatter for high stresses coincide with the F-class data. The parameter $\beta_0$ is increased from 22.48 to 22.60 and $\sigma$ is increased from 0.14 to 0.50. The latter figure is a characteristic scatter for most categories of welded joints. The other parameters are kept constant as a first approximation. The obtained quantile S-N curve ($p = 0.025$) is drawn, together with the F-class design curve (based on logarithmic mean minus two standard deviations) in Figure 8.6. As can be seen from the figure, the fit between the two curves is amazingly good in the high stress region. The curves coincide at stress ranges above 110 MPa. As expected, B = both curves have good safety margins to all the data points in test series 1. As for the median curves, the RFLM curve will predict substantially longer lives than the F-class below 100 MPa. In fact, the RFLM curve is only slightly lower in this stress region compared with the original quantile curve in Figure 8.5. This is due to the fact that it is the parameters that characterize the random fatigue limit that mainly govern the curve in this area. As can be seen from Figure 8.7, the present design curve will approach the Category 71 curve at high and low stress ranges. It is only near the knee point that the discrepancy in fatigue life is significant.

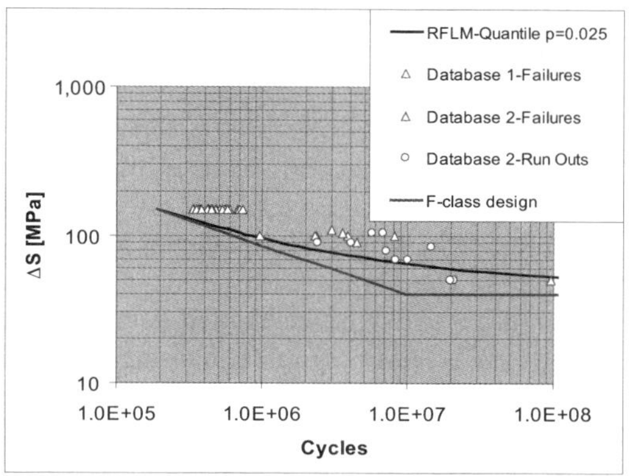

**Figure 8.6.** *Comparison between the RFLM design curve (with $\beta_0 = 22.60$, $\sigma = 0.5$) and the F-class design curve*

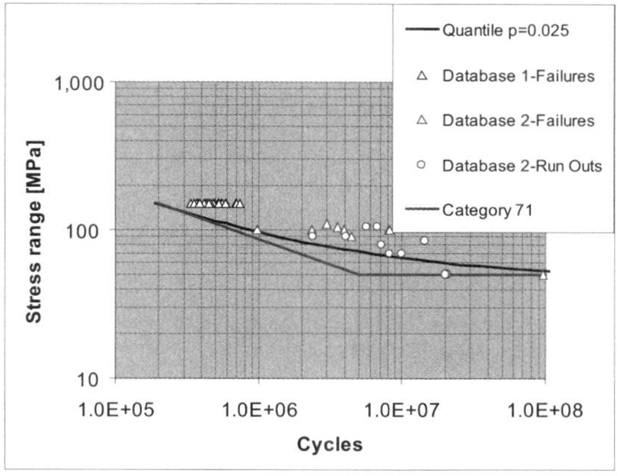

**Figure 8.7.** *Comparison between the RFLM design curve (with $\beta_0 = 22.60$, $\sigma = 0.5$) and the Category 71 design curve*

## 8.6. Conclusions

The statistical behavior of the fatigue life of fillet welded joints has been examined and modeled with reference to conventional S-N curves found in current rules and regulations. An alternative statistical model based on a joint random fatigue life and a random fatigue limit has been applied. Constant amplitude fatigue life data near the "knee point" of the rule-based bilinear S-N curves are assembled to study and corroborate the model. The model has been fitted to experimental fatigue lives and the obtained S-N curve is compared with the traditional bilinear S-N curves given in rules and regulations. The rule-based S-N curves and the RFLM-based curve coincide for stress ranges above 110 Mpa. For stress ranges below 100 MPa, the RFLM curve will predict fatigue lives that are from 2 to 10 times longer than the predictions made by the F-class S-N curve. It appears that the non-linear curve obtained from the RFLM has a much better ability to model fatigue life behavior in this stress region. The abrupt knee point of rule-based bilinear curves does not fit the experimental facts for the assembled data. The fatigue life behavior in this stress regime is obviously more complex than the conventional bilinear S-N curve can describe. The discrepancy between the present RLFM curve and the F-class curve is important as it occurs in a stress region where the majority of the load cycles for a welded detail in service usually occur. The rule-based S-N curves seem overly pessimistic in this regime and this will have a strong bearing on practical fatigue life predictions, fatigue design, and final dimensions of welded details.

## 8.7. References

1    BS 5400, Steel, concrete and composite bridges, 1980. *Part 10: Code of practice for fatigue*, London, British Standard Institution (BSI)

2    EUROCODE 3: Design of steel structures, 1993. Part 1–9: *Fatigue strength of steel structures*, European Norm EN 1993-1–9

3    Manual of Steel Constructions 13th ed. American Institute of Steel Construction, AISC, 2005

4    F.G. Pascual and W.Q. Meeker, "Estimating fatigue curves with the random fatigue-limit model" *Technometrics* 41, 1999, pp 277–302

5    S. Loren and M. Lundstrøm, "Modeling curved S-N curves" *Journal of Fatigue and Fracture of Engineering Materials and Structures* 28, 2005, pp 437–43

6    T. Lassen, P.H. Darcis and N. Recho, "Fatigue behavior of Welded Joints Part 1 – Statistical Methods for Fatigue Life Predictions" *Welding Journal* 84 (12), 2005, 183-s to187-s

7   F.V. Lawrence, S.D. Dimitrakis and W.H. Munse, "Factors influencing weldment fatigue" *Fatigue and Fracture*, Vol. 19, 1996, ASM Handbook, pp 274–86

8   T. Lassen, "The effect of the welding process on the fatigue crack growth in welded joints" *Welding Journal* 69 (2), 1990, 75-s to 85-s

9   G. Lebas and J.C. Fauve, *Collection of Fatigue Data*, 1988, Elf Aquitaine, Pau

Chapter 9

# Physical Modeling of the Entire Fatigue Process

## 9.1. Introduction and objectives

In Chapter 6 we discussed the merits and shortcomings of a fracture mechanics mode (FMM) when it comes to modeling the entire fatigue process in a welded joint. In conclusion, the FMM should not be used to model the entire fatigue process in high-quality welds proven free from detectable initial cracks. Although the model is capable of describing the crack evolution at one given stress level, it will fail to predict the fatigue life as the stress range decreases.

Furthermore, the fatigue limit will be overly optimistic. The FMM should only be applied if cracks are found and sized. In other cases, a crack initiation phase should be modeled before the crack propagation phase is added. This will be addressed in the present chapter.

Let us also bear in mind that in Chapter 8, a statistical model based on a joint random fatigue life and a random fatigue limit (random fatigue-limit model (RFLM)) was applied to predict the fatigue life of welded joints. The S-N curve obtained from this RFLM was non-linear for a log-log scale between the stress range $\Delta S$ and number of cycles N to failure. The S-N curve gradually changes slope as the stress range decreases. A FMM cannot describe such behavior. The model fitted the assembled S-N data in the low stress region far better than the conventional bilinear S-N curve found in rules and regulations. The fatigue behavior is obviously significantly more complex at low stress ranges than the conventional curves are able to describe. The RFLM-based non-linear curves give some new physical insight into the fatigue process itself. This will be addressed in the present chapter by a semi-empirical physical model. The model is a complementary tool to the statistical RFLM developed in Chapter 8. It is important to control the physical parameters that have an influence on the fatigue damage process as it evolves towards final failure. This is important when the following predictions are required:

– predictions of fatigue life under conditions, for which experimental data do not exist;

– predictions of likely crack growth histories leading to failure.

The need for a more a physical model for carrying out the first type of predictions is obvious: fatigue life can only be predicted by a statistical model for joints pertaining to the populations and conditions for which the model was established. If basic fatigue properties such as joint geometry or loading modes are changed it is only a physical model that can predict the effect of such changes on the fatigue life. Hence, the physical model is an important tool in safe life analysis. The second type of predictions is required if inspection planning is to be carried out, i.e. a damage tolerance approach. In this case it is necessary to characterize the fatigue process itself, not only to determine the final fatigue life. This is essential if in-service inspections are to be planned; we must know what crack sizes to look for at different time stages before final failure. This will make the scheduled inspection more efficient and economical. Based on these considerations, a physical model will be established to meet the following criteria:

a) the model should be corroborated by S-N data for the joint in question when these are available (databases 1 and 2 in Chapter 8);

b) the model should predict a crack evolution that coincides with measured crack growth histories before failure (database 1 in Chapters 6, 7 and 8).

With this background we will endeavor to model the fatigue process in fillet welded joints. The total fatigue life is considered to be the sum of the cycles spent in the crack initiation phase and the crack propagation phase:

$$N_T = N_i + N_p .$$
(9.1)

For some time there has been a debate amongst researchers and engineers as to whether or not the crack initiation period is important. It has been a traditional belief that fatigue crack growth often starts from surface-breaking defects in the weld toe region. The initial flaw has often been assumed to have a depth greater than 0.1 mm, sometimes 0.25 mm. In rules and regulations an initial crack depth, even as deep as 0.5 mm, has been recommended; see Ref [1]. These flaws directly start the fatigue crack propagation and the fatigue initiation period can be ignored. However, with advanced fatigue testing where the entire crack depth history is monitored – not only the final fatigue life – it has become obvious that this simple approach does not fit the facts. Several researchers have obtained experimental evidence that supports the existence of a crack initiation period; see Refs [2] and [3]. The conclusion is that rather than speaking of small micro-cracks or inclusions in the vicinity of the weld toe, it is more correct to use the notion of unfavorable surface condition which gives a rather short initiation period under accelerated laboratory conditions. The early damage mechanism is a combination of crack nucleation and micro-crack growth. In welded joints subjected to high stresses in accelerated laboratory condition (typically a stress range of 120-150 Mpa) the initiation period, defined as the time to reach a crack depth of 0.1 mm, is typically 30% of the entire fatigue life; see Ref [3]. This means that if the same joint is subjected to stress levels that are typical of service conditions (equivalent stress range 50-80 MPa) the crack initiation will totally dominate the fatigue life. Due to this fact, a FMM will not be able to meet both of the model criteria A and B, listed above. With this background, a two-phase model (TPM) is developed and investigated. It was Lawrence *et al.* who first suggested the TPM, and a good overview is given in Refs [4] and [5]. As the method now stands, its accuracy depends greatly on the calibration experiments. Our objective in the present work is to elaborate and calibrate the model to fit the various test series presented in Chapters 7 and 8. The predictions made by the model will be compared with the pure fracture mechanics model in Chapter 6 and the predictions made by the RFLM in Chapter 8. The joint used in the analysis is shown in Figure 9.1.

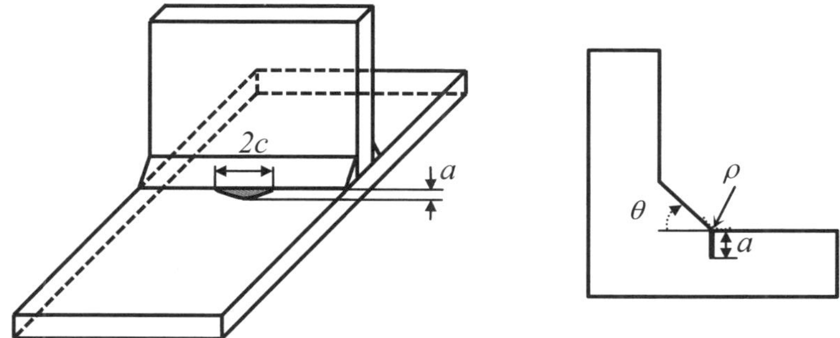


**Figure 9.1.** *Joint configuration with crack shape parameters a and c*

## 9.2. Modeling the fatigue crack initiation period

### 9.2.1. *Basic concept and equations for the local stress-strain approach*

We discussed the local stress-strain approach in Chapters 2 and 3. In the present chapter we will show in more detail how the method is applied for a weld notch. The predictions for the number of cycles to crack initiation, $N_i$, are based on the Coffin-Manson equation with Morrow's mean stress correction (Refs [4] and [5]):

$$\frac{\Delta\varepsilon}{2} = \frac{\left(\sigma'_f - \sigma_m\right)}{E}\left(2\,N_i\right)^b + \varepsilon'_f\left(2\,N_i\right)^c .$$

(9.2)

Here, $\Delta\varepsilon$ is the local strain range and $\sigma_m$ is the local mean stress at the weld toe. The parameters $b$ and $c$ are the fatigue strength and ductility exponents, and $\sigma'_f$ and $\varepsilon'_f$ are the fatigue strength and ductility coefficients respectively. The local stress and strain behavior is given by the Ramberg-Osgood stabilized cyclic strain curve:

$$\Delta\varepsilon = \frac{\Delta\sigma}{E} + 2\left(\frac{\Delta\sigma}{2\,K'}\right)^{\frac{1}{n'}}$$

(9.3)

where $K'$ and $n'$ are the cyclic strength coefficient and strain hardening exponent respectively.

Equation 9.3 is combined with the Neuber rule:

$$\Delta\varepsilon\,\Delta\sigma = \frac{\left(K_t\,\Delta S\right)^2}{E}$$

(9.4)

where $\Delta S$ is the nominal stress range, $E$ is the Young modulus, and $K_t$ is the stress concentration factor at the welded toe. Equation 9.4 is sometimes modified by introducing the fatigue notch factor $K_f$ instead of $K_t$. It is argued that the fatigue notch factor better quantifies the severity of a discontinuity in the fatigue life calculation. Yung and Lawrence conclude (see Ref [4]) that Peterson's equation correctly interrelates the fatigue notch factor with elastic stress concentration factors for welded joints:

$$K_f = 1 + \frac{\left(K_t - 1\right)}{1 + \left(\dfrac{a_p}{\rho}\right)} .$$

(9.5)

Here, $\rho$ is the weld toe root radius and $a_p$ is Peterson's material parameter. The latter may be approximated by the expression $1.087 \times 10^5\,S_u^{-2}$ (in mm, N/mm$^2$) where $S_u$ is the tensile strength of the steel. The following expression for $K_t$ is used (see Ref [8]):

$$K_t = 1 + \left[ 0.5121 \; (\theta)^{0.572} \; \left( \frac{T}{\rho} \right)^{0.469} \right] \tag{9.6}$$

where $\theta$ is the weld toe angle (radians) and $T$ is the plate thickness; see Figure 9.1.

The definition of the local stress-strain variation is illustrated in Figure 9.2. The nominal stress $S$ (left) and the local stress $\sigma$ (right) are shown for the first reversal (0-1) and the stabilized hysteresis loop (1-2-3). The local $\Delta\sigma$ and $\Delta\varepsilon$ and the mean stress, $\sigma_m$, corresponding to the cyclic loading, are determined. The effect of cyclic hardening or cyclic softening is neglected.

In the case of an elastic notch root condition at the weld toe, equation (9.2) reduces to the Basquin equation, ignoring the second term on the right-hand side. This equation can be rearranged:

$$N_i = \frac{1}{2} \left( \frac{K_f \, \Delta S}{2 \, (\sigma_f' - \sigma_m)} \right)^{1/b} = \frac{1}{2} \frac{\left[ 2 \, (\sigma_f' - \sigma_m) \right]^{-1/b}}{\left( K_f \, \Delta S \right)^{-1/b}} . \tag{9.7}$$

From this equation one may construct an S-N curve with slope $b$.

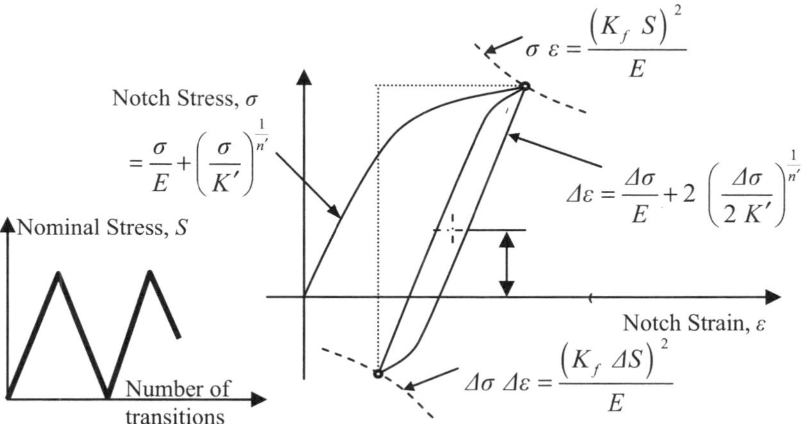


**Figure 9.2.** *Schematic illustration of the local stress-strain hysteresis loop analysis*

### 9.2.2. *Definition of the initiation phase and determination of parameters*

When applying the Coffin-Manson approach to predict time to crack initiation at the weld toe, three questions arise. The first is what stress concentration factor $K_t$ one should choose to characterize the notch effect from the highly variable local toe geometry. The next is the definition of time to crack initiation, i.e. what crack depth is reached at the end of the phase before propagation is considered to take over. This depth will be referred to as the transition depth. If the chosen depth is too large, the initiation phase will not obey equation (9.2) due to a substantial amount of crack growth involved in the process. Several definitions are possible for the transition depth and this is probably one of the reasons why so few TPMs have been applied in practice. The last question is how to determine the cyclic mechanical parameters and the parameters in the Coffin-Manson equation such that they are valid for the weld toe condition. Usually these parameters are determined from tests with small-scale smooth specimens. However, in a full-scale weld there may be a thickness-effect involved and the surface-finish will certainly have an influence on these parameters. Tests with smooth specimens may not be representative for the rough surface conditions found in the vicinity of the weld toe. It is our hypothesis that the actual parameters should be determined directly from full-scale tests with welded joints where the early cracking is measured. Database 1 gives us this possibility. Based on the discussion above, we will emphasize three topics:

- the local toe geometry and stress concentration factor;
- the choice of the transition depth;
- the determination of the parameters in Coffin-Manson law.

### 9.2.3. *Local toe geometry and stress concentration factor*

The statistics for the local geometry for the specimens tested in series 1 (database 1) are given in Table 9.1. If we substitute the mean values in Table 9.1 into equation (9.6), this will give $K_t = 2.9$. The total weld leg length is assumed to be 1.8 times the plate thickness T (see Figure 8.1). This value is corroborated by a refined finite element analysis. However, it is highly likely that the cracks initiate at a more unfavorable geometry. It is above all the random variation in the weld toe radius that gives large variation for $K_t$ with values ranging from as low as 2.9 up to as high as 7.0. To circumvent this problem of variability, Lawrence *et al* suggested using the fatigue notch factor $K_f$ given in equation (9.5) instead of $K_t$. Furthermore, a worst-case notch was defined by setting the toe radius equal to the Peterson constant in equation (9.5). In our case, this will approximately give $\rho = a_p = 0.3$ mm, which corresponds to $K_f = 3.1$. The problem with the $K_f$ concept is that the physical interpretation is not obvious. It has been claimed (see Ref [9]) that the need for a definition of a fatigue notch factor is due to the fact that, in reality, the crack initiation life includes an appreciable amount of crack growth. If the initiation phase were limited to pure nucleation of a micro-crack, the stress concentration factor could have been used directly in the calculations. Furthermore, the smallest radius in a joint is a random variable and its mean value will be a function of the length of the weld seam. This point will be pursued below. Let us set the toe angle constant at the mean value and try to determine the smallest toe radius most likely to be found within one test specimen. The statistics in Table 1 were derived from 300 measurements taken with 5.5 mm spacing along a weld seam of the length of 1,650 mm. It appeared that two neighbor radii with a spacing of 5.5 mm could be very different, whereas at a closer distance there is a correlation. Based on this observation, we assume that at a 5.5 mm spacing the measured radii will be independent. Hence, a weld seam with length $W$ mm contains $k = W/5.5$ totally independent radii. The extremum distribution for the smallest value will then read:

$$1 - F_{\rho\min}(\rho) = \left(1 - F_{\rho}(\rho)\right)^{k}$$

(9.8)

where $F_{\rho}(\rho)$ is the cumulative distribution function (CDF) of the arbitrary radius as given in Table 9.1, whereas $F_{\rho\min}(\rho)$ is the CDF for the smallest value over the length $W$. The mean and peak values for this smallest radius can be obtained from the corresponding probability density function (PDF). The peak value is, by definition, the most likely value for the smallest radius within the length $W$. If we assume that $\rho$ is Weibull distributed (see Ref [10]) and $W$ is set to 1,650 mm ($k = 300$), then our approach will give a peak value for the smallest radius close to 0.1 mm. The smallest radii found in various test series are actually close to 0.1 mm; see Chapter 3 and Ref [10]. Hence, these results support our approach. If we use the same approach within the width of one test specimen ($W = 60$ mm, $k = 60/5.5 = 10$) we get $\rho = 0.42$ mm. The PDFs for an arbitrary and extreme toe radius are shown in Figure 9.3. As can be seen, the mean value and the peak value are not very different for the relatively narrow symmetrical extreme value distribution. The peak value of 0.42 mm will result in a stress concentration factor of 4.5 that will be used in our calculations.

| Weld toe angle $\theta$ (Degrees) | | Weld toe radius $\rho$ (mm) | |
|---|---|---|---|
| Mean | Standard deviation | Mean | Standard deviation |
| 58 | 9 | 1.6 | 0.7 |

**Table 9.1.** *Statistics for local weld toe geometry, database 1*

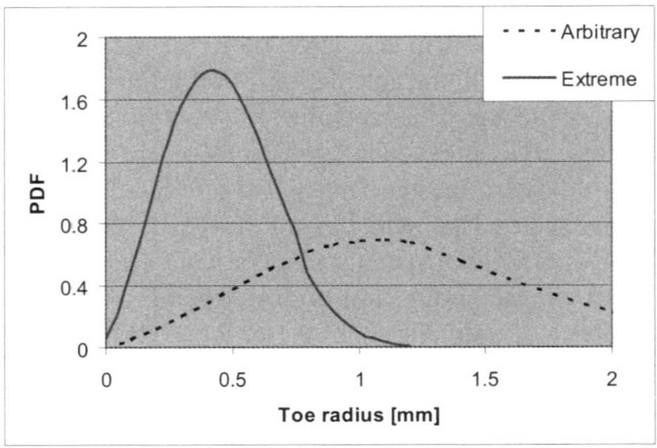



**Figure 9.3.** *Definition of the toe notch geometry in
one specimen by extreme value statistics*

### 9.2.4. *Transition depth*

Often in earlier work the transition depth has arbitrarily been set to 0.25 mm
(0.01 in). However, the time to develop such a deep crack will include a large
amount of crack growth. This has in many cases resulted in models that give
reasonably good predictions at short total fatigue lives (less than $10^6$ cycles), but an
over-prediction of the fatigue lives for low stress levels. The reason is that the
initiation part of the fatigue life at low stress levels will have a life curve with a
slope close to the parameter b $\cong$ -1/10; see equation (9.7). This is only true for pure
initiation, whereas crack growth will have a slope, according to the Paris law,
-1/$m$ $\cong$ -1/3 relative to the applied stress range. Hence, a phase that contains both
nucleation and growth will have a slope between -1/3 and -1/10. If a slope of -1/10
is assumed for such a mixed process it will significantly overestimate the fatigue life
at low stresses. In more recent work, Lawrence *et al.* (see Ref [5]) suggested that the
transition depth should be between 0.05 and 0.1 mm. In the present work we will
investigate the results obtained by setting the transition depth equal to the lower and
upper bound of this given range. The following arguments support a transition depth
of 0.1 mm:

– to apply the fracture mechanics model at crack depths smaller than 0.1 mm
may be dubious because such small crack depths approaches the grain size
(typically 0.01 mm);

– in laboratory tests it is almost impossible to measure any crack less than
0.1 mm with sufficient accuracy without using destructive methods. Hence,
calibration data will not be available;

– cracks with a depth of less than 0.1 mm are not of interest in in-service inspection as no common non-destructive inspections (NDI) method can detect such small cracks. Hence, for inspection planning we do not need the notion of a crack smaller than 0.1 mm.

Our arguments are partly theoretical, partly practical. The actual number of cycles to reach 0.1 mm is given in Table 9.2 for database 1. In contrast to the fictitious initial crack depths obtained from the FMM analysis in Chapter 6, section 6.7.7, this is a measurable quantity. The fatigue initiation life is close to 30% of the entire fatigue life. The entire fatigue life is 10% shorter than the predictions from the F-class curve.

The arguments for a transition depth as low as 0.05 mm are less obvious, apart from the fact that it will exclude any crack growth. It is to be noted that even this shallow a depth is well above the initial crack distribution obtained from the FMM (see Figure 6.21, Chapter 6). The upper bound was found to be 0.03 mm for these cracks. Hence, the transition depth of 0.05 mm is about the smallest transition depth possible based on what may be interpreted as crack size of possible initial flaws. The time to arrive at 0.05 mm crack depth was not measurable for the tests carried out for database 1. The number of cycles to reach this depth can be found by back-calculation from the first measurable crack size (0.1 mm) by applying the FMM in Chapter 6, section 6.7. In this way the number of cycles to reach 0.05 mm was determined to 90,000 cycles, i.e. 20% of total fatigue life.

| $N_i$ is defined as time to reach 0.1 mm | | | |
|---|---|---|---|
| $N_i$ | $N_p$ | $N_t$ | $N_t$ F-class |
| 140 | 330 | 468 | 513 |

**Table 9.2.** *Measured number of cycles (in 1,000) spent in various phases. Accelerated laboratory condition 150 MPa (database 1)*

### 9.2.5. *Cyclic mechanical properties and parameters in Coffin-Manson equation*

It is our hypothesis that parameters determined from small-scale smooth specimens are not directly applicable for weld toe conditions. Hence, we will determine these parameters directly from the time to early cracking as given Table 9.2. We now seek the parameters in equations (9.2) and (9.3) that correspond to the initiation time given in Table 9.2. The calibration is carried out by assuming a dependency between the various parameters with the Brinell hardness (HB) as the key parameter. The following equations are applied (Ref [11]):

$$S_u = 3.45 \; HB \qquad \text{MPa} \qquad\qquad n' = \frac{b}{c}$$

$$S_y' = 0.608 \; S_u \qquad \text{MPa} \qquad K' = S_y \left(0.002\right)^{-n'} \qquad \text{MPa}$$

$$b = -0.1667 \log\left(2.1 + \frac{917}{S_u}\right) \qquad \sigma_f' = 0.95 \, S_u + 370 \qquad \text{MPa} \qquad\qquad (9.9)$$

$$c = -0.7 < c < -0.5 \qquad\qquad \varepsilon_f' = \left(\frac{\sigma_f'}{K'}\right)^{1/n'}$$

where $S_u$ and $S_y$ are the tensile stress and yield stress respectively, and $S_y'$ is the cyclic yield stress. The correlation between the various parameters, using the HB as a master variable, is not exact. In Ref [12] it was found that the relationship might lead to significant overestimation of the initiation life. Furthermore, Dowling (Ref [9]) suggested that the surface condition at the weld toe would primarily alter the fatigue strength exponent $b$. However, in the present work we accept the relationships given in equation (9.9), but not the absolute values. An absolute value for HB is sought such that the time to reach a given transition crack depth coincides with what actually has been measured on the welded joints in database 1. The solution gives a HB close to 202 for the cycles to reach a crack depth 0.1 mm. This HB is very close to the value actually measured in the heat-affected zone (HAZ) of the weld toe. The values measured for the base metal gave an HB = 145 and values of the HAZ gave 213. Hence, our solution HB = 202 is only 5% less than the highest value measured at the potential crack locus. If we had applied HB = 213 directly, the solution would give the time to reach a crack depth of 0.3 mm. Hence, a significant amount of crack propagation would have been included in the initiation phase. When using the search scheme for determining the parameters with a transition depth of 0.05 mm, we obtained a HB of 180, i.e. still within the range measured on the specimens, but 15% less than the value at the HAZ.

The parameters corresponding to HB = 180 ($a$ = 0.05 mm) and HB = 202 ($a$ = 0.1 mm) were used in equations (9.2) and (9.3) to define the first part of the TPM. The propagation phase based on fracture mechanics was subsequently added to calculate the entire fatigue life. The model predicts exactly the mean value for the fatigue lives of database 1 at a stress range of 150 MPa. When the stress range was decreased from 150 MPa to below 100 MPa, the model based on a transition depth of 0.1 mm predicted somewhat longer lives than the median line obtained from the RFLM in Chapter 8, which is representative for database 2 in this stress region. The model based on a transition depth of 0.05 mm predicted results with somewhat shorter lives than figures obtained from the RFLM in this region. At a stress range of

80 MPa, the model with a = 0.1 mm predicts a 28% longer life than the median line of the RFLM, whereas the model with a = 0.05 mm predicts a 20% shorter life than the same median line. These results will be discussed in more detail in the next section. The results indicate that the interval for a transition depth between 0.05 mm and 0.1 mm as proposed by Lawrence *et al.* (Ref [5]) is a reasonable choice. Furthermore, any transition depth in this narrow band will predict fatigue life well within the scatter band of database 2. Based on the arguments stated at the beginning of this chapter, we have selected a transition depth of 0.1 mm in what follows. The corresponding parameters in the Coffin-Manson equation are given in Table 9.3. As we already have shown, the solution given in Table 9.3 is not unique. Other solutions without total dependency between the parameters are possible. However, these solutions will not be far from the one given in Table 9.3. Hence, we regard the solution representative for prediction of time to reach a depth of 0.1 mm in welds made from C-Mn steel with a yield stress close to 345 MPa.

| Parameter, Symbol (units) | Value | |
| --- | --- | --- |
| Cyclic yield stress, $S'_y$ (MPa) | 424 | (61 ksi) |
| Ultimate strength, $S_u$ (MPa) | 697 | (101 ksi) |
| Young modulus, $E$ (GPa) | 206 | (30,000 ksi) |
| Fatigue strength exponent, $b$ | -0.089 | |
| Strain hardening exponent, $n'$ | 1,032 | (150 ksi) |
| Fatigue strength coefficient, $\sigma'_f$ (MPa) | -0.6 | |
| Fatigue ductility coefficient $\varepsilon'_f$ | 0.81 | |
| Cyclic strength coefficient, $K'$ (MPa) | 1,064 | (154 ksi) |
| Fatigue ductility exponent $c$ | 0.148 | |

**Table 9.3.** *Cyclic mechanical properties and parameters in the Coffin-Manson equation calibrated for time to reach 0.1 mm, HB=202*

## 9.3. Constructing the *S-N* curve from the two-phase model

One of our main goals is to construct S-N curves from the TPM that are consistent with the RFLM curves obtained in Chapter 8. As pointed out, the RFLM curve fitted the data points far better than the F-class curve at low stress ranges. Although the TPM is semi-empirical, it has a more physical-theoretical basis than the RFLM which is based on purely statistical methods. The TPM is capable of predicting the influence of, for example, local weld toe geometry, stress ratio, and stress relieving. Thus, high-quality joints will have long fatigue lives, whereas poor-quality joints will be penalized by the model. The TPM model can also be applied to calculate fatigue lives at low stress levels where experimental data do not exist and where it is dubious to extrapolate the statistical RFLM. Let us begin by demonstrating that the model can predict fatigue lives that are in good agreement with the S-N curve obtained from the RFLM under appropriate assumptions of the quality of the joint. For joints that are stress-relieved (database 1), the TPM will predict fatigue lives as given in Table 9.4 at various stress levels. As can be seen from the table, the time to crack initiation at a test stress range of 150 MPa is 30% of the entire fatigue life, whereas it is 88% of the fatigue life at a stress range of 80 Mpa. These results pin point the importance of the crack initiation life at low stress ranges, i.e. in the stress region where service stresses usually occur. The table also lists the fatigue lives predicted by the RFLM and the F-class. At stress ranges below 100 MPa, the TPM predicts somewhat longer lives than the S-N curve based on the RFLM and significantly longer lives than the F-class S-N curve. As can be seen from Table 9.4, the total TPM fatigue life is close to 1.6 times longer than the life obtained from the RFLM and 5.5 times longer than the prediction made by the F-class at 80 Mpa. When comparing these figures we must bear in mind that the figures derived from the TPM correspond to the test series in database 1, i.e. stress relieved (SR) and with an applied stress ratio of $R = 0.3$. This stress relieving has a strong bearing on the time to crack initiation through the Morrow mean stress effect at long lives; see equation (9.2). The RFLM S-N curve is dominated by database 2 in the low stress region. These tests are carried out on non-load-carrying fillet welded joints with thicknesses in the range of 16 to 38 mm. The specimens are all in as-welded (AW) condition and with positive stress ratio, i.e. there may be large residual stresses present in the specimens. Furthermore, the vast majority of tests used to determine the F-class curve are in AW conditions and are often tested at a stress ratio close to $R = 0.1$. Thus, our next step is to simulate these conditions for the initiation part of the TPM by setting the residual stress equal to the actual material yield stress, i.e. 400 Mpa. The results are given in Table 9.5. As can be seen, the TPM results are now almost identical to the non-linear S-N curve obtained from the RFLM, but the model still predicts a fatigue life 2.5 times longer than the F-class at 80 Mpa. These results are illustrated in Figure 9.4 where the F-class and RFLM-based S-N curves are drawn together with the TPM S-N curve and the test results. As can be seen, we have basically two types of curves. The F-class curve is bilinear, whereas the RFLM and TPM curves are both continuously changing slope. All three S-N curves coincide at high stress levels. Hardly any discrepancy in fatigue life (less than 10%) is found above a stress range level of 120 MPa. When the stresses are lowered to under 100 MPa, the RFLM curve and the TPM curve still coincide, but they predict two to nine times longer lives than the F-class curve as long as the stress range is above the F-class fatigue limit of 56 Mpa. It is our judgment that the F-class curve is too conservative in the stress region under consideration, as already discussed in Chapter 8. This is due to the fact that it is a straight line and based on test results that have the center of gravity for the stress ranges between 120 MPa and 150 Mpa. Hence, the curve fails to take into account the increasing fatigue life due to the importance of an initiation phase below 100 MPa. The experimental results plotted in this stress region corroborate the predictions made by the TPM. It has been shown how the TPM is capable of correctly taking into account the effect of residual stresses and loading ratio. This is shown in more detail in Figure 9.5. The life curve obtained for the AW condition coincides with the median curve (i.e. the RFLM curve) for database 2.

| Stress range (MPa) | $N_i$ (cycles) TPM | $N_p$ (cycles) TPM | $N_t$ (Cycles) TPM | $N_i/N_t\%$ TPM | $N_t$ RFLM | $N_t$ F-class |
|---|---|---|---|---|---|---|
| 150 | $1.4\times10^5$ | $3.3\times10^5$ | $4.7\times10^5$ | 30 | $4.6\times10^5$ | $5.1\times10^5$ |
| 120 | $5.6\times10^5$ | $6.5\times10^5$ | $1.2\times10^6$ | 47 | $1.1\times10^6$ | $1.0\times10^6$ |
| 100 | $2.1\times10^6$ | $1.1\times10^6$ | $3.2\times10^6$ | 66 | $2.5\times10^6$ | $1.7\times10^6$ |
| 80 | $1.6\times10^7$ | $2.2\times10^6$ | $1.8\times10^7$ | 88 | $11.0\times10^6$ | $3.4\times10^6$ |
| 60 | $3.7\times10^8$ | $5.1\times10^6$ | $2.9\times10^8$ | 99 | $\infty$ | $8.0\times10^6$ |

**Table 9.4.** *Results derived from the TPM at various stress ranges; SR, R = 0.3*

| Stress range (MPa) | $N_i$ (cycles) TPM | $N_p$ (cycles) TPM | $N_t$ (Cycles) TPM | $N_i/N_t\%$ TPM |
|---|---|---|---|---|
| 150 | $1.3\times10^5$ | $3.3\times10^5$ | $4.6\times10^5$ | 28 |
| 120 | $4.3\times10^5$ | $6.4\times10^5$ | $1.1\times10^6$ | 40 |
| 100 | $1.3\times10^6$ | $1.1\times10^6$ | $2.4\times10^6$ | 54 |
| 80 | $6.6\times10^6$ | $2.2\times10^6$ | $8.8\times10^6$ | 75 |
| 60 | $8.1\times10^7$ | $5.1\times10^6$ | $8.5\times10^7$ | 94 |

**Table 9.5.** *Results derived from the TPM at various stress ranges; AW, R = 0.1*

Finally it should be noted that although the statistically-based RFLM and the physically-based TPM give the same life predictions at almost any stress range level, there is one fundamental difference between them. The RFLM prescribes a fatigue limit, whereas the TPM does not. This is illustrated in Figure 9.5 where the focus is on the lower stress region. The slope of the S-N curve derived from the TPM will not be smaller than b = -1/10 and will never become horizontal as is the case with the RFLM. This gives a discrepancy between the curves at very long lives (longer than $10^8$ cycles) such that the RFLM is more optimistic. The TPM predicts that any joint will eventually fail if the number of cycles is high enough. It is in fact possible to build a fatigue limit into the TPM by assuming that after crack initiation has taken place the crack may stop growing due to the fact that it has a stress intensity factor range (SIFR) below the threshold value. However, there is no data available to corroborate such behavior for shallow surface breaking cracks.

The development of the model carried out thus far is also found in Ref [13]. In what follows we will look into the consequences of the model with regard to damage accumulation and practical results.

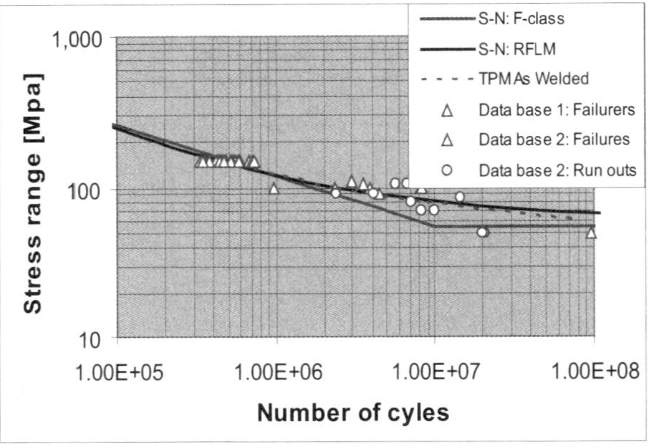


**Figure 9.4.** *S-N curves constructed from the RFLM and TPM together with the F-class median curve and test data*

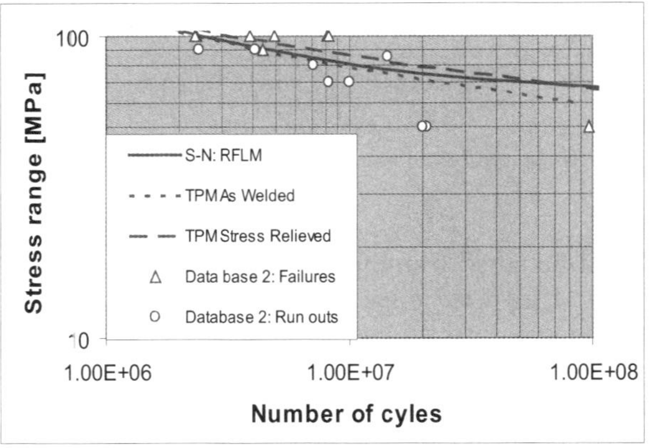


**Figure 9.5.** *S-N curves constructed from the RFLM and TPM in the low stress region*

## 9.4. Damage accumulation using the TPM

In Chapters 5 and 7 we discussed the uncertainty in the Miner's summation rule. The present TPM is an interesting model with respect to damage accumulation under variable amplitude (VA) loading. The model treats crack initiation and crack propagation separately and does not have any fatigue limit. A logical way of accumulating fatigue damage is to first use the Miner's linear damage sum to predict when initiation has taken place, then use the summation rule for the subsequent propagation phase:

$$D_I = \sum_{i=1}^{k} \frac{n_i}{N_{I,i}} = 1.0 \tag{9.10a}$$

$$D_P = \sum_{i=k+1}^{\infty} \frac{n_i}{N_{P,i}} = 1.0 \tag{9.10b}$$

The summation procedure for a given load spectrum is first to carry out the summation of $D_I$ only, until this sum equals 1.0. The subsequent summation of $D_P$ can then start. The fracture criterion will be that both summations are equal to 1.0. Although the summation is linear for both $D_I$ and $D_P$, the total damage sum will be dependent on the sequence of the applied stress spectrum. The S-N curve for initiation and propagation are given in Figure 9.6. This is the S-N curve given in Figure 9.5 for the AW condition, but now split into initiation and propagation life. As can be seen, a stress block with stress range of 70 Mpa will do much less harm in the initiation phase compared to the propagation phase. This is an interesting area of research in order to improve the fatigue failure criterion under variable amplitude loading. As we discussed in section 5.3.2 in Chapter 5, the traditional Miner's linear damage summation rule based on the S-N curve for the total fatigue life is not reliable. Predicting fatigue failure under variable amplitude loading by that approach seams like an illusion. Both equation (9.10a) and (9.10b) must be calibrated against VA test results. Non-linear damage accumulation should also be investigated.

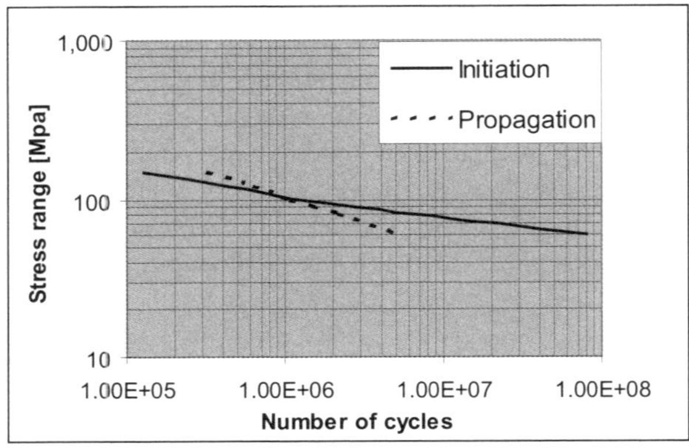


**Figure 9.6.** *The TPM S-N curve (AW) divided into the initiation part and the propagation part*

## 9.5. The practical consequences of the TPM

### 9.5.1. *General considerations*

We have constructed an S-N curve that is non-linear for a log-log scale and that predicts substantially longer lives at stress ranges below 100 MPa than does the F-class linear curve. Furthermore, at these long fatigue lives (close to $5 \times 10^6$ cycles), the initiation life is at least 70% of the entire fatigue life. We will show by a simple example what the consequences of these results are in respect to:

– predicting fatigue life for selecting dimensions for a joint;

– predicting crack growth path for inspection planning.

We will compare our results with the results obtained by the F-class curve and a pure FMM. The latter approach is traditionally used for decision-making regarding inspection planning. We shall use our FMM from Chapter 6, Table 6.4, section 6.7.5, but for simplicity we decrease the growth parameter $C$ by 10%, i.e. $C=1.52 \times 10^{-13}$ (for $da/dN$ in mm/cycle and $\Delta K$ in N/mm$^{3/2}$). This small adjustment will make the predictions made by F-class and by the FMM coincide.

### 9.5.2. *Life predictions and dimensions*

Our goal is to compare results obtained by:

– F-class and the FMM (which are equal when the endurance limit is neglected);

– the present TPM assuming AW conditions, $R = 0.1$.

For simplicity, we assume constant amplitude loading and choose a stress range in the region considered, e.g. $\Delta S$ = 80 MPa. By using equations (9.2) and (9.3) we obtain the results in Table 9.6.

| Stress range (MPa) | TPM | | | F-class |
|---|---|---|---|---|
| | $N_i$ | $N_p$ | $N_t$ | $N_t = N_p$ |
| 80 | $6.6\times10^6$ | $2.2\times10^6$ | $8.8\times10^6$ | $3.4\times10^6$ |
| 58 | | | | $8.8\times10^6$ |

**Table 9.6.** *Median life (cycles) predictions made by the TPM and the F-class*

As can be seen, the TPM predicts 2.5 times longer fatigue life than the F-class, and the initiation part is close to 70% of the entire life at 80 MPa. If dimensions are chosen according to the F-class, the dimensions must be increased by 38% to give the same predicted fatigue life as the TPM, i.e. $8.8 \times 10^6$ cycles. This will correspond to an allowable stress range of 58 Mpa only. These assessments will also be valid for a design curve if the lives predicted by the TPM have the same scatter as the F-class.

### 9.5.3. *Predicted crack evolution and inspection planning*

Let us compare the two alternatives above in relation to inspection planning. The first alternative is the TPM-based design with an allowable stress of 80 MPa; the second alternative is the F-class design with allowable stress range 58 Mpa. The two cases will have the same design life. Our task is to compare the crack evolutions before and up to final fracture based on the TPM and the FMM in each case. We shall more precisely consider the effect of a scheduled inspection program for the two growth histories. The purpose of such a program is of course to detect cracks so that they can be repaired before reaching the final critical crack size. We use the concept of a probability of detection (POD) curve to characterize the performance of the inspection technique; see Chapter 7. The POD is a function of the joint type, environment, and crack size, and is established based on blind tests by inspectors. The POD curve for magnetic particle inspection (MPI) under poor conditions reads (see Chapter 7):

$$POD(a) = 0.9\,[1 - e^{-(a-1)}] \qquad\qquad a > 1 \text{ mm} \qquad\qquad (9.11)$$

The curve is shown to the left in Figure 9.7. The obtained crack histories for the FMM and the TPM are shown to the right. As can be seen, the curve derived from the TPM has a more hidden path that makes the crack more difficult to detect at an early stage. The reliability calculations for an inspection program are carried out using a simple quasi-stochastic approach often used in the aircraft industry; see Chapter 7, section 7.6.4, Figure 7.17. We call the method quasi-stochastic because it does not take into account the randomness of the crack evolution, but applies the mean curves as given to the right in Figure 9.7. A more fully stochastic analysis has been thoroughly presented in Chapter 7. If the strategy is to implement $k$ inspections during the planned service life, the likelihood of all of them failing can be estimated by:

$$P_F = \prod_{i=1}^{i=k} [1 - POD(a_i)] \,. \qquad (9.12)$$

The expression is based on the assumption that each inspection is independent. The reliability of the inspection program is $R=1 - P_F$.

If we are undertaking inspections at a constant time interval corresponding to $5 \times 10^5$ cycles, we will, for the two curves in Figure 9.6, get the figures in Table 9.7. An effective inspection is defined when the current crack depth is larger than $a_B$, i.e. 1 mm. According to the TPM, there will only be two effective inspections. As can be seen, there is an important difference in achieved reliability for the given inspection program for the two different crack evolutions. The FMM predicts a reliability of 0.999, whereas the TPM predicts only 0.935. When comparing the probability of failure, the difference becomes more striking, the probability of failure pertaining to the FMM is in the acceptable range, whereas the one pertaining to the TPM is not. A more refined stochastic analysis should be carried out to corroborate the results. However, if the predictions made by the TPM are accepted as true, one would obtain acceptable reliability if the inspection efforts were concentrated in the last part of the service life with a decreased inspection interval of $2 \times 10^5$ cycles. For a more fully stochastic approach one can introduce the probability of a preexisting crack in the TPM. The difference in reliability will then be less, but the revealed tendency will be the same.

In conclusion, the practical consequences of the applying a TPM will be a reduction in joint dimension of 30% and a scheduled inspection program that is progressive with the time in-service.

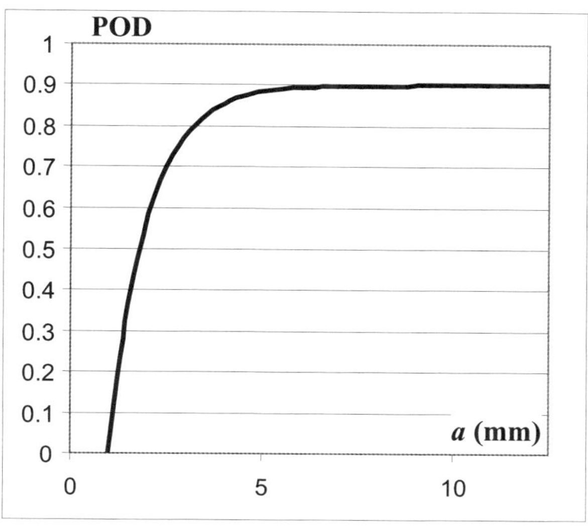


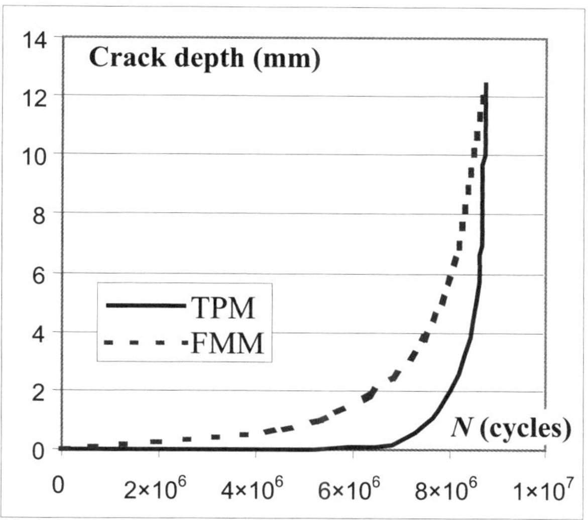


**Figure 9.7.** *Above: POD curve MPI; below: predicted crack evolution for FMM, ΔS=58 MPa and TPM, ΔS=80 MPa*

| Model for prediction | First effective inspection | Last effective inspection | Number of effective inspections | Reliability | Probability of failure |
|---|---|---|---|---|---|
| FMM | $5.5 \times 10^6$ | $8.5 \times 10^6$ | 7 | 0.9999 | $1.0 \times 10^{-4}$ |
| TPM | $8.0 \times 10^6$ | $8.5 \times 10^6$ | 2 | 0.935 | $6.5 \times 10^{-2}$ |

**Table 9.7.** *Reliability calculations for a given inspection program*

## 9.6. Conclusions

Our study in this chapter is primarily concentrates on non-load-carrying fillet welded joints made of C-Mn steel with nominal yield stress close to 345 MPa (50 ksi). A TPM was used to predict the fatigue life. The initiation life was modeled by Coffin-Manson equation, whereas the crack propagation was based on the simple version of the Paris law. The models were validated and calibrated with the use of large databases. The criteria for acceptance of the model were that the model should predict both damage evolution and final fatigue life at any stress level. The FMM model presented in Chapter 6 failed to fulfill these criteria, whereas the TPM in the present chapter gave an excellent fit to both measured crack growth histories and experimental fatigue lives. The following conclusions are drawn.

The fatigue behavior of fillet welded joints is far more complex at typically in-service stresses than fracture mechanics can describe. This is due to the fact that the crack initiation phase dominates the fatigue life at these low stresses.

A TPM is capable of modeling the damage evolution from the initial state to the final fracture provided that the model is accurately calibrated for this purpose. The notch factor at the weld toe is based on extreme value statistics for the toe geometry, and the transition crack depth between the initiation phase and the propagation phase is set to 0.1 mm. The parameters in the Coffin-Manson equation were determined directly from early cracking in full-scale welded joints.

As the TPM has a semi-empirical physical basis, the determining factors, such as residual stresses, global and local joint geometry, and loading mode, are readily accounted for.

The S-N curves constructed from the model are non-linear for a log-log scale and coincide with the curves obtained from the statistical RFLM in Chapter 8. Both models fit experimental data far better than the conventional bilinear S-N curves.

There is a fundamental difference between S-N curves obtained from the RFLM and the TPM in the way that the latter curves do not predict any fatigue limit. At stress ranges below 70 MPa, the RFLM curve will appear flat, whereas the TPM curve will continue to fall with a small slope close to the parameter b in the Coffin-Manson law. At present there is no data to corroborate either one of these curves, but the authors tend to have more confidence in the prediction made by the physical-based TPM than the predictions based on the statistical RFLM when extrapolated outside the range of the data.

The first practical consequence of the present TPM is that it predicts longer lives at low stress ranges than the conventional S-N curves in rules and regulations. With the application of the TPM-constructed S-N curve in the lower stress region it is possible to reduce dimensions by 30-40% and still achieve the same fatigue life as for the F-class S-N curve.

The second practical consequence is that in-service inspection strategy may be optimized. This is due to the fact that the crack path leading to final fracture is quite different from the path calculated by a pure FMM. The TPM with its long initiation phase will give a more hidden path for the crack evolution. Hence, an inspection program with increased inspection frequency at the end of service life is proven to be favorable.

## 9.7. Suggestions for future work

The study in this chapter has been limited to fillet welded joints subjected to constant amplitude loading. Future work should carry out an investigation on other joint configurations such as butt joints and VA loading. As for a butt joint, which is the most common configuration for high load transfer, we already know that the initiation phase will play an even more important role than has been shown for the fillet welded joint. This is due to the significantly lower stress concentration factor at the weld toe. When it comes to variable amplitude loading, the importance and the consequences of an initiation phase are not obvious and an investigation is necessary before drawing conclusions. As the TPM has no fatigue limit, it will be interesting to analyze how the model responds to variable loading by using a damage accumulation law; see equation (9.10a) and (9.10b). For load spectra with the center of gravity in the low stress range area (e.g. exponential distributed stress ranges), the effects revealed for constant amplitude loading will probably prevail.

It is also interesting to determine the role the initiation part plays for welded joints in a seawater environment. Inspection planning on offshore structures is today planned entirely on applied fracture mechanics where the initiation phase is ignored.

In the present chapter the TPM has been used to construct median S-N curves only. Future work should focus on constructing quantile curves for design purposes as well. This can be done by a Monte Carlo simulation treating the main determining factors as random variables as was shown in Chapter 7. The resulting curves should be compatible with the quantile curves obtained from the RFLM in Chapter 8. Finally, the practical consequences of the model in terms of joint dimensions and scheduled inspection programs should be studied in more detail. Other types of joints, such as butt joints, are also of great interest in this regard.

## 9.8. References


1    American Bureau of Shipping, Guide *for the Fatigue Assessment of Offshore Structures*, ABS, April, 2003

2    Y. Verreman and B. Nie, "Early Development of Fatigue Cracking at Manual Fillet Welds" *Fatigue & Fracture of Engineering Materials and Structures* 19 (6), 1996, pp 669–81

3    T. Lassen, "The Effect of the Welding Process on the Fatigue Crack Growth in Welded Joints" *Welding Journal* 96 (2), 1990, pp 75s, 85s

4    JY Yung and FV Lawrence, "Analytical and Graphical Aids for the Fatigue Design of Weldments" *Fatigue Fract. Engn. Mater. Struct.* 8 , 1985, pp 223–41

5    F.V. Lawrence, S.D. Dimitrakis and W.H. Munse, "Factors Influencing Weldment Fatigue" *Fatigue and Fracture*, ASM Handbook, Vol. 19, 1996, pp. 274–86, Materials Park, OH

6    BS7910, *Guidance on Methods for Assessing the Acceptability of Flaws in Fusion Welded Structures*, 2000, London, British Standards Institution

7    Ph.Darcis, *et al.*, "A fracture mechanics approach for the crack growth in welded joints with reference to BS7910" *European Conference on Fracture (ECF 15),* Stockholm 11-13 August 2004

8    X. Niu and G. Glinka, "The Weld Profile Effect on the Stress Intensity Factors in Weldments" *Int. J. of Fracture* 35, 1987, pp 3–20

9    E. Dowling, "Estimating Fatigue Life" *Fatigue and Fracture*, ASM Handbook, Vol. 19, 1996, pp 250–62

10   K. Engesvik and T. Lassen, "The Effect of Weld Toe Geometry on Fatigue Life" The 7th OMAE Conference, Houston, Texas, 1988, pp 441–45

11   R.A. Testin, J.Y. Yung, F.V. Lawrence and R.C. Rice, "Predicting the Fatigue Resistance of Steel Weldments" *Welding Journal* 66, 1987, 93-s to 98-s

12   A. Tricoteaux, F. Fardoun, S. Degallaix and F. Sauvage, "Fatigue Crack Initiation Life Prediction in High Strength Structural Steel Welded Joints" *Fatigue Fract. Engng. Mater. Struct.* 18(2), 1995, pp 189–200

13   Ph. Darcis, T. Lassen and N. Recho, "Fatigue Behavior of Welded Joints Part 2: Physical Modeling of the Fatigue Process" *Welding Journal* 85 (1), 2006, 19-s to 26-s

Chapter 10

# A Notch Stress Field Approach to the Prediction of Fatigue Life

## 10.1. A modified S-N approach

### 10.1.1. *General considerations*

In Chapter 5 the S-N approach for welded joints was thoroughly presented and discussed. The various stress concepts used as keys to fatigue life were explained. We argued that using the weld notch stress range as the key parameter to fatigue life was the most logical choice compared to the nominal stress and geometric stress concept. The theoretical argument was that it is the weld notch stress that is the vehicle for the fatigue damage process. The practical benefit is that we are able to estimate the fatigue life of any joint configuration and weld bead shape. In other words, both detail design and workmanship are accounted for. The method has become popular in rules and regulations for ship structures; see Refs [1] and [2]. At the end of Chapter 5 we discussed some possible improvements of the method and we shall pursue those improvements in the present chapter. The method has the following basis:

1) The reference S-N curve is a modified C-curve based on tests with the base metal.

2) The method uses the weld notch stress as the single parameter to enter the S-N curve.

3) The weld notch stress effect is often determined from average values of the geometry parameters characterizing the weld toe profile.

All these three points are approximations, and the shortcomings are as follows:

1) The reference curve is established for other surface conditions than that which is found in the fusion line of a welded joint.

2) Fatigue life should not be related to the stress range only, but to the entire stress field at the weld toe.

3) When characterizing the irregular profile of the weld toe, extreme value statistics should be used instead of the average values.

The first point is related to the fact that the welding process leaves some sort of fingerprints that reduce the time to crack initiation, compared to the base metal; see Chapter 9 regarding discussion of the initiation period. The second point was explained in Chapter 5 by the fact that the higher the local stress concentration gets, the smaller is the material volume at the weld toe that is actually influenced by this high stress level. Hence, scaling fatigue life directly to the maximum stress range will be overly pessimistic. The conditions are visualized in Figure 10.1. The reference test specimen designated A is subjected to a uniform stress field all over the cross section, whereas for the two fillet welded joints, the stress will just affect a small material volume at the weld toe. As can be seen, the case denoted C with a small radius at the weld toe has a very high stress concentration factor (SCF) and a steep stress gradient. The third point above is related to the fact that the toe angle and radius are highly variable along the weld seam. Cracks are likely to initiate and grow at the worst-case notch within one joint. Hence, extreme value statistics should be used instead of the average values for the geometry variables. The second and third points above must be considered together. The application of the stress concentration factor only is overly pessimistic, whereas the use of an average toe profile is overly optimistic. The assumptions are counteracting each other and will give reasonable good estimates for the fatigue life in most cases, but the assumptions are not physically correct. This will be demonstrated for some extreme cases.

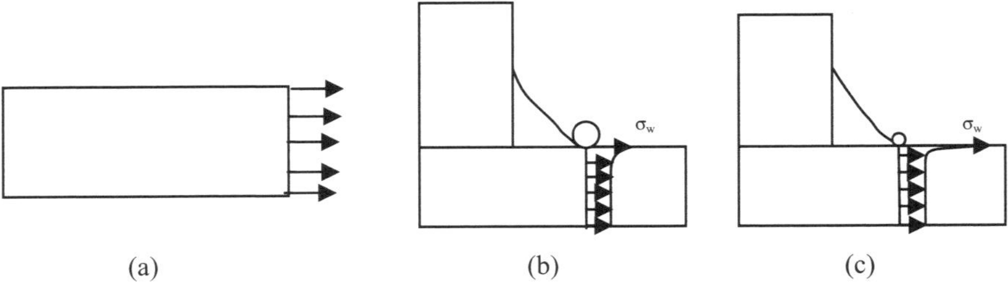


**Figure 10.1.** *(a) reference test specimen, (c) weld with favorable toe profile (larger radius), (c) weld with less favorable toe profile (small radius)*

In this chapter we will present a stress field method as an attempt to improve the weld notch approach:

– the reference curve will be established based on tests with welded joints;

– the entire stress field and not only the stress concentration will be considered;

– extreme value statistics for the weld toe geometry will be applied.

### 10.1.2. *The basic theory for the notch stress intensity factor*

A model that account for these improvements is the notch stress intensity factor (N-SIF) as has been suggested by Lazzarin; see Refs [3] and [4]. He regarded the weld bead as a sharp V-shaped notch with a given angle. The radius was set to zero as a first approximation; see Figure 10.2. From this assumption it is possible to determine the stress intensity factor (SIF) associated with the weld toe notch. The SIF is a measure for the entire stress field and not just the stress concentration. The method has the same basis as the fracture mechanics approach presented in Chapter 6. The difference is that in the present case we do not have the presence of a crack, but a sharp weld notch.

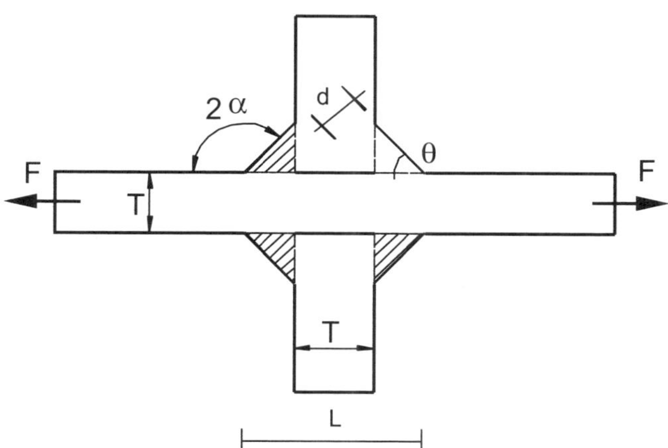


**Figure 10.2.** *Definition of specimen and toe geometry*
*for a fillet weld, Ref [3]*

The suggested approach is an approximation as it assumes that the weld toe radius is zero. We remember that DNV suggested in the ship rules at the end of Chapter 5, section 5.6.6, that the average toe radius should be as large as 2 mm for a butt joint. BV has suggested that a reasonably good weld has a radius of 1 mm; see Refs [1] and [2]. Both these values are based on average values. In Chapter 9 we showed, by extreme value statistics, that the most likely smallest radius within a welded specimen of 60 mm width is close to 0.4 mm. We also showed in Chapter 3 that for large test series the smallest toe radius was actually close to 0.1 mm. In the latter case we are approaching the sharp notch assumption of Lazzarin; see Ref [3]. Based on the approach, the stress intensity factor range (SIFR) can be found on a closed form solution:

$$\Delta K_1^N = k_1 \cdot T^{1-\lambda_1} \cdot \Delta\sigma_{nom}$$

(10.1)

where $k_1$ is a non-dimensional coefficient, T is the plate thickness, and $\lambda_1$ is the Williams eigenvalue which denotes the degree of singularity. The parameters $k_1$ and $\lambda_1$ are the amplitude and the exponent of the first term of the stress field expansion; see Ref [5]. The next step proposed in Ref [4] was to relate the SIFR in equation (10.1) to the elastic strain energy range $\Delta E$ in a defined area A near the crack tip. The averaged elastic energy range $\Delta\overline{W}$ then reads:

$$\Delta\overline{W} = \frac{\Delta E}{A}$$

(10.2)

which can be expressed as:

$$\Delta\overline{W} = \frac{1}{E} \cdot e_1 \cdot \left(\Delta K_1^N\right)^2 \cdot \left(R_C\right)^{2(\lambda_1 - 1)}$$

(10.3)

where $e_1$ is a shape function that depends mainly on the toe angle $\theta$. The expressions are visualized in Figure 10.3. The area A is defined by a radius $R_C$ with a typical size of 0.3 mm. Details are found in Refs [3] and [4].

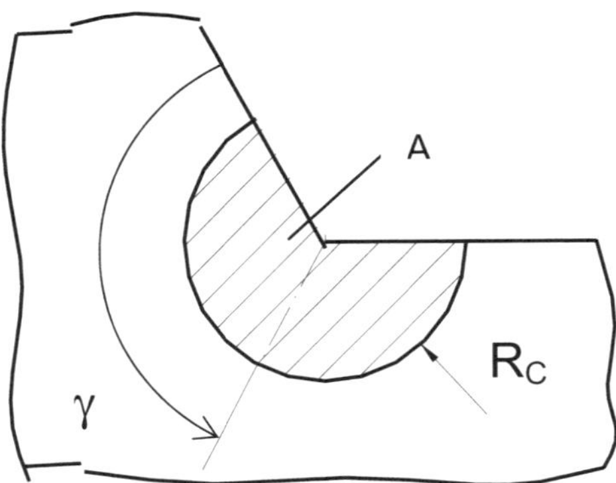


**Figure 10.3.** *Critical volume (area) surrounding the notch tip, Ref [3]*

If we now relate fatigue life to the average strain energy range $\Delta \overline{W}$, we have in fact accounted for the entire stress field in the vicinity of the weld toe. Figure 10.4 shows the fitting from more than 300 fatigue data. The plate thickness ranged from 6 to 100 mm and the flank angles from 30 to 70 degrees. The data plotted as a function of average strain energy range exhibited much less scatter compared to nominal stresses or to the notch stress approach. The curves are valid for fillet welded joints as well as butt joints as the method takes account of both the global geometry of a joint and the local toe profile.

### 10.1.3. *S-N data analysis for fillet welded joints*

To study and compare the rule-based notch stress approach and the present N-SIF approach, we used the two approaches to predict the fatigue life for two large test series carried out on fillet welded joints; see Ref [6]. Both test series contained 42 specimens that were subjected to a constant amplitude (CA) stress range of 150 MPa. The two test series are interesting in the way that series 1 (T=32 mm) has a favorable toe geometry, whereas series 2 (T=25 mm) has a rather unfavorable toe geometry; see Table 10.1. These data were presented in Chapter 3 (see Table 3.3) where the scatter was also listed. The calculations are carried out for the mean toe geometry. The results of the various life predictions are given in Table 10.1.

| Series | Rho Mean [mm] | Theta Mean [Degr.] | SCF Mean Geometry | Shortest Test life | DNV S-N Nominal stress | DNV Weld notch stress | N-SIF Weld notch |
|---|---|---|---|---|---|---|---|
| 1 | 2.7 | 30 | 2.2 | 324,000 | 212,000 | 125,000 | 322,000 |
| 2 | 0.7 | 58 | 3.1 | 189,000 | 212,000 | 49,000 | 226,000 |

**Table 10.1.** *Experimental data and life predictions for fillet welded joint test series with a nominal stress range at 150 Mpa (attachment length L/T=1.8)*

As can be seen from Table 10.1, the shortest fatigue life measured during the two test series is much shorter for series 2 than for series 1, as expected. The predictions based on the DNV nominal stress approach have good margins to the shortest fatigue life for series 1, but this is not the case for series 2. In the latter case the shortest tested life was 189,000 cycles, whereas the prediction is 212,000 cycles, i.e. 12% longer than the tested life. Hence, the nominal stress method can predict overly optimistic lives for joints with an unfavorable toe geometry. As for the DNV weld notch stress method, the predictions underestimate the fatigue life significantly for both series. This is the case even if we have based the SCF values on the average toe geometry of the welds. For series 2, that has an abrupt to geometry, the DNV-based life estimate is close to 4 times less than the experimental fatigue life. The fatigue life predictions would have decreased further if we had used the extreme value statistics for the toe geometry variables instead of the average values. We have explained the reasons for this discrepancy at the beginning of the chapter.

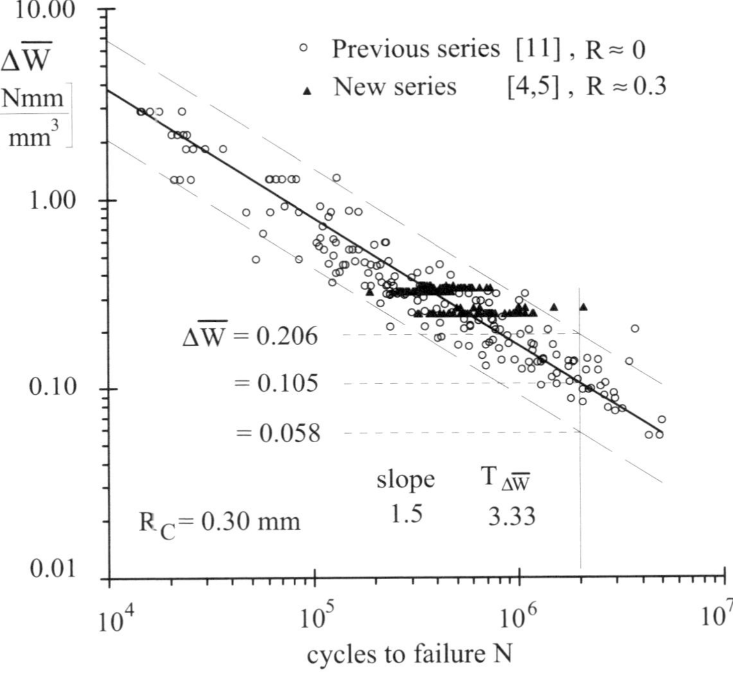


**Figure 10.4.** *Strain energy-based fatigue life curves, Lazzarin et al., Ref [3] and [4]*

The predictions based on the N-SIF method are both quite close to the shortest measured fatigue life for the two series. The N-SIF method overestimates the fatigue life for series 2 by 20%, but this will not be the case if the extreme values for the geometry parameters are applied. The two test series are included in the database in Figure 10.4

For practical applications of the N-SIF method, the average energy $\Delta\overline{W}$ in Figure 10.4 should be related directly to the geometry parameters L/T, $\theta$, and $\rho$/T of the joint. Figure 10.4 can then be used as generic master curve for all types of welded joints with fatigue cracking from the weld toe.

## 10.2. A modified crack growth approach

We would like to add some theoretical developments in determining the N-SIF. We will also propose how the method can be used for predicting fatigue crack growth. The stress field near the notch tip can be written as an expansion series, such as:

$$\sigma_{ij} = A_1 \cdot r^{\lambda_1} \cdot f_{ij}^{(1)}(\lambda_1, \theta) + A_2 \cdot r^{\lambda_2} \cdot f_{ij}^{(2)}(\lambda_2, \theta) + ... \tag{10.4}$$

$(A_1, A_2, .....)$ are the amplitude values and $(\lambda_1, \lambda_2, .....)$ are the degrees of singularity at the *notch tip*. $(f_{ij}^{(1)}(\lambda_1, \theta), f_{ij}^{(2)}(\lambda_2, \theta), .....)$ are proper functions of the notch angle $\alpha$ and depend on the geometry and the boundary conditions.

When the singularity degree of the stress field is high, equation (10.4) can be presented by the first term only. In this case it is only $\lambda_1$ that has a negative value, the other lambdas are positive. Consequently, the higher order terms will approach zero when r gets small. Hence, the stress field can be written as follows near the weld notch ($\lambda_1 = \lambda$):

$$\sigma_{ij} = A \cdot r^{\lambda} \cdot f_{ij}(\lambda, \theta). \tag{10.5}$$

In order to determine the exponent $(\lambda)$ Williams (see Ref [5]) proceeds by expansion and establishes a characteristic equation in the case of V-notch with an angle $(\alpha)$:

$$\sin^2[(\lambda + 1).\alpha] - (\lambda + 1)^2 \sin^2 \alpha = 0 \tag{10.6}$$

In this equation, $(\lambda)$ is the eigenvalue of a set of linear equations.

The crack growth rate of a crack initiated at the tip of a V-notch has to depend on the singularity degree associated to the V-notch. This dependence is limited to a small crack length from the V-notch tip. This crack length can be determined as a function of the material characteristics and the singularity degree. The fatigue life associated with this small crack represents a large amount of the total fatigue life; see the results in Chapter 9.

Going from the stress field expression limited to the first term, one writes the strain energy in the region A concerned by the small crack close to the V-notch tip as follows:

$$W(\varepsilon) = \int_A w(\varepsilon). \, d(A) \text{ with } w(\varepsilon) = A_1^2 \cdot r^{2\lambda_1} \cdot S(\alpha) \tag{10.7}$$

$S(\alpha)$ is a proper function of $\alpha$, the V-notch angle. The energy release rate limited to this region is noted G (see Appendix A for definitions). The crack growth rate could be then a function of the variation of G and of the singularity degree, such that:

$$\frac{da}{dN} = f(\Delta G, \lambda_1).$$

(10.8)

For example, a crack propagation law could be presented as follows:

$$\frac{da}{dN} = C.(\Delta G - \Delta G_s)^{m.\lambda_1}$$

(10.9)

in which $\Delta G_s$ is the threshold value of $\Delta G$. $C$ and $m$ are material parameters. Indeed, such a law has to be verified from experimental results, as was shown for the Paris law in Chapter 3. A work in progress is dealing with this type of crack growth propagation law in mode I and in mixed mode crack.

In fact, the determination of the region affected by the singularity is a major goal for fatigue life estimation. The higher the singularity degree, the smaller the size of this region. As a first estimation, the characteristic length $l_0$ given by Leguillon (see Ref [7]) can be considered:

$$l_0 = \frac{G_C.s^2{}_{\theta\theta}(\theta_0)}{K(a,\theta_0).\sigma_C^2}.$$

(10.10)

$G_C$ and $\sigma_C$ are the critical energy release rate and the ultimate stress, respectively. $S_{\theta\theta}$ and $K(\alpha,\theta_0)$ are specific functions issued from the equivalence between the strain energy in the far field and the strain energy in the asymptotic field; see Ref [7]. In this reference, this equivalence is done in a linear elastic medium under static loading.

It is obvious that the comprehension of the static mechanical fields near the notch tip allows modeling the fatigue behavior for small cracks. When taking into account this singularity degree in the crack propagation law as shown, the predictions of the damage accumulation will be modified compared to conventional fatigue predictions.

## 10.3. References


1   DNV, Fatigue Assessment of Ship Structure Classification note 30.7, Det Norske Veritas, 2003

2   BV, *Fatigue Strength of Welded Ship Structures*, Bureau Veritas, July 1998

3   P. Lazzarin and P. Livieri, "Notch Stress Intensity Factors and Fatigue Strength of Aluminium and Steel Welded Joints" *Int. Journal of Fatigue*, 23, 2001, pp 225–32

4   P. Lazzarin, T. Lassen, and P. Livieri, "A notch stress intensity approach applied to fatigue life predictions of welded joints with different local toe geometry" *Journal of Fatigue and Fracture of Engineering Material and Structure*, 26, 2003, pp 49-58

5   M.L. Williams, "Stress singularities resulting from various boundary conditions in angular corners of plates in extension" *J. Appl. Mech,* 1952, pp 526-528

6   K. Engesvik and T. Lassen, "The Effect of Weld Geometry on Fatigue Life", OMAE, Houston, 1988

7   D. Leguillon, "Strength or Toughness? A Criterion for Crack Onset at a Notch" *European J. Mech. A/Solids,* Vol 21, 2002, pp 61–72

# Chapter 11

# Multi-Axial Fatigue of Welded Joints

## 11.1. Introduction and objectives

Various loading modes and stress situations for details containing a crack are often idealized by pure mode I or pure mode II, see Figure 6.4. The crack tip stress field is determined by linear elastic fracture mechanics (LEFM) for these cases, see equation (6.1). In Chapter 6 we argued that fatigue cracks at the weld are often subjected to pure load I mode. This is of course not always the case; combinations of loading modes may appear. The combinations of loading that involve more than one stress crack tip mode are referred to as mixed mode. For mixed-mode loading, crack growth behavior is related to given the loading condition. Furthermore, a mixed mode may involve bifurcation, i.e. the crack changes direction during the propagation. In order to study the crack growth process that occurs under mixed-mode loading, a series of mixed-mode experiments have been carried out under static loading condition [Refs 1–5].

When a crack is subjected to the cyclic loading, the study of fatigue crack growth is of great interest to predict the structure and component life. Traditional applications of fracture mechanics due to fatigue have been concentrated on cracks growing under an opening or mode I mechanism. A number of crack growth laws were developed in order to evaluate the fatigue crack growth rate, for example, the Paris law [Ref 6], the empirical formula of Forman *et al.* [Ref 7] or the empirical relation of Erdogan and Ratwani [Ref 8], etc. These laws were established for a crack subjected to mode I fatigue loading as outlined in Chapter 6. The effect of the loading angle on the propagation was not considered in these laws.

However, many service failures occur from cracks subjected to mixed-mode loadings. Under mixed-mode loading conditions, not only the crack growth direction is of importance, but also the fatigue crack growth rate. Several criteria have been proposed regarding the crack growth direction under mixed-mode loadings, such as the maximum circumferential criterion [Ref 9], the minimum strain-energy density criterion [Ref 10], the maximum energy release rate criterion [Ref 11], and the crack tip opening displacement criterion [Ref 12], etc. Also, several parameters have been suggested to correlate mixed mode to fatigue crack growth rates; for example, Tanaka [Ref 13] used the equivalent stress intensity factor in Paris' law to evaluate the crack growth rate; Yan *et al.* [Ref 14] proposed another formula of the equivalent stress intensity factor; Socie *et al.* [Ref 15] presented the equivalent strain intensity factor for small cracks; Hoshide and Socie [Ref 16] developed an equation by using a J-integral to predict the crack growth rate, and also for small crack; and Tong *et al.* [Ref 17] proposed a model of Paris's type to predict the crack growth rate in which the branch crack tip local mode I stress intensity factor was used.

In addition to the effect of the loading angle on the crack growth rate, the residual stresses due to welding may influence the crack propagation. When a crack exists in the metallic welded structure, considerable work has been carried out on the assessment of welding residual stresses [Refs 18 and 19]. In order to evaluate the influence of residual stresses due to the weld on the crack propagation, Parker [Ref 20] made a review of the various methods in order to evaluate the level of the residual stress intensity factor $K_{res}$. From these methods, the Green's function was usually applied.

In [Ref 28], experiments of a fatigue crack under mixed-mode loading are performed with compact-tension-shear (CTS) specimens. The effect of the loading angle on the crack growth rate and crack growth direction is analyzed. Moreover, the welded specimens are introduced in the experiments in order to examine the influence of a fillet weld on the crack growth rate. Furthermore, on the basis of the experimental results, a crack growth model is proposed in order to numerically evaluate a fatigue crack growth rate, and the effects of loading mode and of the residual stresses due to the weld are taken into account in the model. In addition, the effect of the welding residual stresses on the kinking angle is examined.

The final aim of the present study is the development of a method that allows a better estimate of the fatigue life of welded structures (armored vehicles, ships, floating production storage off-loading, wrecking cranes, cars, etc.) when subjected to various loading modes.

## 11.2. Overview of theory and crack extension criteria

In order to determine the crack growth path under mixed-mode loading, one can use different criteria to calculate the crack extension angle. For example, the maximum circumferential stress $\sigma_{\theta\theta max}$ criterion (Erdogan and Sih [Ref 9]), the maximum energy release rate criterion-MERR (Palasniswamy and Knauss [Ref 6]), the stationary strain energy density criterion (Sih [Ref 10]), the $J_{II}=0$ (Pawliska *et al.* [Ref 11]) and $K_{II}=0$ (Cotterell and Rice [Ref 26]) criteria ($J_{II}$ is the value of the J-integral corresponding to pure mode II, and $K_{II}$ is the value of the stress intensity factor corresponding to pure mode II), the crack tip opening displacement (or angle) criterion (Sutton *et al.* [Ref 12]), etc. Recently, we have developed the $J$-$M_p$-based criteria (Li, Zhang and Recho [Ref 27]) to assess the propagation of a crack in elastic-plastic material under mixed-mode loading.

In the case of a crack in elastic material, the elastic crack extension criterion ($\sigma_{\theta\theta max}$ criterion) is most often used. According to this criterion, the crack always propagates in the direction of the maximum circumferential stress at $\theta=\theta_0$. Consider the equation of the circumferential stress $\sigma_{\theta\theta}$ as follows:

$$\sigma_{\theta\theta} = \frac{1}{4\sqrt{2\pi r}}\left[ K_I(\cos\frac{\theta}{2} + \cos\frac{3\theta}{2}) - 3K_{II}(\sin\frac{\theta}{2} + \sin\frac{3\theta}{2}) \right] \qquad (11.1)$$

where $r$ and $\theta$ are the polar coordinates from the crack tip.

The crack extension angle $\theta_0$ can be determined after calculating the values of the stress intensity factors $K_I$ and $K_{II}$ by $\dfrac{\partial\sigma_{\theta\theta}}{\partial\theta} = 0$ :

$$\text{tg}(\frac{\theta_0}{2}) = \frac{1}{4}\,(\frac{K_I}{K_{II}}) \pm \frac{1}{4}\sqrt{(\frac{K_I}{K_{II}})^2 + 8}\;. \qquad (11.2)$$

The numerical simulation of a crack growth is made in this work by using this criterion. Furthermore, the maximum energy release rate criterion (MERR) and the $K_{II}=0$ can also be considered.

## 11.3. The crack box technique

### 11.3.1. *General considerations for finite element analysis and element mesh*

In industrial complex structures, such as welded structures, the prediction of the crack path is necessary to determine the fatigue life and failure mode. Crack extension criteria are also needed to predict the crack path. Their choice depends on the mechanical material characteristics and the loading levels. For industrial purposes, two methods of finite element analysis (FEA) are commonly used. Either step-by-step complete re-meshing of the global model is performed or a very simplified determination is made by choosing the crack path that is perpendicular to the maximum of the principal stresses [Ref 29]. The first approach is very time consuming and cannot be used for industrial purpose unless an optimized meshing is performed. Nevertheless it is necessary when the whole stiffness is affected by the crack path and in large-scale plasticity.

Close to the crack tip, the local stress field is determined by the use of asymptotic analysis. This enables the prediction of the critical loading level to crack propagation and the determination of the crack extension angle. But, this local asymptotic stress field presents a very high gradient, which is why a specific and regular finite element mesh is required in this zone.

Also, during the crack propagation this mesh has to move with the crack tip. Apart from crack failure and crack extension criteria, two major problems remain, i.e. what are the mesh characteristics of the crack tip region and how is it connected to the overall structure?

### 11.3.2. *Methodology*

In order to use this technique, one has to automatically create a transition zone between the "crack box" and the whole structural unchanged mesh (see Figure 11.1) [Ref 30]:

– Zone (A): crack box (Figures 11.2 and 11.3). It contains a specific and regular mesh. It is affected by the asymptotic solution at the crack tip. For elastic calculations, few elements are needed. The crack tip is modeled with degenerated quadratic elements with one side collapsed and mid-side nodes are moved to the quarter point nearest the crack tip to create a strain singularity in $r^{-0.5}$ (r is the distance from the crack tip). For plastic calculations, more elements are needed to determine precisely the J-integral. To introduce a $r^{-1}$ singularity for perfectly plastic material strains, degenerated quadratic elements are also used, but crack tip nodes are allowed to move independently and mid-side nodes remain at the mid-side point.

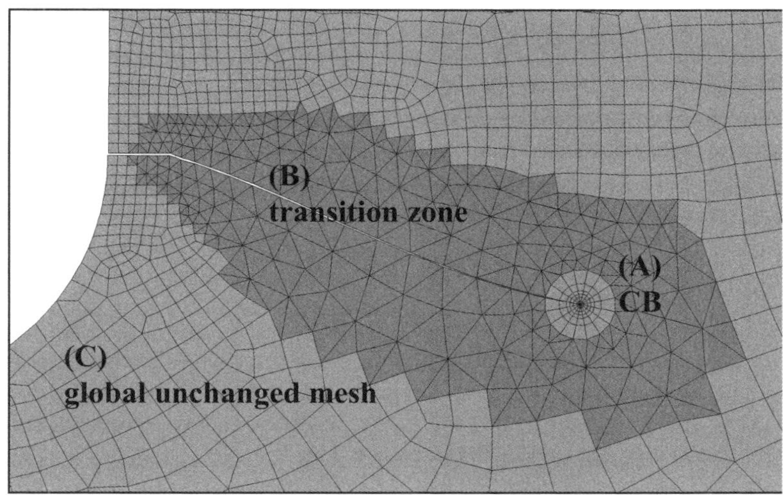



**Figure 11.1.** *Crack box in a structure (regions A, B and C)*

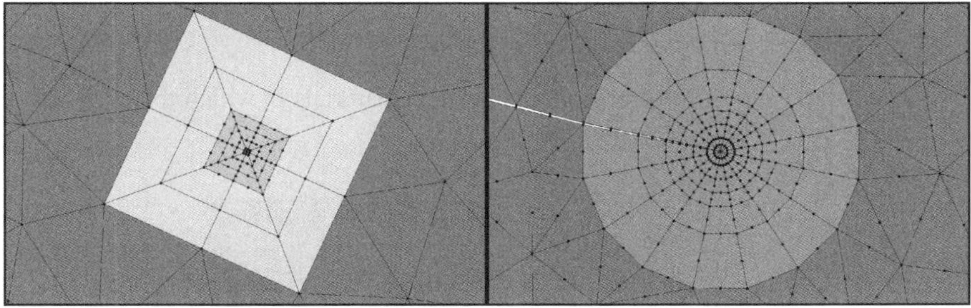

**Figure 11.2.** *Refined crack box*          **Figure 11.3.** *Coarse crack box*

– Zone (B): transition region. It contains an optimized linear (for elastic calculations) or quadratic (to increase precision for elastic calculations and for elastic-plastic calculations) triangular mesh obtained by using the Delaunay triangulation procedure from NAG [Ref 31]. This enables the specific crack box to be connected with the whole ABAQUS [Ref 32] model, which can be a 2D plane strain or stress and a 3D shell model.

– Zone (C): whole model. It represents a usual finite element mesh. It is to be noted that this mesh is unchanged during the crack propagation.

The automatic crack-box technique used is developed using the ABAQUS code and consists of the following steps, [Ref 30]:

– Meshing of the three regions for the initial crack.

– Performing FEM calculations associated with crack extension criterion in order to determine the crack extension angle.

– Taking a crack growth increment in the direction corresponding to the crack extension angle.

– Updating of local crack tip region mesh and connecting it by the use of region (B) to the whole structure.

Note: region (B) works such as a moving contour around the crack tip. It looks like a static condensation of the structural behavior to the crack tip region. This technique is almost similar to the one based on the boundary integral equations in which the contour is replaced by the transition zone.

### 11.3.3. *Examples*

Two simple examples are shown as to how to apply this crack box technique.

*Crack growth near two holes in elastic material*

This example shows the crack path for elastic materials in which the crack grows between the both holes (see Figure 11.4).

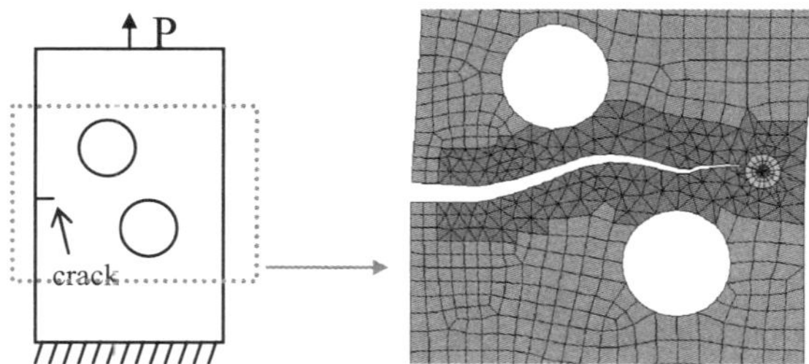


**Figure 11.4.** *Crack growth numerical results using crack box technique (CBT): two-hole specimen*

*Crack from a fillet in elastic material*

This example shows the capability of the CBT to predict crack paths in various geometry and loading conditions. It is important to note that only the crack zone is re-meshed during iterative calculations. The bending stiffness is modified by varying the size of the bottom I-beam: h. For low h values, there is a large bending component so the crack direction will change significantly. For high h values, the crack will tend to be straight (see Figure 11.5).

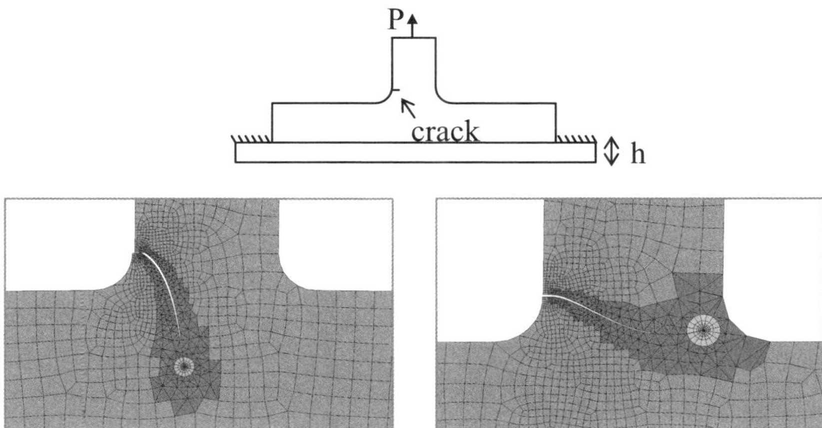

**Figure 11.5.** *Crack growth numerical results using CBT: fillet specimen (left: h low values; right: h high values)*

## 11.4. Tentative mixed-mode model to crack propagation in welded joints

This section is based on the work done by [Ref 28], in which series of experiments are done in order to analyze the effect of mixed-mode on crack growth rate; see Figure 11.6. A mixed-mode crack propagation model is developed on the basis of these experiments.

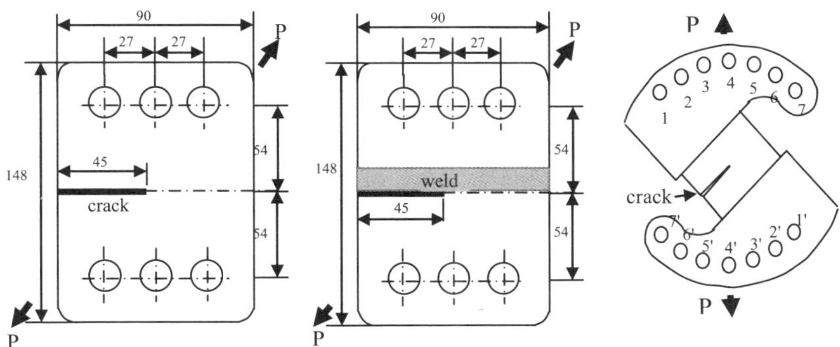

**Figure 11.6.** S*pecimens and loading device (left-hand-side) non-welded specimen (middle) welded specimen (right-hand side) loading device*

The fatigue tests are conducted on the machine MTS-810 (material test system) at room temperature. The CTS specimens are tested with two loading levels and three loading angles, 90°, 60° and 30°, with respect to the crack axis, as shown in Figure 11.7. The 90° loading corresponds to the pure mode I test. Two specimens are used for each loading condition. During the tests, the load ratio $R(= \sigma_{min}/\sigma_{max})$ for all loading angles and for all loading levels is kept constant at 0.5. Loads are applied sinusoidally at a frequency of 25 Hz.

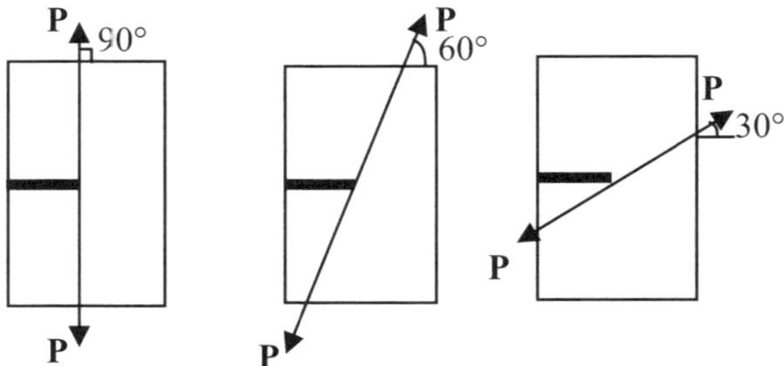


**Figure 11.7.** *Loading angles*

In general, when a crack is subjected to a fatigue loading of small amplitude, it is commonly accepted that the small-scale yielding condition is satisfied. Therefore, in this section of the chapter, the maximum circumferential stress $\sigma_{\theta\theta max}$ criterion can be used to predict the crack extension angle for a mixed-mode crack under fatigue loading of small amplitude.

In order to evaluate numerically the influences of loading mode and of the residual stress due to the weld on the crack growth rate, a numerical model is proposed and calibrated the experimental results. This model is established on the basis of the Paris law with the equivalent stress intensity factor $K_{eq}$:

$$\frac{da}{dN} = C(\Delta K_{eq})^{m}. \tag{11.3}$$

For a mixed-mode loading, the variation of the equivalent stress intensity factor $\Delta K_{eq}$ is used, which can be obtained according to the formula proposed by Tanaka [Ref 13]:

$$\Delta K_{eq} = \left[ \Delta K_I^4 + 8\Delta K_{II}^4 \right]^{0.25} \tag{11.4}$$

In the following experimental results, the crack growth rate $da/dN$ is presented as a function of the variation of the equivalent stress intensity factor $\Delta K_{eq}$ which is calculated by FEM according to equation (11.4).

Two factors are taken into account in the model, one is the degree of mixed fracture modes (often referred to as mixity), and the other is the residual stress due to the weld. In fact, these two factors have influence on the material resistance.

## 11.4.1. *Modeling the effect of the loading mode on the crack growth rate*

Although the equivalent stress intensity factor $\Delta K_{eq}$ is the combination of $K_I$ and $K_{II}$, the influence of mixed-mode loading on the crack growth rate cannot be explained completely by $\Delta K_{eq}$. It appears from the experimental observation and the numerical simulation that a kinking crack subjected to a mixed-mode fatigue load grows almost perpendicularly to the loading angle. Consequently, $K_{II}$ is close to 0 except in the first step. $\Delta K_{eq}$ of a kinking crack is almost equal to $K_I$ after the first step. So $\Delta K_{eq}$ is not enough to reflect the influence of mixed-loading on $da/dN$. According to experimental results [Ref 28], the coefficient $m$ remains constant when the crack propagates under mixed-mode loading. However, the coefficient $C$ varies as function of the fracture mode (Figures 11.8 to 11.13). Therefore, the influence of the mixed mode on the crack growth rate can be taken into account by an approximate representation of the coefficient $C$. Therefore we use a coefficient $C^*$ which is a function of the elastic mixity parameter $M^e$:

$$C^* = C\,f(M^e) \tag{11.5}$$

where $C$ is the coefficient of pure mode I, whereas $C^*$ is the coefficient of mixed mode:

$$M^e = \frac{2}{\pi}\tan^{-1}\left|\frac{K_I}{K_{II}}\right| \tag{11.6}$$

$M^e$ varies between 1 and 0. For pure mode I, $M^e=1$; for pure mode II, $M^e=0$.

According to the experimental results, the coefficient $C^*$ is written as follows:

$$C^* = C\,f(M^e) = C[1+\beta(M^e-1)^2] \tag{11.7}$$

$\beta$ is a constant is a constant obtained by fitting the equation to experimental data. In this work, $\beta=3$ seems to fit the experimental data for all CTS specimens. For the pure mode I, $M^e=1$, $C^*=C$.

By using the shape of the Paris law, a model is proposed to evaluate the crack growth rate under a mixed-mode loading. It is expressed as follows:

$$\frac{da}{dN} = C^* (\Delta K_{eq})^m \tag{11.8}$$

In this equation, the constants $C$ and $m$ are measured from the pure mode I fatigue test. For each crack length, $K_I$ and $K_{II}$ are calculated by FEM. $\Delta K_{eq}$ (equation (11.4)) and $M^e$ (equation (11.6)) can be determined from the values of $K_I$ and $K_{II}$.

### 11.4.2. *Modeling the effect of the residual stress due to the weld on the crack growth rate*

The experimental data (Figure 11.8 to Figure 11.11) show that for the same loading condition, the crack growth rate in welded and non-welded specimen is different, i.e. the residual stress due to the weld has influence on the crack growth rate. According to the experimental results, the coefficient $C$ varies as a function of the magnitude and the distribution of the residual stresses in the welded specimens. Therefore, the effect of the welding residual stress on the crack growth rate can be considered by an approximate representation of the coefficient $C$. Hence, a new coefficient $C^R$ is proposed, which is expressed as follows:

$$C^R = C \frac{1 + \dfrac{K_{res}}{K_{max}}}{1 + \beta \dfrac{K_{res}}{K_{min}}} \tag{11.9}$$

where we have:

 – $K_{min}$: minimum of stress intensity factor;

 – $K_{max}$: maximum of stress intensity factor;

 – $K_{res}$: stress intensity factor due to the welding residual stress;

 – $\beta$ is the same experimental constant as in equation (11.7). For the type of CTS specimen in this work, $\beta$=3. In the case of without weld, $K_{res}$=0, so $C^R$=$C$ is just the Paris law. The stress intensity factor due to the welding residual stress $K_{res}$ can be determined by Green's function [Ref 22]:

$$K_{res} = 2 \sqrt{\frac{a}{\pi}} \int_0^a \frac{\sigma_r(x) dx}{\sqrt{a^2 - x^2}} \tag{11.10}$$

where $a$ is the crack length, $\sigma_r(x)$ is the distribution of the residual stress due to the weld. The expression of $\sigma_r(x)$ was proposed by Masubuchi and Martin [Ref 23]. When the welded residual stress is horizontal to the fillet of the weld:

$$\sigma_r(x) = \sigma_{r-\max}\left[1-(\frac{x}{b})^2\right] \times e^{-\frac{1}{2}(\frac{x}{b})^2} \tag{11.11}$$

where $\sigma_{r-max}$ is the maximum residual stresses, $b$ is the width of the compressive residual stress region, and $x$ is the distance to the weld. In our experiments, the initial crack is parallel to the fillet weld in each welded specimen. According to the experimental data, the residual stress due to the weld decreases the crack growth rate, i.e. it is compressive residual stress near the crack tip in the direction perpendicular to the initial crack axis. This tendency is consistent with that measured in the same direction by Lieurade [Ref 24]. In these tests, the shape of the distribution of the residual stress in the direction perpendicular to the fillet of the weld is similar to that in the direction horizontal to the fillet of the weld. Therefore, it is assumed that the shape of the initial residual stress distribution corresponds to the distribution $\sigma_r(x)$ described in equation (11.11) can be adopted. In equation (11.11), the width of the compressive residual stress region $b$ represents one-third of the specimen width, which was evaluated by YB Lee *et al.* [Ref 25] for the same shape of residual stress distribution. The maximum residual stress $\sigma_{r-max}$ is estimated from experimental data to fit the test values of the crack growth rate $da/dN$. Consequently, the distribution of the residual stress can be determined.

Hence the model proposed to evaluate the crack growth rate is expressed as equation (11.3). In this model, the effect of the residual stress due to weld is taken into account in the term of $C^R$:

$$\frac{da}{dN} = C^R(\Delta K_{eq})^m \tag{11.12}$$

### 11.4.3. *Measured effect of the loading angle on the crack growth rate*

Both aluminum alloy specimens and steel specimens, with or without weld, are tested under mixed-mode loading conditions (shown in Figure 11.6). Figures 11.8 and 11.9 give us the results of crack growth rate in aluminum specimens; Figures 11.10 and 11.11 show the experimental results in steel specimen. In each figure, the specimens are subjected to 90°, 60°, and 30° loads respectively. The mixed-mode loading levels are listed in Table 11.1. To obtain the same initial intensity of stress field near the crack tip, the same initial $\Delta K_{eq}$ for different loading angles is preferred for the comparison of the crack growth rate. For the aluminum specimens,

$\Delta K_{eq}$=6.2 $MPa\sqrt{m}$ , and for the steel specimens, $\Delta K_{eq}$=15.6 $MPa\sqrt{m}$ . The loading levels of steel specimens are higher than that of aluminum specimens because of the higher fracture toughness of steel.

From Figure 11.8 to Figure 11.11, we get the same tendency: that for the same initial $\Delta K_{eq}$, the crack grows faster in the case of 30° loading than in the case of 60° loading, and the crack grows faster under 60° loading condition than that under 90° loading conditions. To obtain the same $\Delta K_{eq}$, the closer to pure mode II, the greater the load is needed. Therefore, because of the lowest load, the crack growth rate is also the lowest when the crack is subjected to pure mode I load. When the crack is subjected to 30° loading, the amplitude of the load is greater than that of 60° and 90° loading, so the crack growth rate is the fastest among these three loading conditions whatever the material is, whatever with or without weld.

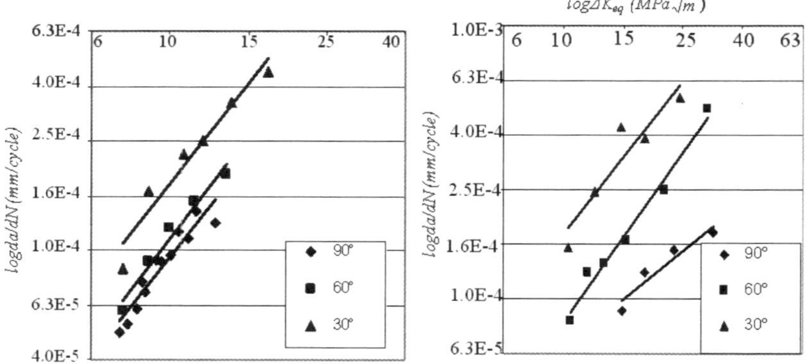

**Figure 11.8.** *Non-welded aluminum specimen*

**Figure 11.9.** *Welded aluminum specimen*

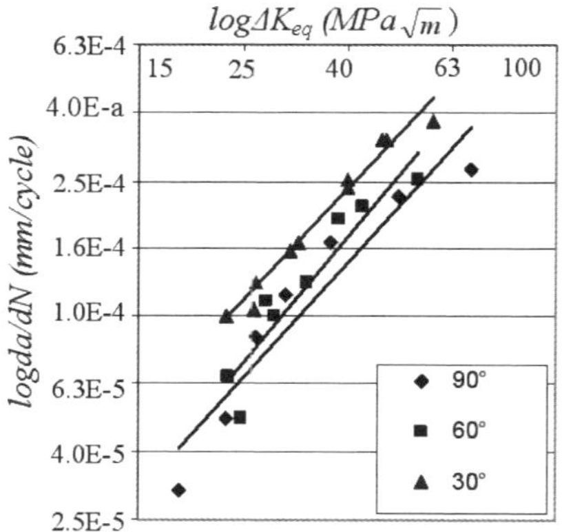

**Figure 11.10.** *Non-welded steel specimen*

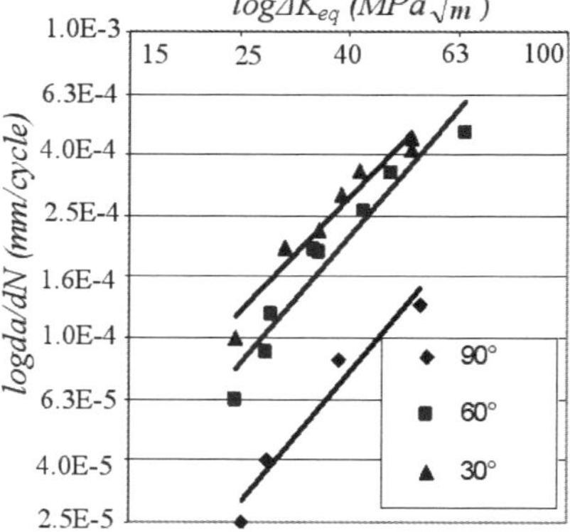

**Figure 11.11.** *Welded steel specimen*

#### 11.4.4. *Measured effect of weld on the crack growth rate*

In this part of the experiment, the welded and non-welded specimens are subjected to the same loading angle and to the same loading level (see Table 11.1).

| Loading angle | Aluminum specimen | | Steel specimen | |
|---|---|---|---|---|
| | $F_{min}(kN)$ | $F_{max}(kN)$ | $F_{min}(kN)$ | $F_{max}(kN)$ |
| 90° | 6.0 | 12.0 | 8.0 | 16.0 |
| 60° | 6.0 | 12.0 | 9.0 | 18.0 |
| 30° | 7.0 | 14.0 | 10.4 | 20.8 |
| $\Delta K_{eq\text{-}initial}\ (MPa\sqrt{m}\ )$ | 6.2 | | 15.6 | |

**Table 11.1.** *Mixed-mode loading*

Figures 11.12 and 11.13 show the results of crack growth rate in the aluminum specimen subjected to 90° and 30° loading. From these figures, it can be noted that for the same loading level, the crack growth rate is greater in non-welded specimens than that in welded specimens. This means that the weld process introduces the compressive residual stress near the fillet of the weld in the direction perpendicular to the initial crack axis. This residual stress decreases the crack growth rate. When the specimens are subjected to pure mode I loading conditions (Figure 11.12), this effect is more obvious than in the case of 60° and 30° loading because of the different crack growth rate path. Here, only the welding residual stress in the direction perpendicular to the fillet weld is considered.

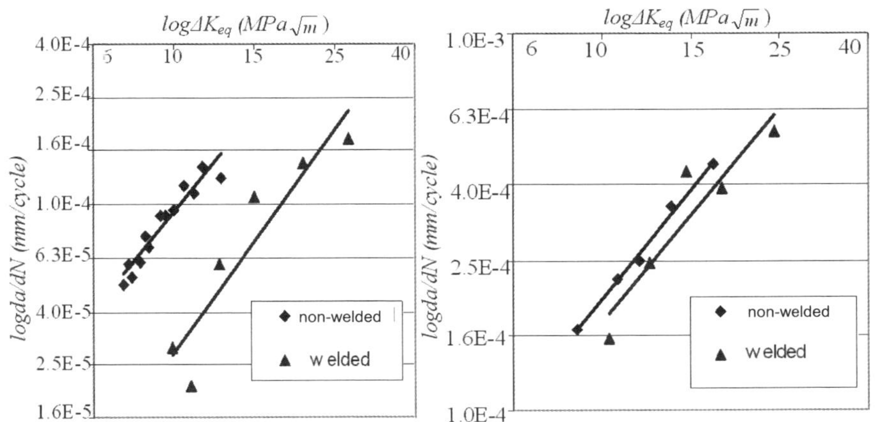


**Figure 11.12.** *90° loading for aluminum specimen*

**Figure 11.13.** *30° loading for aluminum specimen*

When the crack is subjected to mode I loads, the direction of the welding residual stress is horizontal to the loading direction and perpendicular to the direction of the crack propagation; in this case, the crack grows along the fillet of the weld and it is always in the effective zone of the residual stress until fracture. Therefore, the effect of the residual stress on the crack growth is important. However, when the crack is subjected to mixed-mode loads, for example 60° and 30° loading, the angle between the welding residual stress is not consistent with the loading direction; in these cases the crack grows far from the weld zone. Therefore, the influence of the residual stress on the crack growth decreases rapidly.

Figure 11.13 shows a mixed-mode example of 30° loading. The crack growth rate in the welded specimen is similar to that in the non-welded specimen. The effect of the residual stress on the crack growth rate is not significant. The steel specimens with and without weld are then tested under 90°, 60°, and 30° loading conditions. The same effect of the residual stress on the crack growth rate is observed as the experimental data from steel specimen shown in Figure 11.14 and Figure 11.15. The crack grows faster in the non-welded specimen than in the welded specimen. The residual stress due to weld retards the crack propagation, especially in the case of mode I loading. When loading angle is 60° or 30°, the crack propagates far from the weld zone, so the effect of the residual stress due to weld is not significant (see Figure 11.15).

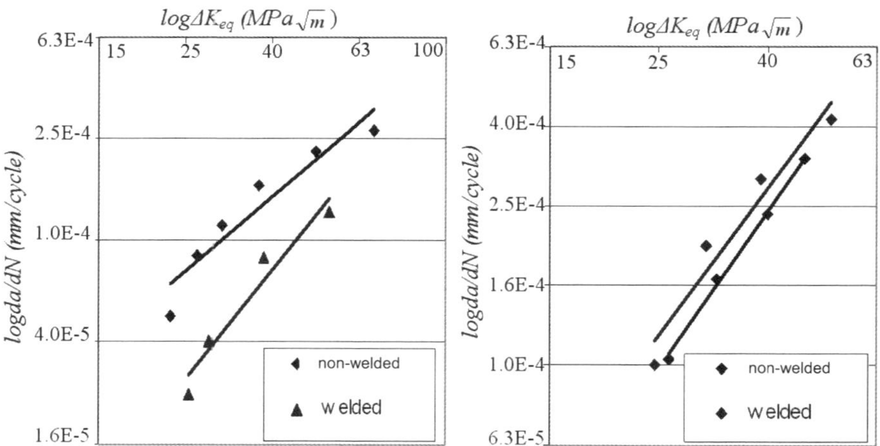


**Figure 11.14.** *90° loading for steel specimen*    **Figure 11.15.** *30° loading for steel specimen*

### 11.4.5. *Measured crack extension angle under mixed mode loading*

Table 11.2 lists the crack growth angle of different specimens and under different loading angles. Figures 11.16 and 11.17 show the photos of the crack growth path in aluminum and steel specimens under 30° loading conditions. According to Table 11.2 and the photos, it can be observed experimentally that there is no great difference of crack growth path between welded specimens and non-welded specimens; that is to say, the residual stresses due to welding have no obvious effect on the crack extension angle.

| Materials | Specimens | 60° loading | 30° loading |
|---|---|---|---|
| Aluminum alloy | Non-welded | -32° | -47° |
| | Welded | -33° | -47° |
| Steel | Non-welded | -26° | -45° |
| | Welded | -29° | -46° |

**Table 11.2.** *Mixed-mode crack extension angle (θ)*

**Figure 11.16.** *Crack growth path of 30° loading of aluminum specimen (a) non-welded; (b) welded*

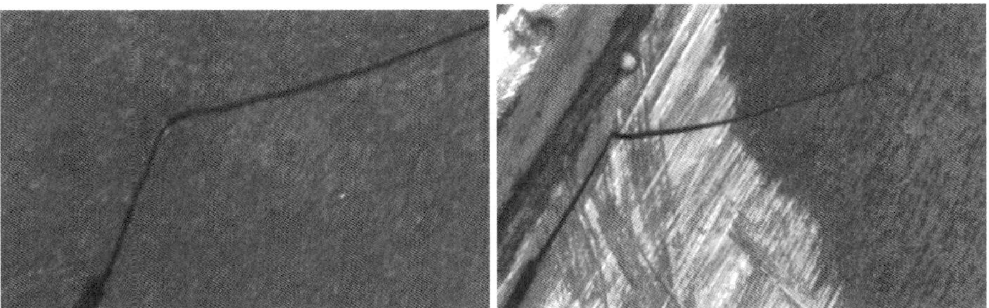

**Figure 11.17.** *crack growth path of 30° loading of steel specimen (a) non-welded; (b) welded*

## 11.5. Validation of the model

This numerical model is proposed on the basis of the experimental results of specimens in aluminum alloy given in [Ref 28]. In order to verify this model, the numerical calculations are performed on the steel specimen tested in the same work.

### 11.5.1. *Verification of the models for non-welded steel specimens under mixed-mode loading*

The specimen, which has 6mm thickness, is shown in Figure 11.6. The loading angles $\alpha$ are 90°, 60° and 30° (Figure 11.7). The loads are listed as follows:

$$\alpha=90°,\ F_{max}=16KN,\ F_{min}=8KN$$

$$\alpha=60°,\ F_{max}=18KN,\ F_{min}=9KN$$

$$\alpha=30°,\ F_{max}=20.8KN,\ F_{min}=10.4KN$$

For the different loading angles, the initial equivalent stress intensity factor $\Delta K_{eq}$ is constant as $15.6\mathrm{MPa}\sqrt{m}$ in this example:

– First, the coefficients $m$ and $C$ are measured in pure mode I test. $m$ is equal to 2,1, and $C$ is about $7\times10^{-8}(mm/cycle/MPa\sqrt{m})^{m}$.

– Secondly, $\Delta K_{eq}$ is calculated by using equation (11.14). Then the mixity parameter $M^{p}(=M^{e})$ and the parameter $C^{*}$ are determined.

– Finally, for 90° loading (mode I), $da/dN$ is calculated by using the Paris law, and for 60° and 30° loading, $da/dN$ is evaluated with equation (11.18).

The numerical results are listed in Table 11.3. Figure 11.18 shows the comparison between the numerical evaluation and the experimental results.

| a | 90° | | 60° | | 30° | |
| (mm) | $K_{eq}$ $(MPa\sqrt{m}\,)$ | $da/dN(\times10^{-5})$ (mm/cycle) | $K_{eq}$ $(MPa\sqrt{m}\,)$ | $da/dN(\times10^{-5})$ (mm/cycle) | $K_{eq}$ $(MPa\sqrt{m}\,)$ | $da/dN(\times10^{-5})$ (mm/cycle) |
|---|---|---|---|---|---|---|
| 45 | 15.9 | 0 | 15.7 | 2.7 | 15.6 | 0 |
| 47 | 17.5 | 2.9 | 18.7 | 3.9 | 18.4 | 5.3 |
| 49 | 19.2 | 3.5 | 20.5 | 4.7 | 20 | 6.3 |
| 51 | 21.2 | 4.3 | 22.6 | 5.8 | 21.9 | 7.7 |
| 53 | 23.5 | 5.3 | 25 | 7.2 | 24 | 9.3 |
| 55 | 26.2 | 6.7 | 27.7 | 8.9 | 26.4 | 11 |
| 57 | 29.2 | 8.4 | 30.7 | 11 | 29 | 14 |
| 59 | 32.8 | 11 | 34.3 | 14 | 32 | 17 |
| 61 | 37.1 | 14 | 38.5 | 18 | 35.3 | 21 |
| 63 | 42.1 | 18 | 43.4 | 23 | 39.2 | 26 |
| 65 | 48.3 | 24 | 49.3 | 30 | 43.6 | 33 |

**Table 11.3.** *Numerical results of the non-welded steel specimens under mixed-mode loading*

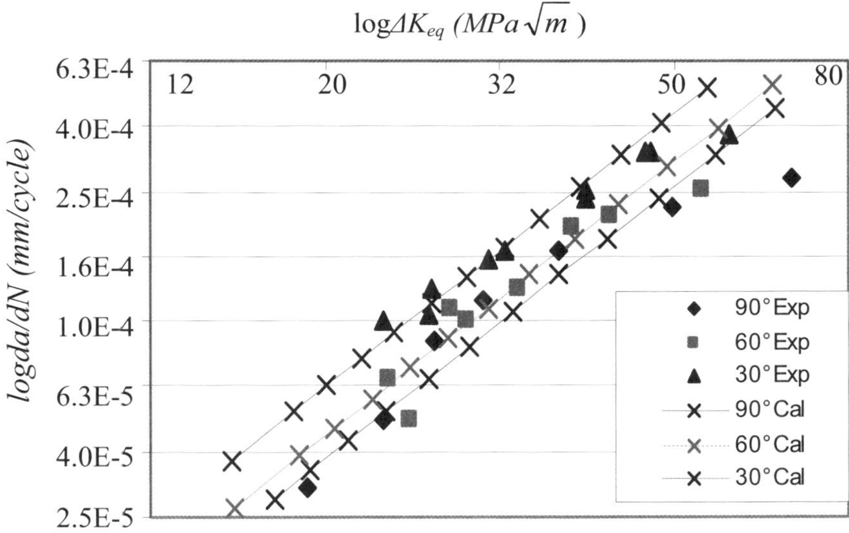



**Figure 11.18.** *Crack growth rate in non-welded steel specimens under mixed-mode loading*

It can be noted, from Figure 11.18, that the numerical evaluations are close to the experimental results except for the results of 90° loading.

### 11.5.2. *Verification of the models for non-welded and welded steel specimens under mode I loading*

The same material is used, so $m = 2.1$ and $C = 7 \times 10^{-8} (mm/cycle/MPa\sqrt{m})^m$. When the crack is subjected to mode I loading, we have $K_{eq}=K_I$. In the case of "without weld", $da/dN$ is calculated by using Paris' law. In the case of "with weld", $C^R$ is determined by FEM using equation (11.9), then $da/dN$ is evaluated according to equation (11.12).

Table 11.4 lists the results of the cases "without and with weld". The comparison is shown in Figure 11.19. As can be seen, the numerical prediction of crack growth rate in the welded specimen is in good agreement with the experimental observations. It proves that the effect of the welded residual stresses is well considered in the model.

| $a$ (mm) | $K_{eq}$ (MPa $\sqrt{m}$) | $da/dN(\times 10^{-5})$ (non-welded) (mm/cycle) | $da/dN(\times 10^{-5})$ (welded) (mm/cycle) |
|---|---|---|---|
| 45 | 15.9 | 0 | 0 |
| 47 | 17.5 | 2.9 | 1.4 |
| 49 | 19.2 | 3.5 | 1.6 |
| 51 | 21.2 | 4.3 | 1.9 |
| 53 | 23.5 | 5.3 | 2.3 |
| 55 | 26.2 | 6.7 | 2.8 |
| 57 | 29.2 | 8.4 | 3.5 |
| 59 | 32.8 | 11 | 4.5 |
| 61 | 37.1 | 14 | 5.9 |
| 63 | 42.1 | 18 | 7.7 |
| 65 | 48.3 | 24 | 10 |

**Table 11.4.** *Numerical results of the non-welded and welded steel specimens under mode I loading*

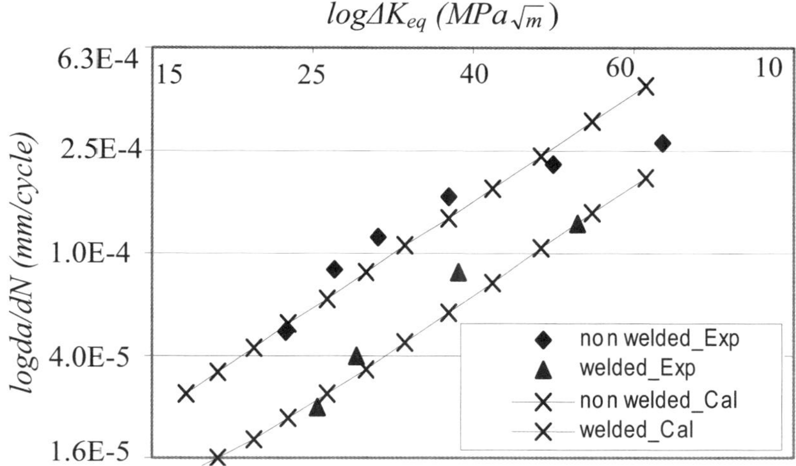



**Figure 11.19.** *Crack growth rate in non-welded and welded steel specimens under mode I loading*

### 11.5.3. *Verification of the models for welded steel specimens under mixed-mode loading*

The same loading angles and the same material are selected.

It is explained above that the effect of the welded residual stresses on the crack growth rate is important only in the case of mode I loading. When the crack is subjected to a mixed-mode loading, the growth path is far from the weld zone. Therefore, the influence of the weld is ignored in the calculations in the case of mixed-mode loading.

For the welded specimen under 90° loading, equation (11.12) is used to evaluate the crack growth rate. For the welded specimen under 60° and 30° loadings, equation (11.3) is used, so only the influence of the loading angle is considered.

Figure 11.20 shows the numerical and experimental results.

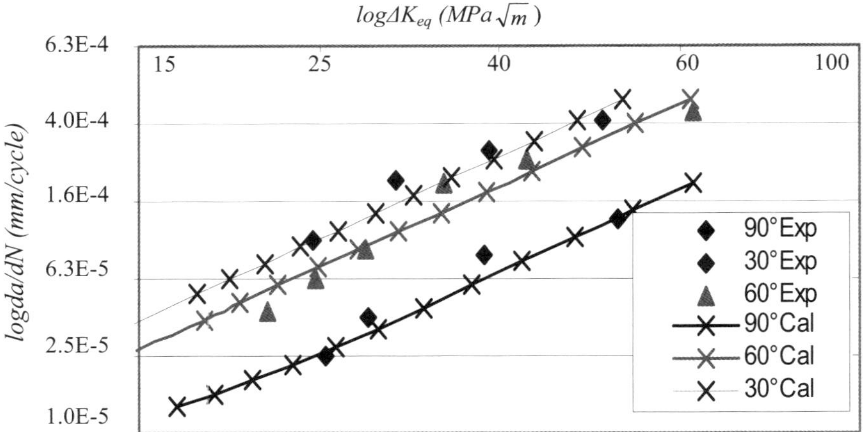



**Figure 11.20.** *Crack growth rate in welded steel specimens under mixed-mode loading*

In Figure 11.20, the numerical evaluations are similar to the experimental results. It can be said that the crack propagation models proposed in this work are valuable for the welded metallic specimen under mixed-mode loading.

### 11.5.4. *Verification of the effect of the welded residual stress on the fatigue life*

From the results of the crack growth rate $da/dN \sim \Delta K_{eq}$, the fatigue life in terms of number of cycles N to failure can be evaluated. For the non-welded specimen, Paris' law is adopted to determine the value of N; and for the welded specimen, the proposed model (equation (11.12)) is used. Figures 11.21 and 11.22 show the obtained a-N curves in aluminum specimens and in steel specimens, respectively.

The period of the beginning of the crack propagation is not considered in the proposed model, therefore it is not included in the numerical results of welded specimens (Figures 11.21 and 11.22). It should be noted that the comparison shows a good correlation between the numerical calculations and the experimental results. The results prove that the proposed model can give predictions that are in good agreement with experimental fatigue lives. When the crack length is more than 65 mm, the fracture will occur, and the crack becomes unstable. Therefore, there is little difference between the numerical and experimental results.

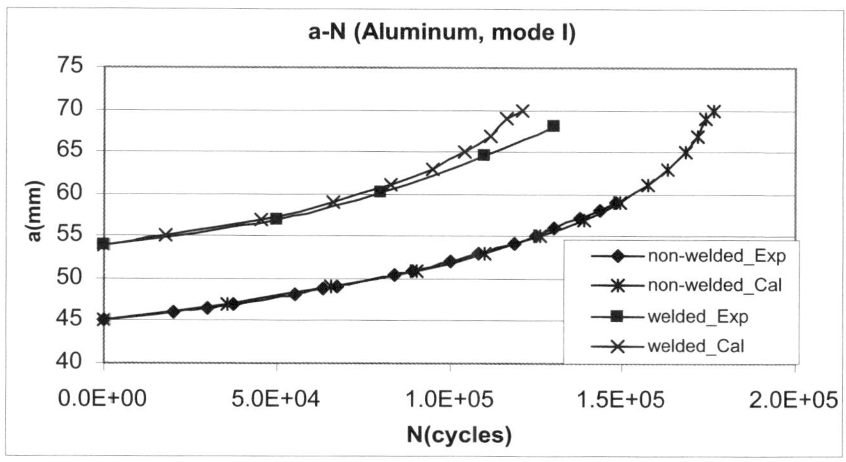

**Figure 11.21.** *Results of cycle number for aluminum specimens*

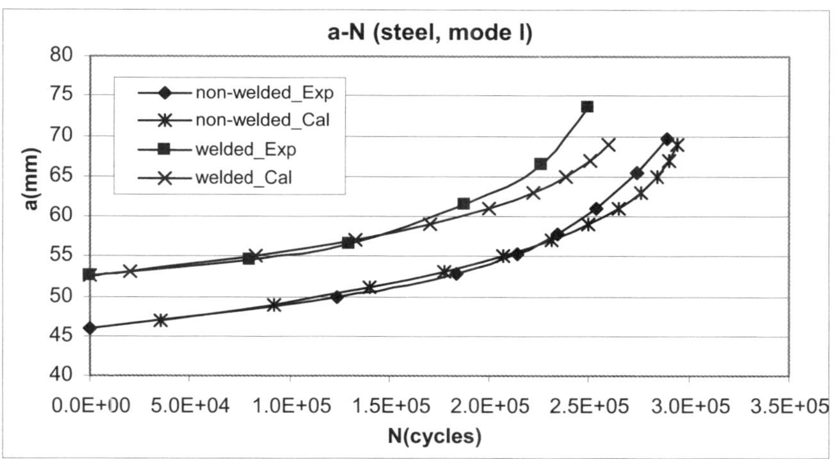

**Figure 11.22.** *Results of cycle number for steel specimen*

### 11.5.5. *Discussion and conclusions*

The crack propagation and the crack extension in CTS specimens under mixed-mode loading conditions are studied experimentally and numerically. Two types of specimens and two ductile materials are used. The results indicate the following.

The crack growth rate is related to the loading angle. For the same initial $K_{eq}$, 60° and 30° loading angles need higher loading levels than 90° loading. Therefore, the closer to pure mode II, the faster the crack growth rate is.

When a crack is parallel to the fillet of the weld, for the same loading level, the compressive residual stress due to weld decreases the crack growth rate in the case of pure mode I loading, but this effect is weak as the crack grows far from the weld zone in the case of mixed-mode loading conditions.

The fillet weld has little influence on the crack growth direction.

According to the experimental results, a numerical model is proposed to evaluate the effect of loading mode and of the residual stress due to weld on the crack growth rate under cyclic loading. The equivalent stress intensity factor $K_{eq}$ is used in the model. The coefficient $C$ in Paris' law is modified when a crack is subjected to mixed-mode loading in welded specimens. This model is validated in the non-welded and welded steel specimens under mode I loading and mixed-mode loading. The numerical evaluations are in good agreement with the experimental data. Nevertheless, it is felt that as no convenient analytical solution exists, the approximate method provides a simple procedure to evaluate the crack growth rate, considering the effect of the loading mode and the residual stress due to weld. It should be pointed out that the present study is a first step that has to be completed by experimental measurements of the residual stress field and its redistribution.

## 11.6. Extension to full test

The final aim of the present study is the development of a method that allows better estimation of the fatigue life of welded structures (armored vehicles, ships, floating production storage off-loading, wrecking cranes, cars, etc.) when subjected to various loading modes. The complexity of these structures has lead to the adoption of a multi-scale approach, based on the use of finite elements codes associated with various levels of modeling, going from the global cartography of damaged zones to the local calculation with cracks inserted in the models. The aim is to develop an industrial procedure, avoiding successive re-meshing, that is an efficient and easy tool to apply. It is also to be open enough to provide tools to allow the engineer to assess crack initiation and propagation until failure. Coupled with an extension of the line spring method, multi-initiation of fatigue cracks in welds and through crack growth are considered in order to calculate the stress intensity factors for various loading modes and geometries. Furthermore, a set of tools has been developed to predict the crack extension and to take into account the influence of the loading history on fatigue crack growth, such as crack growth retardation effect, as a result of overloads (see Chapter 12). This approach is then applied to an overall aluminum welded structure experiment, which has been designed to allow several cracks to initiate and propagate. Local micro-geometries and residual stresses have been measured at weld toes, as needed for local stress calculations. Furthermore, a complete instrumentation of this welded structure enabled the precise determination of crack initiations and allowed one to follow crack propagations. The results are in good agreement with calculations and show that it is an industrial necessity to measure the local characteristics of welds and to control the quality for fatigue design. This work was part of a global study performed among a research partnership between three industrial partners and a research laboratory [Ref 35].

### 11.6.1. *Modeling methodology*

The failure of an element of a structure may be explained by the presence of a crack which is first initialized and then propagates under complex and random repeated loads. These loads due to the environmentally induced fatigue phenomenon that depend on the joint geometry, the material microstructure, and the nature and the size of the defects. The aim of the proposed methodology is to predict crack initiation and crack growth in industrials structures until failure. For these structures, what is particularly important to be aware of is that under in-service conditions, the load combinations and related stress situations may be very complex. These loadings, which can be static or dynamic, can be decomposed in combination of unitary loadings or modes if the response of the structure is linear elastic; but a crack represents a non-linearity and if we want to take into account the effect of the crack on the stiffness of the global structure, one has to insert them in a finite element model and make them propagate, which means re-meshing. Re-meshing is very time consuming, but can be easily used for industrial purpose only with 2-D models [Ref 30]. For part-through cracks, the problem is much more difficult, because modeling the crack tip requires tri-dimensional refined elements. Furthermore, as the crack propagates in complex geometries, under multi-axial loadings, analytical solutions can not be used. Considering the structure response being linear elastic, the use of the line spring method makes it possible to take into account the crack stiffness during growth without re-meshing [Refs 33, 34].

### 11.6.2. *Global calculation scheme*

The global scheme of the proposed method allows one to study the crack from its initiation until the failure occurs. Various tools using either damage determination or crack growth calculation have been combined to define a global tool called Vericrack.

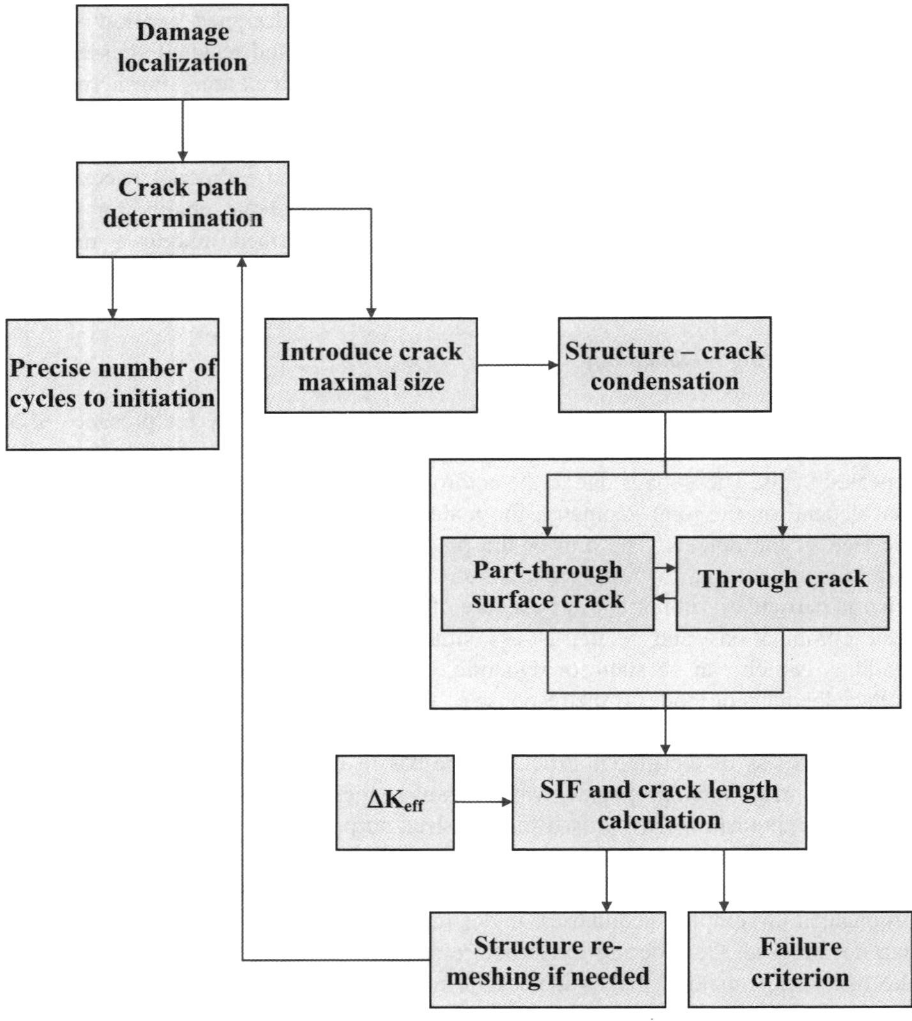



**Figure 11.23.** *Global calculation scheme of Vericrack*

The first step is to carefully carry out an accurate stress analysis. It is this stress field that governs the crack behavior in the material. From the knowledge of the stress field, we are able to predict where the most-damage zones located, and which path the crack will take. Then a more precise estimation of the crack initiation duration is made. A crack of the maximum length is inserted in the global mesh and all loadings are applied on the fully-open crack. Structure is then condensed on the crack. From the crack stiffness one can compute the displacements of the crack nodes for a part-through (Figure 11.24b) or through crack (Figure 11.24c) by determining the loads on the crack nodes necessary to close the crack until we obtain the desired crack shape.

So during the simulation, after a crack initiates for nodes that have accumulated enough damage (Figure 11.24a), the stress intensity factors (SIF) are calculated for the crack. Moreover, as the stiffness of the structure evolves when the crack grows (Figure 11.24b), stress field is updated and damage is cumulated for the nodes of the crack that have not already initiated. This allows one to take into account multi-initiation crack phenomenon for welded structures. All the different steps of the global scheme are introduced below.

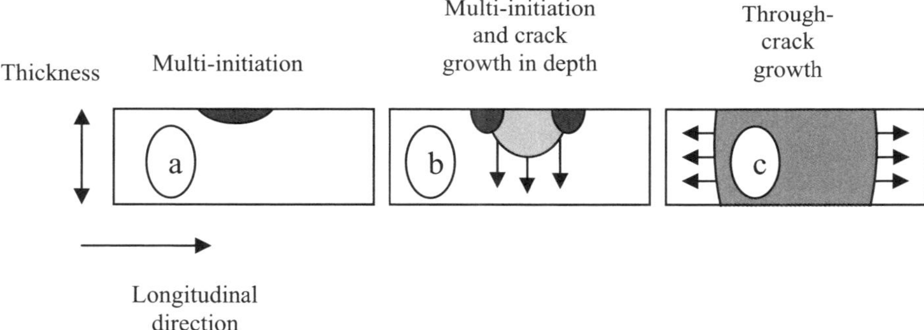


**Figure 11.24.** *Different crack front shapes*

### 11.6.3. *The crack box technique*

It is important to emphasis that SIF calculated based on the Tada coefficient is important for cracks propagating in the depth of a plate. When the crack front is straight in the depth of the plate, the Line Spring assumption can no longer be used as the crack now grows in the longitudinal direction as it is represented in Figure 11.24c.

So the idea is to use the displacements calculated at the neighbor nodes, as has been done for stresses, to determine the SIF for the through crack. These displacements are the applied on a tri-dimensional sub-model of the structure, including the crack tip which is meshed using fracture mechanics elements. In order not to carry out finite element calculations during Vericrack simulation, the SIF in the depth of the sub-model is previously calculated for unitary displacement at its border. Then, during Vericrack simulation, SIF are combined with the displacements. Figure 11.25 represents the sub-model with a unitary translation applied, all the other nodes of the border being clamped.

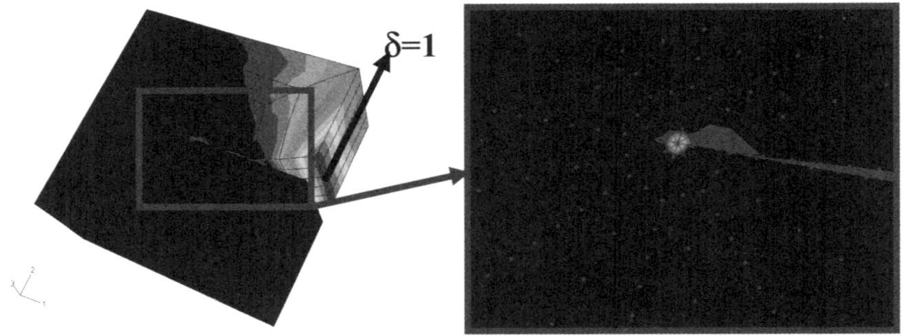


**Figure 11.25.** *Sub-model of the structure: 3-D crack box (left); zoom on crack tip (right)*

The accuracy of the method depends on where (which nodes) the displacements are imposed on the sub-model. It is particularly these nodes distance from the crack mouth that is important, see Figure 11.25. If we use the first line of neighboring nodes, errors on SIF are between 7 and 13%, depending of the crack length; but when we use the second line of neighbor node, errors on SIF are between 1 and 5%. Error can also depend on the sub-model refinement and on the type of elements chosen for the global model (linear or quadratic). In the future, the aim of this technique is to include in the sub-model geometry details as welds.

### 11.6.4. *Crack propagation rate*

The crack growth is controled by the Paris law, using the SIF amplitude calculated from Tada equations or CBT. It can be corrected using the stress intensity factor necessary to open the crack (Kop) to determine the effective range of stress intensity factor $\Delta K_{eff}$. The Kop can be fixed, determined using Elber's equation; see Chapter 12.

### 11.6.5. *Description of experiments carried out*

This approach is then applied to an overall aluminum welded structure experiment (Figure 11.26) [Ref 35]. A complete set of strain gauges has been used to precisely determine the stress field in the structure and the moment when a crack initiates. They have been placed at 10 mm from the welds. Dye penetrant has been used to detect crack initiation. At the back floor of the structure, the structure was bolted on the ground. A hydraulic jack has been placed on the sides of the structure to allow four welds to be loaded, so there are eight possible cracks at the weld toe. The stress ratio of the applied loading is 0.1 and the maximum loading is 15,000 N.

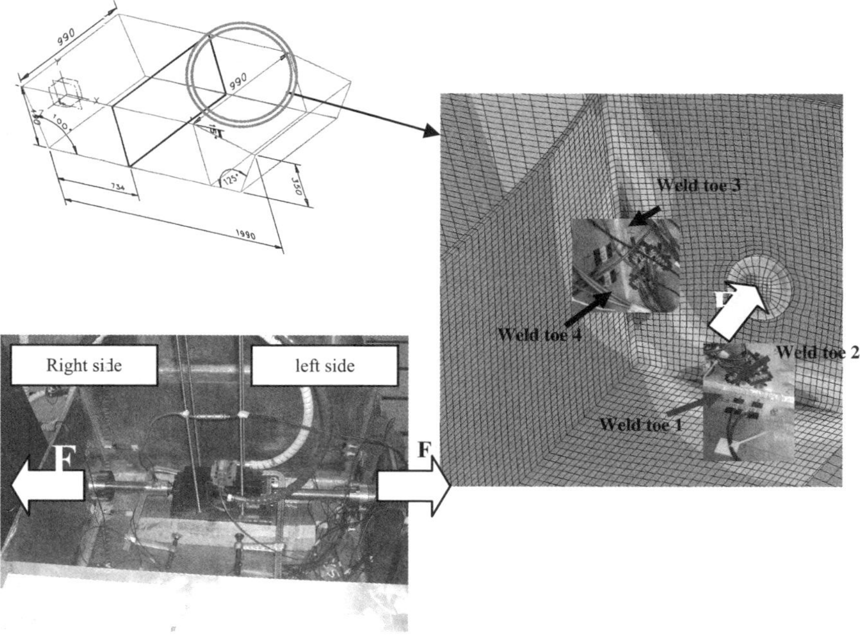


**Figure 11.26.** *Structure dimensions (mm), crack gauge position and picture of the experimental assembly*

### 11.6.6. *Results*

During the test three cracks have initiated: at weld toes 3 and 2 on the right side (Figure 11.27) and at weld toe 3 on the left side. The first crack to initiate was on weld toe 3 on the left side (between 50,000 and 280,000 cycles) and then the crack on the right side initiated (between 300,000 and 400,000 cycles). Later, a crack initiated on the weld toe 2 on the right side (between 400,000 and 550,000 cycles).

The crack was difficult to observe when its length was less than 10 cm as a multi-initiation phenomenon seems to be responsible for the crack initiation. The case of the left side was different as the crack had been observed at 280,000 cycles; its length was about 25 cm, whereas evolution of gauges seem to give 50,000 cycles; but the crack on the left side did not propagate in the direction perpendicular to the crack surface, contrary to other cracks, but in a parallel direction, maybe as a consequence of a material hardness in the heat-affected zone (HAZ). Therefore, all calculations on crack growth will be carried out on cracks on the right side.

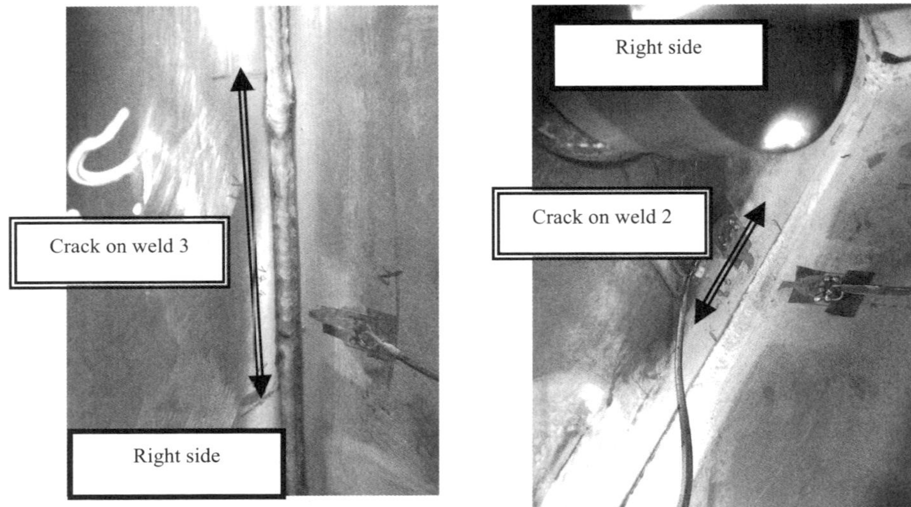


**Figure 11.27.** *Picture of crack on weld toes 2 and 3 (right side)*

### 11.6.7. *Weld toe geometry*

Considering the results given in the previous paragraph, the difference between the left side and the right side may be explained considering local parameters (residual stresses and weld toe radii). Residual stresses were measured at the surface and at 100 μm depth by X-ray diffraction. Measurements of the residual stresses at 100 μm depth were performed to avoid the very superficial residual stresses introduced by wire brushing after welding. Local geometry of weld toe was measured. Profile recordings were performed, with a laser and a CCD Camera, and were used to obtain the radius distribution all along the weld toe (Figure 11.28). Results for all weld toes are presented in Table 11.5. By taking into account the value of transverse stress at 10 mm from the weld toe, one can explain why only three cracks have initiated. More precise calculations are done below.

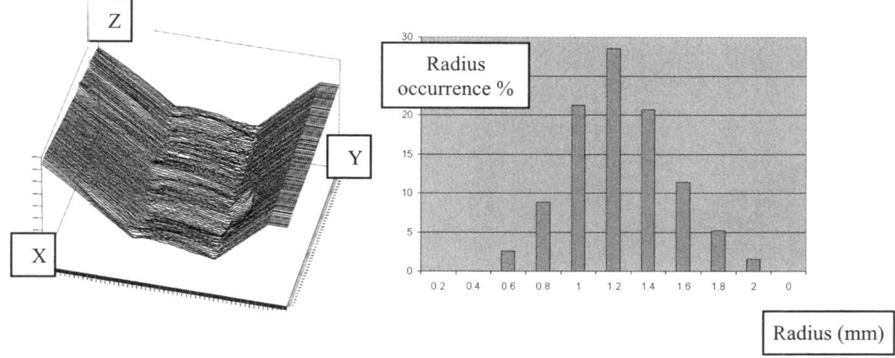



**Figure 11.28.** *Crack profile measurement (left) and radius distribution for weld toe 1 on the right side (right)*

| side | Weld toe | Transverse stress at 10 mm (MPa) (FEM) | mean radius (mm) |
|---|---|---|---|
| | Weld toe 1 | 82 | 1.31 |
| left side | weld toe 2 | 78 | 0.97 |
| | weld toe 3 | 95 | **0.59** |
| | weld toe 4 | 50 | 1.23 |
| | weld toe 1 | 82 | 1.26 |
| right side | weld toe 2 | 78 | **0.42** |
| | weld toe 3 | 95 | **1.05** |
| | weld toe 4 | 50 | 0.9 |

**Table 11.5.** *Weld toe mean radius and transverse stress at 10 mm from the weld toe*

### 11.6.8. *Numerical calculations*

#### 11.6.8.1. *Crack initiation*

To determine crack initiation, it is necessary to calculate the local stress at the weld toe, which depends on the residual stresses and on the local toe radius. The global model used is a shell modeling of the structure, with linear elements. 3-D calculations have been performed using either a 3-D sub-model or a 2-D sub-model under plane strain assumption, driven by the global model displacements (Figure 11.29). When the crack initiate in the middle of the weld, both the geometry and the loading mode support the assumption of plain strain condition. The 2-D modeling was used as it is simple enough to perform calculations for radii varying between 0.2

to 1.2 mm. Results are shown in Figure 11.30 and compared with experimental results. It shows that consideration of the residual stresses at 100 µm depth and the mean radius of the weld toe gives a good prediction.

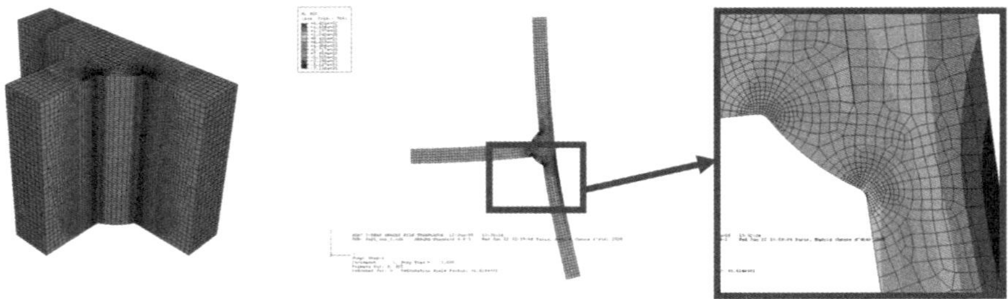

**Figure 11.29.** *Zoom of the weld toe in 3-D and in 2-D modeling*

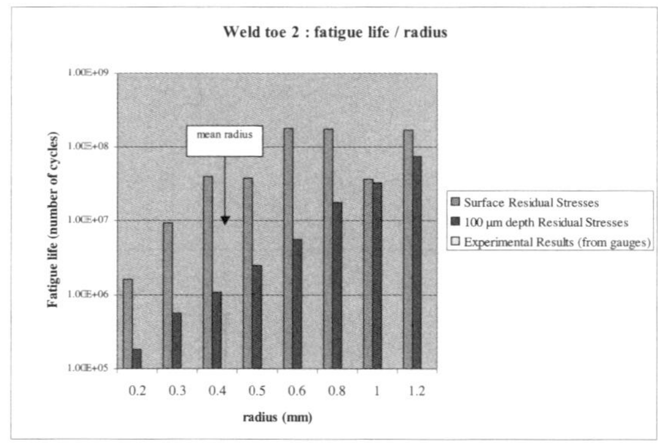

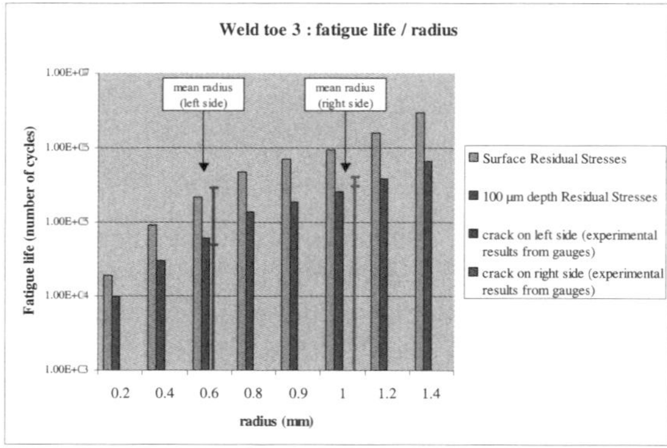

**Figure 11.30.** *Crack initiation determination for weld toe 2 and weld toe 3 for both sides*

### 11.6.8.2. *Crack growth*

Crack initiation and growth is then simulated in Vericrack for the right side of the structure. Material data for crack initiation are based on experiments where the local approach has been applied with mean radius. Material data for crack growth were obtained from crack growth experiments on MIG welded joints. Different assumptions have been made while carrying out a simulation with Vericrack.

A crack initiates when the damage accumulation is 1.

When the damage accumulation is 1 at a node, a crack of 1 mm depth is introduced.

When the stress in the ligament is greater than the yield stress at a node, the crack is supposed to have fully grown in the thickness of the plate.

When the crack front is almost straight, the crack can propagate by virtue of the crack box.

There is a competition between crack growth, thanks to fracture mechanics and thanks to damage accumulation at nodes.

The radius of the weld toe 2 is considered as the minimum values of radii instead of mean value.

Results are shown in Figure 11.31, where the crack shape during simulation is presented. Simulations are compared with experiments for crack 3 (weld toe 3) and crack 2 (weld toe 2).

With the assumptions made, the agreement seems to be good, but many sensitivity tests have to be carried out.

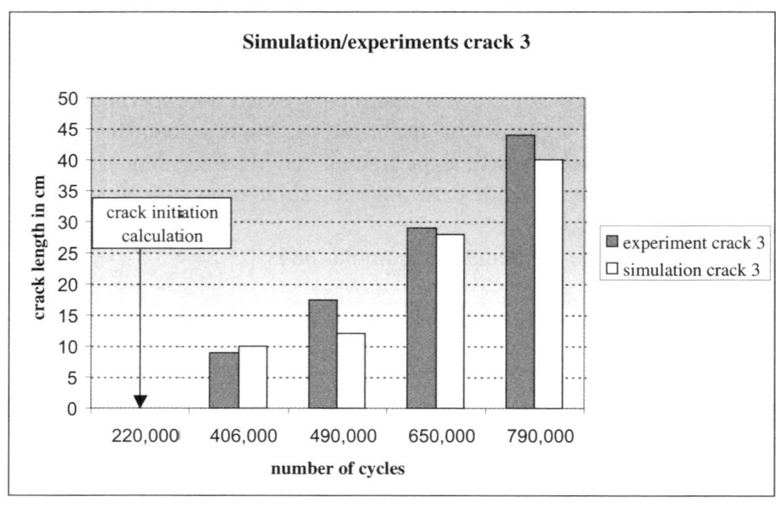


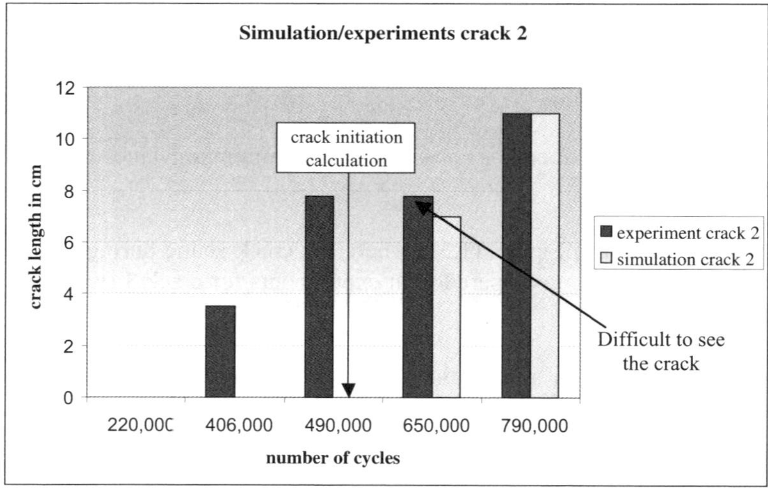


**Figure 11.31.** *Crack growth simulation with Vericrack for weld toe 2 and 3 on the right side of the structure*

In this work [Ref 35], a global methodology has been introduced to allow crack initiation and propagation without re-meshing. All calculations are linear elastic in order to decompose multi-axial loadings into unitary loadings (for static response) or modes (for dynamic response) that are combined in the Vericrack code. This program can determine crack initiation in welds and make part-through and through cracks grow in a 3-D shell structure. This method has been applied on a welded aluminum structure which has a geometry and a load level that will make cracks appear. Results of crack initiation duration are in good agreement with experiments if we consider the local geometry of the weld toes (radius) and the residual stress at 100 µm depth. Crack growth is then simulated and the results are close to experimental results. For the simulation, different assumptions have to be carefully studied as the crack growth is very sensitive to the involved parameters and failure criteria. Typical failure criteria are the ligament failure criterion and the straight front criterion.

## 11.7. References


1   B.E. Amstutz, M.A. Sutton, D.S. Dawicke and J.C. Newman Jr, "An experimental study of COD for mode I/II stable crack growth in thin 2024-T3 aluminum specimen" *ASTM STP 1256 on Fracture Mechanics* 26, 1995, pp 256–71

2   B.E. Amstutz, M.A. Sutton, D.S. Dawicke and M.L. Boone, "Effects of mixed mode I/II loading and grain orientation on crack initiation and stable tearing in 2024-T3 aluminum" *ASTM STP 1296 on Fatigue and Fracture Mechanics*, 27, 1997, pp 105–26

3   K. Tohgo and H. Ishii, "Elastic-plastic fracture toughness test under mixed mode I-II loading" *Engineering Fracture Mechanics*, 41, 1992, pp 529–40

4   S. Aoki, K. Kishimoto, T. Yoshida, M. Sakata and A. Richard, "Elastic-plastic fracture behavior of an aluminum alloy under mixed-mode loading" *J. Mech. Phys. Solids*, 38, 1990, pp 195–213

5   X.B. Zhang, S. Ma, N. Recho and J. Li, "Bifurcation and propagation of a mixed-mode crack in a ductile material" *Engineering Fracture Mechanics*, reference: EFM-0865-05-KHS (to appear)

6   P.C. Paris and F.A. Erdogan, "A critical analysis of crack propagation laws", transactions of the American Society of Mechanical Engineers, Series D, 85(4), 1963, pp 528–34

7   A.G. Forman *et al.*, *Journal of Basic Engineering*, 89, 1968, pp 459–64

8   F. Erdogan and M. Ratwani, "Fatigue and fracture of cylindrical shells containing a circumferential crack" *International Journal of Fracture Mechanics*, vol. 6, issue 4, 1970, pp 379–392

9   F. Erdogan and G.C. Sih, "On the crack extension in plates under plane loading and transverse shear", transaction of the ASME, *J. Basic Eng.*, 85, 1963, pp 519–27

10  G.C. Sih, "Strain energy density factor applied to mixed-mode crack problem" *Int. J. Fractures*, 10, 1974, pp 305–21

11  K. Palaniswamy and W.G. Knauss, "On the problem of crack extension in brittle solids under general loading" in S Memat-Nasser (ed), *Mechanics Today*, 4, 1978, pp 87–148

12  M.C. Sutton, X.F. Deng, S. Ma, J.C. Newman Jr and M. James, "Development and application of a crack tip opening displacement-based mixed-mode fracture criterion" *Int. J. Struct. Solids*, 37, 2000, pp 3591–618

13  K. Tanaka, "Fatigue crack propagation from a crack inclined to the cyclic tensile axis" *Engineering Fracture Mechanics*, 6, 1974, pp 493–507

14  X.M. Yan, D. Du and Z.L. Zhang, "Mixed-mode fatigue crack growth prediction in bi-axially stretched sheets", *Engng. Fracture Mech.*, 43, 1992, pp 471–75

15  D.F. Socie, C.T. Hua and D.W. Worthem, "Mixed-mode small crack growth" *Fatigue Fracture Engng. Mater. Structures*, 10, 1987, pp 1–16

16  T. Hoshide and D.F. Socie, "Mechanics of mixed-mode small fatigue crack growth" *Engineering Fracture Mechanics*, 26, 1987, pp 841–50

17  J. Tong, J.R. Yates and M.W. Brown , "The formation and propagation of mode I branch cracks in mixed- mode fatigue failure" *Engineering Fracture Mechanics*, 56(2), 1997, pp 213–31

18  K. Masubuchi, "Models of stresses and deformations due to welding – a review" *Journal of Metals*, 1981, pp 19–23

19  J. Lu et al., "Study of residual welding stress using the step-by-step hole drilling and X-ray diffraction method" *Welding in the World*, 33(2), 1994, pp 118–28

20  A. Parker, "Linear elastic fracture mechanics and fatigue crack growth – residual stress effects" in E. Kula and V. Weiss Saga (eds), *Residual Stress and Stress Relation*, Army Materials Research Conference, Plenum Press, New York, 1982

21  H.A. Richard and K. Benitz, "A loading device for the creation of mixed-mode in fracture mechanics" *Int. J. Fractures*, 22, 1983, R55

22  O.O. Green, *Trans. American Society for Steel Treating*, 18, 1930, p 369

23  K. Masubuchi and D.C. Martin, "Investigation of residual stresses by use of hydrogen cracking" *Welding Journal*, 40(12), 1961, pp 553–63

24  H.P. Lieurade, "Fatigue in welded constructions" *Welding in the World*, 26(7/8), 1988, pp 158-187

25  Y.-B. Lee, C.-S. Chung, Y.-K. Park et al., "Effects of redistribution residual stress on the fatigue behaviour of SS330 weldment" *International Journal of Fatigue*, 20(8), 1998, pp 565–73

26  B. Cotterell and J.R. Rice, "Slightly curved or kinked cracks" *Int. J. Fractures*, 16, 1980, pp 155–69

27  J. Li J, X.B. Zhang and N. Recho, "*J-M²* based criteria for bifurcation assessment of a crack in elastic-plastic materials under mixed-mode I-II loading" *Engineering Fracture Mechanics*, Vol 71 (3), February 2004, pp 329–43

28  S. Ma, X.B. Zhang, N. Recho and J. Li, "The mixed-mode investigation of the fatigue crack in CTS metallic specimen" *Int. Journal of Fatigue*, 28(12), December 2006, pp 1780–1790

29  D. Lebaillif, S. Ma, M. Huther, H.P. Lieurade, E. Petitpas and N. Recho, "Fatigue crack propagation and path assessment in industrial structures", *International Conference on Fatigue Crack Paths/FCP 2003*, 18–20 September 2003, Parma, Italy

30  D. Lebaillif and N. Recho, "Crack box technique associated to brittle and ductile crack propagation and bifurcation criteria", *11$^{th}$ International Conference on Fracture (ICF 11)*, March 2005, Turin, Ital

31  NAG, The Numerical Algorithms Group Ltd, Oxford, 2001, www.nag.co.uk

32  ABAQUS Software Version 6.4.1. www.hks.com

33  J.R. Rice and N. Levy, "The part-through surface crack in an elastic plate", *Journal of Applied Mechanics*, March 1972, pp185–94

34  D.M. Parks, "The inelastic line-spring: estimates of elastic-plastic fracture mechanics parameters for surface-cracked plates and shells" *Journal of Pressure Vessel Technology*, 103, August 1981, pp 246–54

35  D. Lebaillif, I. Huther, M. Serror and N. Recho, "Fatigue crack initiation and propagation: a complete industrial process compared with experiments on industrial welded structure", *International Conference on Fatigue Design*, 1$^{th}$ ed., 16-18 November 2005, Cetim, Senlis, France

Chapter 12

# The Effect of Overloads on the Fatigue Life

## 12.1. Introduction and objectives

In Chapter 5 the shortcomings of the Miners linear summation rule as a criterion for fatigue fracture were discussed. One particular phenomenon that makes the damage accumulation dependent on the sequence of the stress time series is the crack growth retardation effect due to an overload. The delay phenomenon of the crack growth in fatigue following an overload, although discovered by Schijve [Ref 1] in 1962, remains only partially understood. In the technical literature, this delay was analyzed as a consequence of various physical mechanisms, such as crack tip blunting [Ref 2], crack tip strain hardening [Refs 3 and 4], branching [Refs 5 and 6], crack closure (induced by plasticity [Ref 7], oxidation [Refs 8 and 9] or roughness [Refs 10 and 11]), residual compressive stresses ahead of the crack tip [Refs 12 and 13], etc. Each of these mechanisms has its own importance. According to the loading conditions, material mechanical properties and environment can be considered as secondary mechanisms during the overload application. According to Fleck [Ref 14], the delay calculated by considering crack tip blunting is smaller than that measured in experimental tests, and can therefore be ignored. The same observation was made for crack tip stain hardening according to [Refs 15 and 16]. Also, according to Suresh [Ref 17], the crack branching remains a secondary mechanism and, finally, the crack closure induced by oxidation, which generally takes place on crack lips, cannot be considered as a principal mechanism according to [Ref 18]. That is the reason why we consider two principal mechanisms, namely the crack closure (induced by plasticity and roughness) and the residual compressive stresses ahead of the crack tip, as predominant factors of the problem.

The concept of $\Delta K_{eff}$, proposed by Elber in 1970 [Ref 19], has been generally accepted by the scientific community in order to describe the overload effect. This concept consists of introducing a stress intensity factor range into the Paris law, namely:

$$\frac{da}{dN} = C.\left(\Delta K_{eff}\right)^m \tag{12.1}$$

with $\Delta K_{eff} = K_{max}\text{-}K_R$ $\qquad$ (12.2)

where $C$ and $m$ are the Paris law constants and $K_R$ ($R$ for $R$estriction) is a critical value of the stress intensity factor from which $\Delta K$ contributes to the crack growth. In the technical literature, two different arguments are used to explain this restriction (as shown in papers [Refs 20, 21, 22, 23]). Historically, over 20 years, the Elber model, dealing with the crack closure concept, was generally accepted. It was only in the 1990s that certain researchers, basing themselves on the earlier work of Schijve [Ref 1] and Marci [Ref 24], reintroduced other phenomena, such as the residual compressive stresses ahead of the crack tip in the restriction of the stress intensity factor range. Consequently, the $\Delta K_{eff}$ concept remains correct [Refs 7 and 19]; nevertheless, the capacity of the crack closure concept seemed to be insufficient to explain the delay phenomena [Refs 25 and 26].

After the above introduction, and in order to better model the delay phenomena with $\Delta K_{eff}$, we propose, in this chapter, a new approach based on taking into account the residual compressive stresses. This approach allows us to determine the crack growth dependence following an overload. First, the crack growth retardation is modeled until a minimum level is reached for the growth rate. Then a polynomial law is suggested in order to model in order to model the increase of the crack growth rate until the total restoration of the initial rate before overload.

It is well known that during the fatigue service under constant amplitude loading, an overload involves a delay of the crack growth rate, measured by $N_D$ (Figure 12.1(a)), which increases with the overload amplitude [Refs 26, 27, 28 and 29]. To model this phenomenon it is crucial to understand the mechanism of the crack growth evolution during the period ($N_D$).

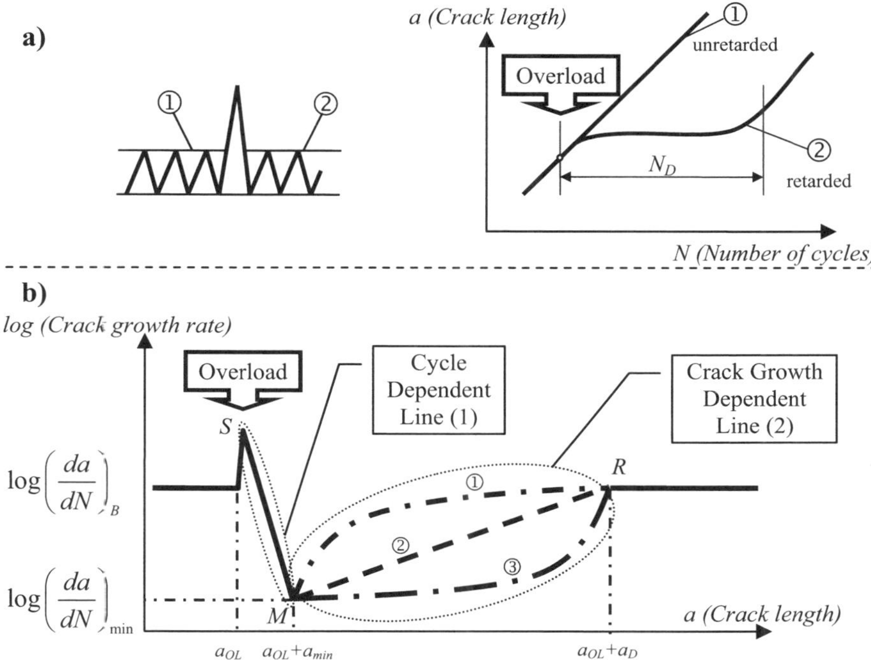


**Figure 12.1.** *a) Effect of an overload application, b) two separate phases during the delay phenomenon (case of constant ΔK)*

Before analyzing the phenomenological aspects of the present model, let us reconsider the delay aspect, which, as we noted in Figure 12.1(b), can be divided into two separate phases:

– The first one is due to the loading cycle dependence effect (cycle dependent) during the overload. For this part, one has to calculate the minimum crack growth rate, $(da/dN)_{min}$, and the minimum crack length, $a_{min}$.

– The second part is due to the crack growth dependence effect (crack growth dependent). For this part, one has to determine of the crack length affected by the delay, $a_D$, and describe the path of the total restoration of the crack growth rate (before overload).

The principal difficulty is related to $(da/dN)_{min}$ calculation.

The calculation of this minimum crack growth rate depends on the calculation of $K_R$ (equation (12.2)) which may have several definitions, as noted in Figure 12.2 (according to [Ref 30]):

– The first definition corresponds to the opening stress intensity factor KOP (OP for OPening) [Ref 19]. Although this term has been employed for more than 30 years, its definition remains imprecise. We could define it as being the limit between the linear part and the curved part of the compliance curve [Ref 31] (see Figure 12.2b).

– The second definition used is also based on the restrictive action of the crack lips. Donald [Ref 32] shows that the crack tip action affecting the crack growth rate must take place below opening load, i.e. after the first contact between the crack lips. Donald interpreted that as follows: the unloading of the cracked specimen, exposed to the crack closure phenomenon, is similar to the process of unloading with a "natural wedge" between the crack surface, but which appears after the KOP level. Donald and Paris introduced a stress intensity factor due to the wedge effect Kw (w for wedge).

– The third definition used was introduced by Lang and Marci [Refs 30, 33 and 34], in order to distinguish the principal mechanism of the crack growth propagation. This parameter, named KPR (PR for PRopagation), was established following the development of a test methodology, named CPLM (crack propagation load measurement) [Ref 30], which enables the delimitation of the effective part of the propagation. Lang and Marci explain that KPR is identified as being the expression of the compressive residual stresses ahead of the crack tip.

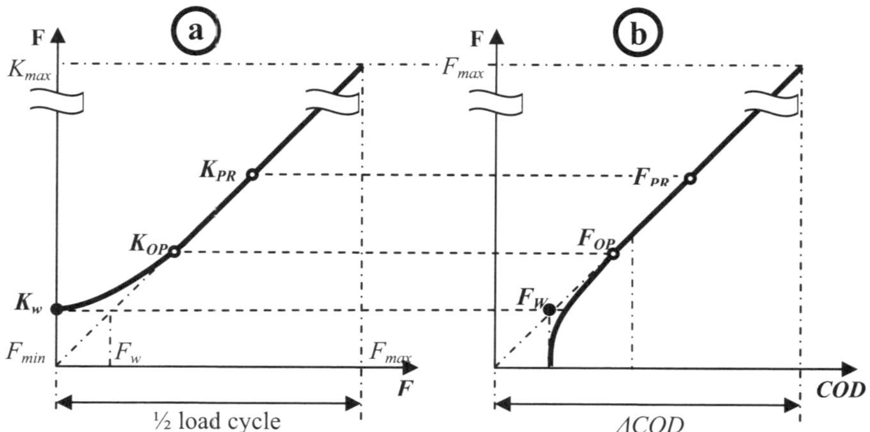


**Figure 12.2.** *a) Crack closure influence on the stress intensity factor, K, b) relation between COD and the applied loading F for a specimen (compliance curve)*

Even if the closure concept remains widely used in load interaction modeling [Refs 35 and 36], the physical aspects are not well understood and remain somewhat illusive, [Refs 1, 30 and 37]. One example is the observation of the lack of fatigue crack surface interference at the crack tip, even when compressive loads are applied (Bowles [Ref 38]). Moreover, several authors [Refs 31, 39 and 40] consider that the stress intensity factor due to the natural wedge $K_w$ remains very small.

That is why we have decided to take into account the residual stresses as being responsible only for the retardation effect. We have developed a numerical methodology enabling us to determine the restriction of the effective stress intensity factor range $\Delta K_{eff}$, which we name $K_{RCS}$ (RCS for Residual Compressive Stress) [Ref 53]. Moreover, we also developed a methodology, based on experimental results, enabling us to better model the crack growth dependence [Ref 60].

## 12.2. Residual stress opening approach at the crack tip following an overload during fatigue

Before performing the $K_{RCS}$ calculation, we will deal with the determination of $a_{min}$, which is the second parameter to be defined in this section. For the $a_{min}$ calculation, several authors [Refs 41 and 42] found that this distance is equivalent to one-quarter of the overload monotonic plastic zone size (following Irwin's calculation [Ref 43]). This is confirmed by comparison with experimental results [Refs 30, 41 and 44]. For the material and the specimen used in [Ref 30], the experimental method of CPLM [Ref 30] enables us to obtain the $K_{PR}$ value as function of the unloading cycle following the overload:

$$K_{PR,OL} = (0.322 + 0.57.U_R + 0.23.U_R^2 - 0.145.U_R^3).K_{max,OL} \left[-0,7 < U_R < 1\right] (12.3)$$

where $U_R$ is the load ratio following an overload, which can take three values:

$K_{ul}/K_{max}$, if there is no crack closure effect (where $K_{ul}$ is the unloading cycle following the overload),

$K_w/K_{max}$, if there is a crack closure effect,

$\sigma_p/\sigma_y$ when the cycle is in full compression ($\sigma_y$ being the yield stress in compression).

However, the second value of $U_R$, taking the natural wedge effect into account, is very small [Refs 30, 33 and 34], and we can ignore it and replace $U_R$ by $R_U$, which is expressed as follows:

$$R_U = \frac{\sigma_p}{\sigma_y} \text{ under compression load and } R_U = \frac{K_{ul}}{K_{max}} \text{ under tensile load} \quad (12.4)$$

Equation (12.3) is represented in Figure 12.3.

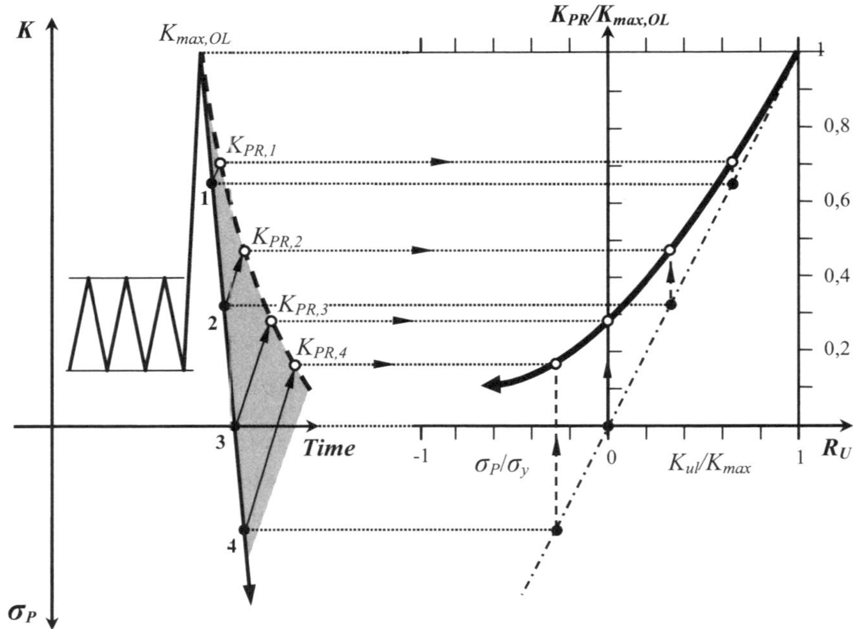


**Figure 12.3.** *Evolution of $K_{PR}$ for an overload (equation (12.3))*

The modeling of $K_{RCS}$ is performed by means of finite element analysis [Ref 53]. As an extension to the models of Newman [Ref 35] or McClung [Ref 45], which consider the crack closure induced by plasticity only, we will consider the residual compressive stresses. In order to avoid problems due to crack closure, the crack surfaces are considered as free of "mechanical restrictions". The present methodology, based on the decoupling between the reversals of the overload, can be described following three steps that are represented in Figure 12.4. In this work, a bilinear behavior law is used to represent the stress-strain relationship, namely:

$$\varepsilon = \frac{\sigma}{E} + \left(\sigma - \sigma_y\right)\left(\frac{E - E_T}{E_T E}\right) \tag{12.5}$$

where $E_T$ is the Young's tangent modulus and $\sigma_y$ is the yield stress.

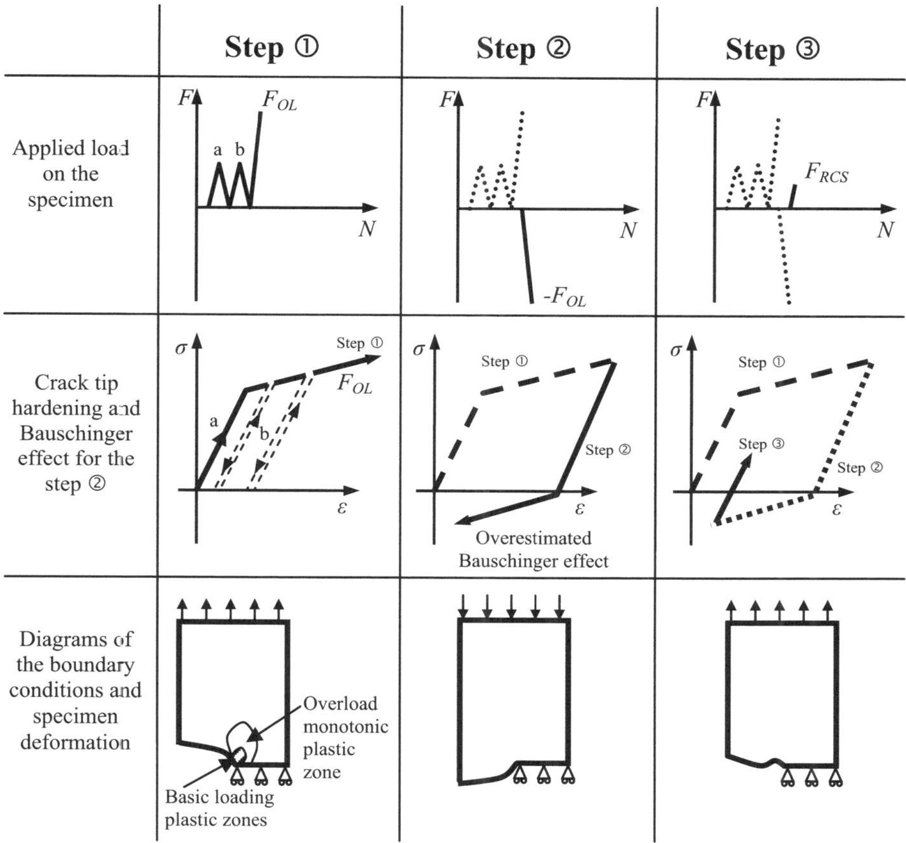



**Figure 12.4.** *Decoupling method for obtaining the stress intensity factor variation due to the residual compressive stresses, $K_{RCS}$: 1) creation of the plastic zones and application of the overload 2) unloading after overload by application of a reversal compressive loading, 3) determination of the loading level corresponding to the residual compressive stresses ahead of the crack tip which removes the crack lips interpenetration*

**Step ①** consists of creating the monotonic and cyclic plastic zones for basic loading, and the monotonic plastic zone for the overload. We make the assumption that a hardening occurs in the material at the crack tip, introduced by the stress cycling which tends to increase the compressive residual stresses ahead of the crack tip. By comparison of the evolutions of the plastic zones radius, created by the basic loading, we found that two cycles are sufficient to obtain a stable plastic zone size at the crack tip (this is also showed by [Refs 36 and 46]). This hardening behavior is represented in Figure 12.4 by the letters a and b in step ① (a for the first cycle and b for the second cycle).

**Step** ② introduces a decoupling of the reversals overload by a jump of the loading level, as shown in Figure 12.4. This jump consists of taking the same displacements, strains, and stresses fields of step ① and applying a compressive reversal loading ($-F_{OL}$).

**Step** ③ is to carry out a jump, as in the previous step, by using the same displacements, strains, and stresses fields as in step ②. Thus, the load making it possible to remove the crack surface interpenetration, created during step ②, can be determined (see Figure 12.4, step 3). This load corresponds to the contact compressive forces existing between the crack lips created by the residual stresses. Thus, $K_{RCS}$ is obtained in a purely geometrical way. It is evident that this geometrical way does not consider the wedge effect.

## 12.3. Numerical modeling

Finite element analysis has been carried out with the Abaqus code [Ref 47] in order to derive the effective stress intensity factor range, namely the residual stress crack opening phenomenon ($K_{op}$ equivalent to $K_R$ in equation (12.2)).

### 12.3.1. *Modeling aspects*

In recent years, the crack opening modeling has become the main subject of investigation for many researchers. In particular, works from Ellyin and Wu [Ref 48], Pommier [Ref 49], and Solanki *et al.* [Ref 50] have been released. Indeed, computational means have been greatly improved in order to accurately determine the stress gradient at the crack tip in elastic-plastic materials. However, many assumptions have to be made to model the complexity of fatigue crack growth. So various modelings have been developed in the literature and have covered the following issues:

– Plane stress or plane-strain conditions, depending on the geometry and on the $K_{op}$ measurements location.

– Two-dimensional or three-dimensional finite element modeling; the latter is much more time consuming, but enables one to consider the plastic zone size at the surface and in the depth of specimens.

– Specimen geometry: compact tension (CT) and others.

– Element type and element size at the crack tip. It is important that the number of elements in the plastic zone is sufficient.

– Crack surface contact modeling.

– Crack advance scheme. Generally, the methods are based on the following principle: releasing nodes, assuming a growth of one element size, and then applying cyclic loadings. But we can de-bond the crack surfaces at the maximum load after one or two cycles. The main difficulty is to define a sequence of crack advances and loadings that refers to the crack growth phenomenon; but it is worth noting that this decoupling has no physical considerations, and no stress or strain criteria drive the crack tip advance in this model.

– The $K_{op}$ crack opening stress intensity factor (SIF) assessment. Either displacement or stress criteria at the crack tip can be considered.

– The model is based on assumptions and choices described in the following sections (see Ref [53]).

### 12.3.2. *Finite element modeling choices*

Refined meshes are needed at the crack tip to model the crack growth by FEM in order to precisely calculate the stresses and strains in the monotonic and in the reverse plastic zones. Furthermore, the loading cycles have to be finely discretized to determine $K_{op}$. Calculations are non-linear, on the one hand by geometrical aspects as the contact between the crack surfaces is considered, and on the other hand by material aspects, i.e. the material response is elastic-plastic. Only two-dimensional calculations have been performed on a common CT specimen in which plane-strain conditions are fulfilled.

*Element type:* eight noded biquadratic quadrilateral elements have been chosen, with reduced integration. Plane-strain locking is avoided due to hybrid elements from Abaqus, which enable one to calculate the hydrostatic stresses only at the center of the element and so prevent numerical singularities when the Poisson's ratio tends to 0.5 (especially at the crack tip).

*Material behavior law:* cyclic plasticity occurs at the crack tip. The model is based on the assumption that this behavior governs the crack opening level. To better represent the material, we use the Chaboche constitutive law which combines a non-linear kinematic hardening and an isotropic hardening.

*Element size:* size of the elements is determined, thanks to the Irwin's plastic radius $r_p$:

$$r_p = \frac{1}{6\pi}\left(\frac{K_{max}}{S_y}\right)^2 \tag{12.6}$$

where $S_y$ is the initial yield strength. Solanki *et al.* [Ref 50] suggest at least three to four elements in the reverse plastic zone and at least 10 elements in the monotonic plastic zone. In our model the element size is about 10 microns, which corresponds to about five elements in the reverse plastic zone after an overload (see Figure 12.5).

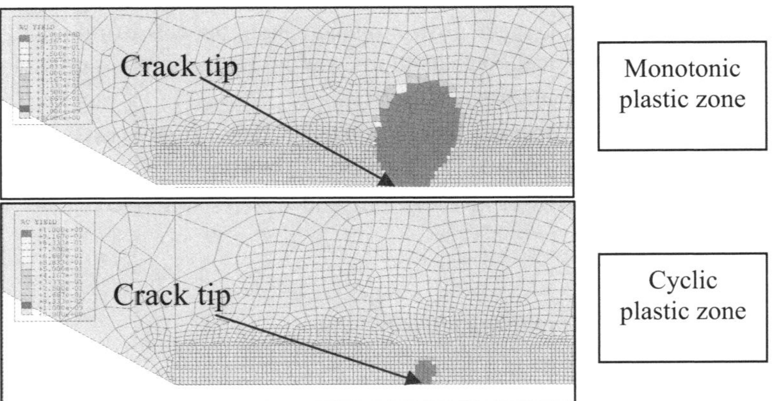


**Figure 12.5.** *Monotonic and cyclic plastic zones*

*Crack tip de-bonding:* the Abaqus procedure includes the definition of two initially bonded surfaces between which the crack will propagate. The first one is based on the faces of the crack elements and the other one is created with a rigid surface, as only half of the specimen is modeled. De-bonding is linked with a reduction of the transmitted force between both surfaces at specified nodes during computations. Convergence difficulties can occur for large-strain problems, typically at the crack tip and a specific de-bonding curve is required. Furthermore, to model fatigue crack growth, elastic-plastic stress (associated to cyclic behavior law) have to be stabilized before de-bonding the nodes.

*Crack opening criteria*

At least two types of criteria can be defined:

– Displacement criterion: based on the displacement of crack surface nodes near to the crack tip. Pommier [Ref 49] determines the displacement variation ratio of the second node behind the crack tip (CTOD) with the maximum displacement of this node. This criterion can be expressed as: $(U_{op} - U_{min})/(U_{max} - U_{min})=1.5\%$, where $U_{max}$ and $U_{min}$ are respectively the maximum and minimum displacements of that node, and $U_{op}$ is the calculated crack opening parameter.

– Stress criterion: based on the transition between compressive and tensile stresses, either at the crack tip [Ref 48], behind the crack tip, or in the elements surrounding it. This criterion is defined on the assumption that the crack can grow only when the compressive stress field at the crack tip is released.

*Specimen used*

A CT specimen has been chosen for application (width: 50 mm, thickness: 10 mm). This specimen was used in Lang and Marci tests. Modeling, meshing, and boundary conditions are presented in Figure 12.6.

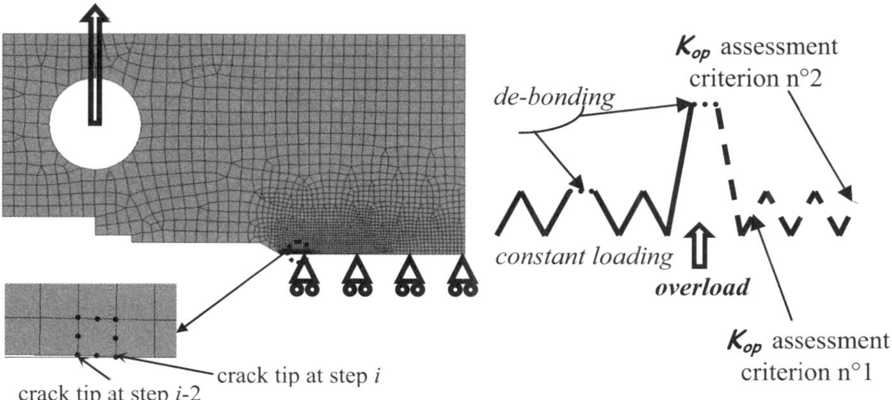


**Figure 12.6.** *Meshing of CT specimen – loading cycles and $K_{op}$ assessment location*

To prevent any short crack phenomenon or failure conditions, our calculations have been carried out with a 6-mm crack. The material used is an aluminum alloy Al7475-T7351 with the following characteristics: $S_y$=350 MPa, $E$=71,500 MPa and $v$=0.3. The constitutive behavior of the material is well described by using a Chaboche plasticity model (with kinematic hardening parameters: $C$=290 MPa, $C_A$=50,000 MPa, and isotropic cyclic hardening parameters: $Q$=0, $b$=15) [Ref 51].

*Loading sequence*

Considering a kinematic cyclic hardening behavior of the material, it has been checked that only two loading cycles are needed to stabilize the crack tip plastic zone. As a consequence, these two cycles are applied before each de-bonding in order to simulate a fatigue crack growth.

The following sequence has been considered for the simulation: 2 cycles at Constant loading ($K_{max}$ = 8.1 MPa.m$^{1/2}$, $K_{min}$ = 0), 1 element (2 nodes) de-bonding ($\leftrightarrow$), 2 cycles at constant loading, 1 **O**verload of 3x8.1 i.e. ($K_{max}$ = 24.3 MPa.m$^{1/2}$,

$K_{min} = 0$), 1 element de-bonding, a **Decrease** at loading to $K_{ul}$ and then $N$ cycles at constant loading. The sequence could be noted as $_2C_0^{8.1} \leftrightarrow {_2C_0^{8.1}O_0^{24.3}} \leftrightarrow D_{Kul}{_N}C_{Kul}^{8.1}$ (see Figure 12.6 with $K_{ul}=0$).

*Comparison of displacement and stress criteria*

The stress criterion has been split into two criteria (see Figure 12.6).

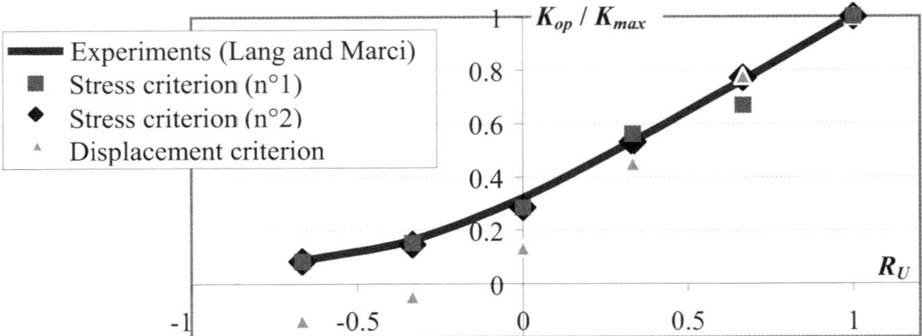


**Figure 12.7** *Comparison between FEM calculations and experiments by Lang and Marci*

– Stress criterion n°1: crack opening assessment at the cycle following the overload.

– Stress criterion n°2: crack opening assessment after two cycles following the overload. This criterion has been considered with reference to Lang and Marci tests. In this criterion a crack opening assessment procedure predicts when the plastic zone is stable.

Results obtained with these different criteria are reported in Figure 12.7. Stress criterion n°2 best fits the experimental results, whereas the displacement criterion seems to underestimate $K_{op}$.

## 12.4. Proposed deterministic approach to fatigue crack growth following an overload

This approach considers that the total fatigue life duration $N_{Total}$ is determined as being the sum of four parts: the number of cycles to crack initiation $N_i$ corresponding to an initial crack length $a_0$, the number of cycles $N_1$ from $a_0$ until $a_{pic}$ corresponding to the overload, the number of cycles $N_2$ from $a_{pic}$ to $a_D$ which is the crack length corresponding to the total restoration of the crack growth rate, and $N_3$, the number of cycles from $a_D$ to the critical crack size until failure.

*Fatigue life initiation: $N_i$*

The number of cycles to crack initiation $N_i$ may be estimated on the basis of a local strain-life approach as described in Chapter 9, equation (9.2).

*The first period: $N_1$*

From the fatigue life initiation until the overload, a crack propagation law such as Paris law is used:

$$\frac{da}{dN} = C\,(\Delta K)^m$$

where $C$ and $m$ are two material constants ($C$ will be considered as function of $m$ in our study). Thus, the first period, $N_1$, is given by integration as follows:

$$N_1 = \frac{1}{C} \int_{a_0}^{a_{pic}} \frac{da}{(\Delta K)^m} \tag{12.7}$$

where $a_0$ is the initial crack length and $a_{pic}$ is the crack where the overload is applied.

*The second period: $N_2$*

The second period is the most important part to be determined when overload occurs.

Before analyzing more in detail our modeling choices concerning $N_2$, we will discuss the retardation effect, which can be divided, as seen in the introduction, into two parts (Figure 12.1), i.e. cycle dependence and crack growth dependence.

*Cycle dependence* – the two characteristic aspects to be determined are the minimum crack growth rate $(da/dN)_{min}$ and the crack length $a_{min}$ at this minimum. As seen before, $a_{min}$ is taken as equal to one-quarter of the plastic zone. On the other hand, the calculation of the lowest crack growth rate remains more difficult. Indeed, this calculation consists of understanding the retardation effect. The concept of $\Delta K_{eff}$, established by Elber [Ref 19] in the 1970s, is generally admitted; but the calculation of $\Delta K_{eff}$ is still discussed because various physical phenomena can explain it.

Two aspects are related to Elber's concept (equations (12.1) and (12.2)):
– the first one consists of the total restoration of the initial crack growth rate (before overload), and
– the second one concerns the physical description of the retardation effect.

Concerning the first aspect, almost of authors agree with this calculation on the basis of the use of $\Delta K_{eff}$ associated with a classical crack propagation law such as Paris law. Nevertheless, the second aspect is still very contentious. Despite the great use of the Elber's concept, the principle of the crack closure was not admitted by Schijve [Ref 1] and Marci [Ref 24]. In fact, they explain the retardation effect by studying the residual stress field near the crack tip following the overload. Many experimental observations [Refs 38 and 52] confirmed Marci's studies.

The numerical method described before allows us to take into account the residual compressive stress field ahead of the crack tip when overload occurs [Ref 53]. The $(da/dN)_{min}$ is then determined for welded joints. The equation of the Line (1) is given by (Figure 12.1):

$$\frac{da}{dN} = \left(\frac{da}{dN}\right)_0 + \xi.a \tag{12.8}$$

where, $\xi$ is the slope calculated with the points $S$ and $M$ as follows:

$$\xi = \frac{y_M - y_S}{x_M - x_S} = \frac{(da/dN)_{min} - (da/dN)_0}{a_{min} - 0} \tag{12-9}$$

where $(da/dN)_0$ is the crack growth rate before the overload.

*Crack growth dependence* – in this part, we have to determine the crack length $a_D$ and the physical behavior representing the total restoration of the crack growth rate (before overload).

Many studies [Refs 54, 55 and 56] recommend setting $a_D$ equal to the monotonic plastic zone due to the overload, although the size of the plastic zone can vary according to authors. For example, [Refs 54 and 56] propose to take the Irwin meaning of the plastic zone size and [Ref 55] the Rice meaning. Nevertheless, recent papers [Refs 54 and 36] recommend a crack length equal to twice the overload monotonic plastic zone. This is considered in our work and checked for tests with constant $\Delta P$ [Ref 18] and with constant $\Delta K$ [Ref 41]. Concerning the modeling of the total restoration of the crack growth rate, between $a_{min}$ and $a_D$, the technical literature distinguishes three manners (Figure 12.1): convex manner ① [Refs 30 and 57], straight manner ② [Ref 41] and concave manner ③ [Refs 58 and 59].

According to the work by [Ref 60], we found that the convex manner corresponds to the tests of [Refs 41 and 44] as confirmed also by Ranganathan [Ref 57]. Thus, the Line (2), which describes the increase of $da/dN$, can be given by (see Figure 12.1):

$$\frac{da}{dN} = \alpha + \beta\,a + \gamma.\sqrt{a} + \left(\frac{da}{dN}\right)_{S'} .\left[\exp\left\{\eta.\left(a - a_{pic}\right)\right\} - 1\right] \tag{12.10}$$

where $\alpha$, $\beta$, and $\gamma$ are obtained from the points $M$ and $R$, and $\eta$ is the slope between the points $S$ and $S'$ for the test at $\Delta P$ constant, as we can see in Figure 12.8.

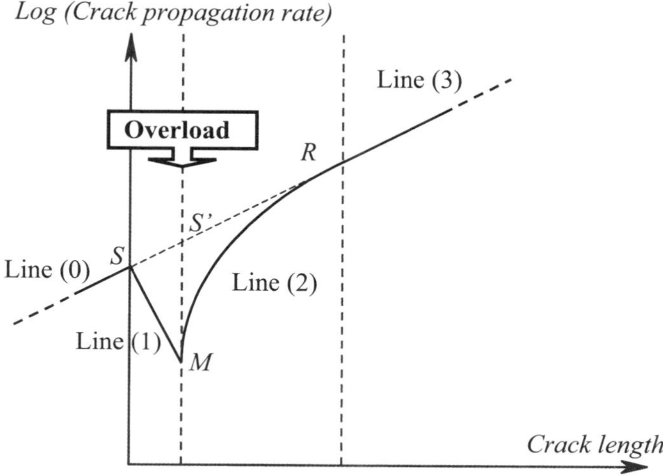


**Figure 12.8.** *Model of fatigue crack growth following an overload at $\Delta P$ constant*

Finally, the crack growth dependence is described by the reduction of the effective stress intensity factor variation $\Delta K_{eff,D}$ from the point $M$ to the point $R$ deduced from the Line (2). We then have:

$$N_2 = \frac{1}{C'} \int_{a_{pic}}^{a_{pic}+a_{\min}} \frac{da}{\left(\Delta K_{eff}\right)^{m'}} + \frac{1}{C'} \int_{a_{pic}+a_{\min}}^{a_{pic}+a_D} \frac{da}{\left(\Delta K_{eff,D}\right)^{m'}} \tag{12.11}$$

where $C'$ and $m'$ are the two material parameters which could be different from $C$ and $m$. Generally, $m'$ is given as equal to $m$.

*The third period: $N_3$*

This period is not influenced by the overload. It can be determined as follows:

$$N_3 = \frac{1}{C} \int_{a_{pic}+a_D}^{a_c} \frac{da}{\left(\Delta K\right)^m} \tag{12.12}$$

where $a_c$ is the critical crack size to failure.

The total fatigue life duration is given then by:

$$N_{Total} = N_i + N_1 + N_2 + N_3 \tag{12.13}$$

This equation is the basis of the limit state function in the following reliability analysis.


## 12.5. Reliability modeling including the effect of an overload

In this section we will develop an approach that accounts for the overload effect on the reliability level. A random variable model based on fracture mechanics is employed. The aim is to study the probability of failure as a function of time, particularly after the occurrence of the overload. The basic random variables are the initial crack size, the material parameters of the crack propagation law, the radius of the weld toe, and the size of the plastic zone. Four phases of the fatigue process are studied as outlined in the previous section. The first one concerns the crack initiation, the second phase deals with crack propagation period from the initial crack until the application of the overload, the third phase takes into account the retardation effect. The last phase corresponds to the restoration of the initial crack growth rate. The results of the reliability analysis show that a gain in reliability is obtained when the overload occurs. This analysis is based on the deterministic model of the retardation effect following overloads, as described in the foregoing section. Material parameters related to the plastic zone are now considered as random variables. The following work is based on [Ref 64] and introduces two new aspects:

– the first one is the taking into account of the fatigue life initiation as a random variable; see Chapters 7 and 9;

– the second one is the development of a new reliability analysis associated to fatigue overloads.

From equations (12.1), (12.7), (12-11), (12.12), and (12.13), the total fatigue life is written as follows:

$$N_{Total} = N_i + \frac{1}{C} \int_{a_0}^{a_{pic}} \frac{da}{(\Delta K)^m} + \frac{1}{C'} \int_{a_{pic}}^{a_{pic}+a_{min}} \frac{da}{\left(\Delta K_{eff}\right)^m} + $$
$$\frac{1}{C'} \int_{a_{pic}+a_{min}}^{a_{pic}+a_D} \frac{da}{\left(\Delta K_{eff,D}\right)^m} + \frac{1}{C} \int_{a_{pic}+a_D}^{a_c} \frac{da}{(\Delta K)^m} \tag{12.14}$$

The probability to failure of a given crack length at a certain number of cycles $N$ can be modeled by the following limit state function:

$$g(N) = N_{Total} - N .$$ 
(12.15)

Using the definitions presented in equations (12.14) and (12.15), the failure criterion is written as a limit state function $g(z_1,z_2...z_n)$ for reliability analysis (see Chapter 7, equation (7.3)):

$$g'(z_1,z_2...z_n) = \left( \begin{array}{c} N_i + \dfrac{1}{C} \displaystyle\int_{a_0}^{a_{pic}} \dfrac{da}{(\Delta K)^m} + \dfrac{1}{C'} \displaystyle\int_{a_{pic}}^{a_{pic}+a_{min}} \dfrac{da}{(\Delta K_{eff})^m} + \\[2em] \dfrac{1}{C'} \displaystyle\int_{a_{pic}+a_{min}}^{a_{pic}+a_D} \dfrac{da}{(\Delta K_{eff,D})^m} + \dfrac{1}{C} \displaystyle\int_{a_{pic}+a_D}^{a_c} \dfrac{da}{(\Delta K)^m} \end{array} \right) - N$$
(12.16)

The failure occurs when $g(z_1,z_2...z_n) < 0$.

The random variables $(z_1,z_2...z_n)$ in this equation are:

– $a_0$, the initial crack length,

– $m$, the Paris law exponent,

– $\rho$, the radius of the weld toe (used in the calculation of $N_i$), and

– $a_D$, the crack length that takes into account the plastic zone as a random variable (for this case we multiply this quantity by a bias factor $\gamma_{a_D}$ which takes into account the variable random characteristic).

The use of the first-order reliability method (FORM) associated to the limit state function equation (12.16) enables us to determine the reliability index $\beta$ for a given structural detail; see section 7.2.5. This $\beta$ index can be compared with a target reliability index in order to consider the structure as safe.


## 12.6. Application of the reliability model to a fillet welded joint

The approach developed in section 12.5 was applied to the cruciform welded joint tested by Lassen [Ref 61] and studied by Grous, Recho, Lassen and Lieurade [Ref 62]. The database 1 for this test series was presented in Chapters 3, 6, and 7. The joint is shown in Figure 12.9 and we shall confine the analysis to constant amplitude loading with $\Delta\sigma_\infty$=150 Mpa. Material data are given in Table 12.1.

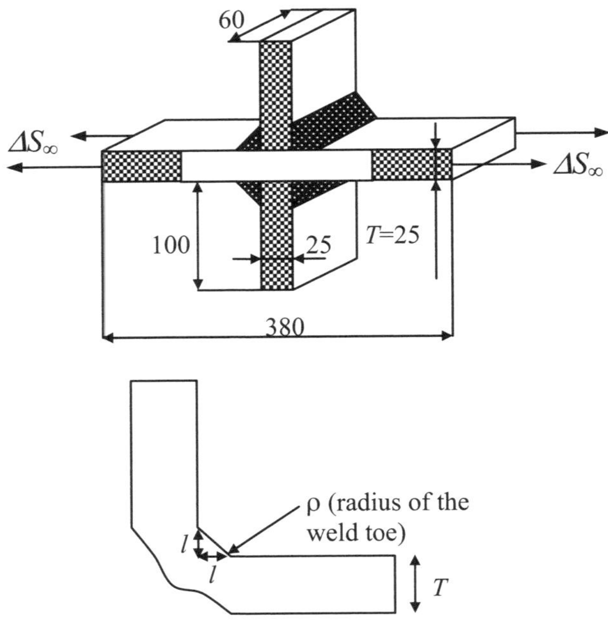



**Figure 12.9.** *Considered welded joint specimen (Lassen [Ref 61])*

In order to evaluate $N_i$ we used the "joint of the G4 type" that is referenced by Yang and Lawrence [Ref 63] as the basis. It consists of a cruciform welded joint with full penetration subjected to an axial loading. The elastic stress concentration factor $K_t$ is described by the following equation:

$$K_t = \beta.\left[1 + \alpha.\left(\frac{T}{\rho}\right)^\lambda\right] \tag{12.17}$$

where $\alpha$, $\beta$ and $\lambda$ are obtained as follows:

$$\alpha = 0.2\left(2 - \frac{l}{T}\right)^{0,5}$$
$$\beta = 1 \tag{12.18}$$
$$\lambda = 0.5$$

where $\rho$, $l$, and T, used in calculations (equations (12.17) and (12.18)) are defined in Table 12.2.

| Mechanical properties | |
| --- | --- |
| Yield strength (in MPa) | 416 |
| Tensile strength (in MPa) | 501 |
| Elongation | 26 |
| Chemical composition % | |

| C | Si | Mn | P | S | Cu | Ni | Cr | Mo | Nb |
| --- | --- | --- | --- | --- | --- | --- | --- | --- | --- |
| 0.08 | 0.15 | 1.40 | 0.006 | 0.002 | 0.01 | 0.02 | 0.02 | 0.01 | 0.008 |

**Table 12.1.** *Mechanical and chemical properties of CLC steel [Ref 61]*

| Toe length $l$ | Plate thickness of the welded joint $t$ | Notch-root radius of the welded joint $\rho$ |
| --- | --- | --- |
| 8 mm | 25 mm | 1.75 mm |

**Table 12.2.** *Geometrical parameters of a welded joint*

The determination of $K_t$ enables us to define the fatigue notch factor $K_f$ from equation (9.5); see the modeling of the fatigue crack initiation period in Chapter 9. In the present reliability analysis we have used the concept of the $K_f$ instead of the direct use of $K_t$ as developed in Chapter 9.

From the tests, we can define the deterministic and random parameters used in our reliability model. The deterministic parameters are as follows:

– the geometrical parameters of plate and weld ($l$, T) (only the weld toe radius is random),

– the nominal stress variation ($\Delta S_\infty$),

– the mechanical properties of the material (Young modulus, yield strength, tensile strength, the Ramberg-Osgood material, etc.).

The random parameters are:

– Parameter $m$ (exponent of the Paris law); we used a Gaussian rule with an average of 3 and a standard deviation of 0.03.

– From considered tests, Grous *et al.* [Ref 62] found a relationship between the parameters $C$ and $m$ of the Paris law:

$$C = \frac{6.069.10^{-8}}{24.64^{m}} \; ; units : \Delta K (daN.mm^{3/2}), \frac{da}{dN}(mm/cycle) \qquad (12.19)$$

$$\text{or } C = \frac{6.069.10^{-8}}{7.791^{m}} \; ; units : \Delta K (MPa\sqrt{m}), \frac{da}{dN}(mm/cycle)$$

This relationship means that $C$ is also considered as a random variable.

– Parameter $a_0$: Grous *et al.* take a statistical representation of the initial crack length $a_0$ as a Weibull law with two parameters with an average of $7.267.10^{-3}$ mm and a standard deviation of $3.112.10^{-3}$.

– Parameter $\rho$: a statistical representation of the radius [Ref 61] of the weld toe $\rho$ as a log-normal with an average of 1.75 mm and a standard deviation of 0.75.

– Parameter $\gamma_{a_D}$ : in order to take the plastic zone as a random variable, we use a statistical representation of the bias $\gamma_{a_D}$ as a Gaussian rule with an average of 1 and a standard deviation of 0.2.

The impact of an overload applied to the cruciform joint is illustrated in Figure 12.10 where the cumulative probability of failure is given as a function of number of cycles. The case is principally equal to the sketch in Figure 7.15 in Chapter 7 where the reliability function was drawn. The curve in Figure 12.10 is based on the information that:

– the structure survives the overload;

– the overload decreases the growth rate of the crack.

Based on the first information the failure curve makes a step down, whereas the second information makes the slope of the curve less than the original slope before the overload.

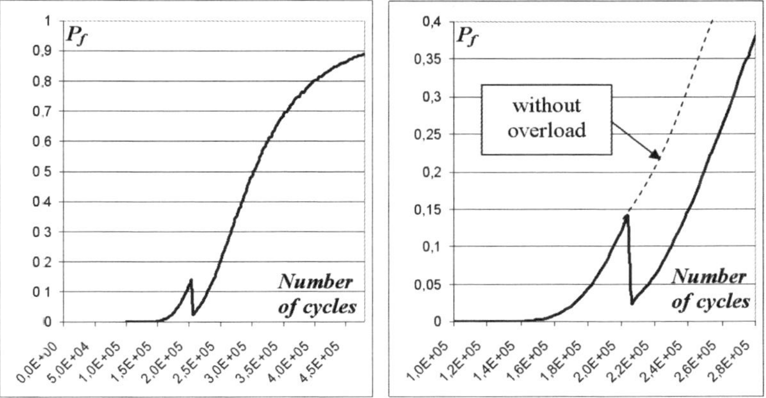


**Figure 12.10.** *Results of the reliability analysis of the fatigue crack growth, after an overload*

This reliability analysis is performed with the FORM technique; see section 7.2.5, Chapter 7. The following deterministic variables are considered:

– the stress ratio: $R$=0,

– the overload ratio: $R_{pic}$=3,

– the crack length where the overload occurs: $a_{pic}$=1.2 mm,

– the critical crack length: $a_c$=0.8 × $t$=20 mm, and

– the nominal stress: $\Delta S_\infty$=150 MPa.

It is clearly demonstrated that the retardation effect of an overload reduces the probability of failure in the time domain after the overload has occurred.

## 12.7. References

1   J. Schijve, "Fatigue crack propagation in light alloy sheet material and structure" in *Advances in Aeronautical Sciences*, vol. 3, Oxford: Pergamon, 1962, pp 387–408

2   R.G. Christensen, "Fatigue crack, fatigue damage and their detection" *Metal Fatigue*, New York: McGraw-Hill, 1959

3   R.E. Jones, "Fatigue crack growth retardation after single-cycle peak overload in Ti-6Al-4V titanium alloy" *Engng. Fract. Mech*, 5, 1973, pp 585–604

4   J.F. Knott and A.C. Pickard, "Effect of overloads on fatigue crack propagation – aluminum alloys" *Metal Science*, 11, 1977, pp 399–404

5   J. Schijve "Fatigue damage accumulation and incompatible crack front orientation" *Engng. Fract. Mech.*, 6, 1974, pp 245–52

6    S. Suresh, "Crack deflection: implications for the growth of long and shorts cracks" *Metall. Trans. A.*, 14A, 1983, pp 2375–385

7    W. Elber, "The significance of fatigue crack closure" in *Damage Tolerance in Aircraft Structures*, ASTM STP 486, 1971, pp 230–42

8    P.C. Paris, R.J. Bucci, E.T. Wessel, W.G. Clark and T.R. Mager, "Extensive study of low fatigue crack growth rates in A535 and A508 steels" in *Stress Analysis and Crack Growth*, ASTM STP 513, 1972, pp 141–76

9    D. Benoit, H.P. Lieurade, R. Namdar-Irani and R. Tixier, "Oxydation des surfaces de rupture par fatigue des aciers aux basses vitesses de fissuration" in *Mémoires et Etudes Scientifiques Revue de Métallurgie*, 1981, pp 569–83

10   K. Minakawa and A.J. McEvily, "On crack closure in near-threshold region" *Scr. Metall.*, 15, 1981, pp 633–36

11   S. Suresh, "Crack growth retardation due to micro-roughness: a mechanism for overloads effects in fatigue" *Scripta Metall*, 16 1982, pp 995–99

12   J. Schijve and D. Braoek, "The result of a test program based on a gust spectrum with variable amplitude loading" *Aircraft Engng.*, 34, 1962, pp 314–16

13   W. Sun and H. Sehitoglu, "Residual stress fields during fatigue crack growth" *Fatigue Fract. Engng. Mater. Struct.*, 15(2), 1992, pp 115–28

14   N.A. Fleck, "Influence of stress state on crack growth retardation" *Questions in Fatigue*, ASTM STP 924, vol. 1, 1988, pp 157–83

15   J. Schijve, "The effect of pre-strain on fatigue crack growth and crack closure" *Engng. Fract. Mech.*, 8, 1976, pp 575–81

16   L. Legris, M.H. El Haddad and T.H. Topper, "The effect of cold rolling on the fatigue properties of SAE 1010 steel" Material Experimentation and Design in Fatigue (Proceedings), Society of Environmental Engineers Conference, F. Sherratt and J.B. Sturgeon (eds), Mach 1981, pp 24–27

17   S. Suresh, "Micro-mechanisms of fatigue crack growth retardation following overloads" *Engng. Fract. Mech.*, 18, 1983, pp 577–93

18   S. Suresh, G.F. Zamiski and R.O. Ritchie, "Oxidation and crack closure. An explanation for near-threshold corrosion fatigue crack growth behaviour" *Metall. Trans. A.*, 13A, pp 1435–443

19   W. Elber, "Fatigue crack closure under cyclic tension" *Engng. Fract. Mech.*, 2, 1970, pp 37–45

20   A.K. Vasudevan and K. Sadananda, "Classification of fatigue crack growth behaviour" *Metall. Mater. Trans. A.*, 26A, 1995, pp 1221–234

21   F.O. Riemelmoser and R. Pippan, "Crack closure: a concept of fatigue crack growth under examination" *Fatigue Fract. Eng. Mater. Struct.*, 20, 1997, pp 1529–540

22   F.O. Riemelmoser and R. Pippan, "Discussion of error in the analysis of the wake dislocation problem" *Metall. Mater. Trans. A.*, 29A, 1998, pp 1357–359

23  K. Sadananda and A.K. Vasudevan, "Reply to Riemelmoser and Pippan" Metall. Mater. Trans. A., 29A, 1998, pp 1359–360

24  G. Marci, "Effect of the active plastic zone on fatigue crack growth rates" ASTM STP 667, 1979, pp 168–86

25  C. Robin, C. Chehimi, M. Louah and G. Pluvinage, "Influence of overloads on the subsequent crack growth of a fatigue crack in a E36 steel" *Proceedings of the 4th European Conference on Fracture*, 1982, pp 488–94

26  R.W. Hertzberg, C.H. Newton and R. Jaccard, "Crack closure: correlation and confusion" ASTM STP 982, 1988, pp 139–48

27  R.C. Rice and R.I. Stevens, "Overload effects on sub-critical crack growth in austernitic manganese steel" ASTM STP 536, 1973, pp 95–114

28  R.P. Wei, T.T. Shih and J.H. Fitzgerald, "Load interaction effects on fatigue crack growth in Ti-6Al-4V alloy" NASA Report CR-2239, 1973

29  L.G. Vargas and R.I. Stephens, "Sub-critical crack growth under intermittent overloading in cold-rolled steel" *3rd International Conference on Fracture*, Munich, Germany, 1973

30  M. Lang and G. Marci, "The influence of single and multiple overloads on fatigue crack propagation" *Fatigue Fract. Engng. Mat. Struct.*, 22, 1999, pp 257–71

31  M. Lang and X. Huang, "The influence of compressive loads on fatigue crack propagation" *Fatigue Fract. Engng. Mat. Struct.*, 2, 1998, pp 65–83

32  J.K. Donald, "Introducing the compliance ratio concept for determining effective stress intensity" *Int. Jnl. of Fatigue*, 19(1), 1997, pp 1–195

33  M. Lang, "Quantitative Analyse von Reihenfolgeeinflüssen auf Ermudungsrißfortschritt" A quantitative investigation of load interaction effects on fatigue crack propagation, PhD Thesis, University of Karlsruhe, Germany, 1996

34  M. Lang and G. Marci, "Reflecting on the mechanical driving force of fatigue crack propagation" *Fatigue Fract. Mech.*, 29, 1998, pp 474–95

35  J.C. Newman Jr, "A crack closure model for predicting fatigue crack growth under aircraft spectrum loading" in *Methods and Models for Predicting Fatigue Crack Growth Under Random Loading*, ASTM STP 748, 1981, pp 53–84

36  S. Pommier "Plane strain crack closure and cyclic hardening" *Engng. Fract. Mech.*, 69, 2002, pp 25–44

37  W. Yisheng and J. Schijve, "Fatigue crack closure measurements on 2024-T3 sheet specimens" *Fatigue Fract. Engng. Mater. Struct.*, 18, 1995, pp 917–21

38  C.Q. Bowles, "The role of environment, frequency and shape during fatigue crack growth in aluminum alloys" Doctoral Dissertation, Delft University, 1978

39  M. Lang, "Description of load interaction effects by the $K_{eff}$-concept" *Advances in Fatigue Crack Closure Measurement and Analysis*, ASTM STP 1343, 1999, pp 207–23

40  M. Lang and J.M. Larsen, "Fatigue crack propagation and load interaction effects in a titanium alloy", *Journal of Fatigue and Fracture* (30), ASTM STP 1360, 2000, pp 201–13

41  X. Decoopman, "Influence des conditions de chargement sur le retard à la propagation d'une fissure de fatigue après l'application d'une surcharge" PhD Thesis, University of Sciences and Technologies of Lille (France), 1999

42  M. Lang, "A model for fatigue crack growth, part I: phenomenology" *Fatigue Fract. Engng. Mat. Struct.*, 23, 2000, pp 587–601

43  G.R. Irwin, "Plastic zone near a crack and fatigue toughness" Mechanical and Metallurgical Behaviour of Sheet Material, Proceedings of the 7th Sagamore Ordinance, Materials Research Conference, Section IV, Syracuse University Research Institiute, 1960, pp 63–71

44  R Kumar, A Kumar and S Kumar, "Delay effects in fatigue crack propagation" *Int. Jnl. Pres. Ves. & Piping*, 67, 1996, pp 1–5

45  R.C. McClung, "Finite element modelling of fatigue crack growth" Theoretical Concepts and Numerical Analysis of Fatigue, Proc. Conf., University of Birmingham, UK, 1992, pp 153–71

46  S. Pommier and Ph. Bompard, "Bauschinger effect of alloys and plasticity-induced crack closure: a finite element analysis" *Fatigue Fract. Engng. Mat. Struct.*, 23, 2000, pp 129–39

47  ABAQUS Software Version 6.3.1. www.hks.com

48  F. Ellyin and J. Wu, "A numerical investigation on the effect of an overload on fatigue crack opening and closure behaviour" *Fatigue and Fracture of Engineering Materials and Structures*, vol 22, 1999, pp 835–47

49  S. Pommier, "Plane strain crack closure and cyclic hardening" *Engineering Fracture Mechanics*, vol 69, 2002, pp 25–44

50  K. Solanki, S.R. Daniewicz and J.C. Newman Jr, "Finite element modeling of plasticity-induced fatigue crack closure: an overview" *Engineering Fracture Mechanics*, vol. 71, 2004, pp 149–71

51  F. Labesse-Jied, B. Lebrun, E. Petitpas and J.-L. Robert, *Multi-axial Fatigue Assessment of Welded Structures by Local Approach, Biaxial/Multiaxial Fatigue and Fracture*, Andrea Carpinteri (ed), Elsevier Science Ltd, 2003, pp 43–62

52  P.C. Paris, Paper presented at the International Congress of Applied Mechanics, Delft, 1976

53  D. Lebaillif, P. Darcis, and N. Recho, "A new residual stress opening approach at the crack tip following an overload during fatigue" *11th International Conference on Fracture (ICF 11)* , March 2005, Turin

54  M. Lang, "A model for fatigue crack growth, part I: phenomenology" *Fatigue Fract. Engng. Mat. Struct.*, 23, 2000, pp 587–601

55 P.J. Cotterill and J.F. Knott, "Overload retardation of fatigue crack growth in a 9%Cr 1%Mo steel at elevated temperatures" *Fatigue and Fracture of Engineering Materials and Structures*, vol 19, No 2/3, 1996, pp 207–16

56 Y. Lu and K. Li, "A new model for fatigue crack growth after a single overload" *Engineering Fracture Mechanics*, vol. 46, No 5, 1993, pp 849–56

57 N. Ranganathan, M.C. Lafarie-Frenot and J. Petit, "Effect of overloads on the evolution of plastic zones and its significance" *8<sup>th</sup> Congress on Material Testing, Scientific Society of Mechanical Engineers*, Omikk-Technoinform, Budapest, 1982, pp 309–13

58 O.E. Wheeler, "Spectrum loading and crack growth" *Journal of Basic Engn.,* Trans ASME, vol 4, 1972, pp 181–86

59 J.D. Willenborg, R.M. Engle and H.A. Wood, "A crack growth retardation model using an effective stress concept" AFFDL-TM-FBR-71-1, USAF Flight Dynamics Lab, 1971

60 M. Filippi, Ph. Darcis and N. Recho, "Modeling of crack growth retardation due to plastic zone following an overload" *Symposium on Fatigue Testing and Analysis Under Variable Amplitude Loading*, Tours, 29–31 May 2002

61 T. Lassen, "The effect of the welding process on the fatigue crack growth" Agder College of Engineering, Grimstad, Norway Weld, Research Supplement 76-s, 1990

62 A. Grous, N. Recho, T. Lassen and H.P. Lieurade, "Caractéristiques mécaniques de fissuration et défaut initial dans les soudures d'angles en fonction du procédé de soudage" *Revue Mécanique Industrielle et Matériaux*, vol 51, no. 1, April 1998

63 J.Y. Yang and F.V. Lawrence, "Analytical and graphical aids for the fatigue design of weldments" *Fatigue Fract. Eng. Mater. Struct.*, 8, 1985, pp 223–41

64 Ph. Darcis and N. Recho, "Fatigue reliability analysis of overload effects in welded joints including crack initiation and plastic zone as random variables, *Fatigue Testing and Analysis Under Variable Amplitude Loading Conditions*, ASTM STP 1439, PC McKeighan and N Ranganathan (eds), 14 pages, American Society for Testing and Materials, West Conshocken, PA, 2003

# Appendix A

# Short Overview of the Foundations
# of Fracture Mechanics

## A1. Introduction

Fracture mechanics analysis is related to the appearance of a crack in an engineering material. In such a case, the following two questions are crucial:

– How can the deformation, stress and strain fields close to the geometrical singularity of the crack tip be determined?

– How can failure criteria be established?

Griffith introduced the concept of fracture mechanics in about 1920. The objective was to characterize the failure behavior of the material using quantifiable parameters within the concepts of engineering analysis, in particular the stress field, the crack size and the resistance to failure of the material. Westergaard carried out the first theoretical developments for the analysis of the displacements, strains and stresses fields in the vicinity of a crack about 1940. Irwin started the extension of the discipline in about 1960. Since this date the development of the fracture mechanics has extended to nonlinear problems both with regard to material behavior and geometrical changes. The problem of crack bifurcation in mixed loading modes has also developed vigorously over the last decade. More recently, fracture mechanics has been applied to composites and for the dimensioning of various complex structures. In the latter case, numerical approaches and the application of software play an important role.

The crack propagation is the creation of a surface discontinuity. It is the ultimate phase of a tensile test and sometimes the only response to a test of deformation. The solid mechanics tackles the problems of structural analysis where we speak about an element of volume whose dimensions are of 1 mm up to structures with 10 to $10^3$ mm of size. In this range, we deal with problem of the crack propagation.

When the damage mechanism is brittle fracture, the failure occurs by cleavage without warning, generally without plastic deformation, and the specimen can be assembled perfectly after failure. The failure is either intercrystalline, or intra crystalline.

Cleavage is the mechanism of deformation that brings into play the rupture of the atomic links in the dense plan in which the deformation must occur. It is a question of considering that each link is split up one after the other rather than all at the same time. Hence, a slit that increases is created; i.e. we have crack propagation (Figure A1a). This mechanism does not suppose perfect crystallization of material. In fact, cleavage occurs as well in perfectly crystallized materials as in partially crystallized materials like numbers polymers, or primarily amorphous materials like glass.

The failure mechanism can occur according to two types of cracking:
– brittle cracking: for the solids, or materials with very high strength, the working stresses are very high and a considerable potential energy is thus created. The presence of small cracks can then lead to a brittle fracture often without macroscopic plastic strains as a consequence of the very low ductility of material;
– successive cracking: a succession of mechanisms (fragile-ductile) which, under repeated stress, involves successive cracking, called fatigue failure or stable fracture.

The factors influencing the crack propagation behavior of materials are of two types: metallurgical and mechanical. Mechanical factors relate to the state of displacements, strains and stress, as well as the environmental conditions such as the temperature.

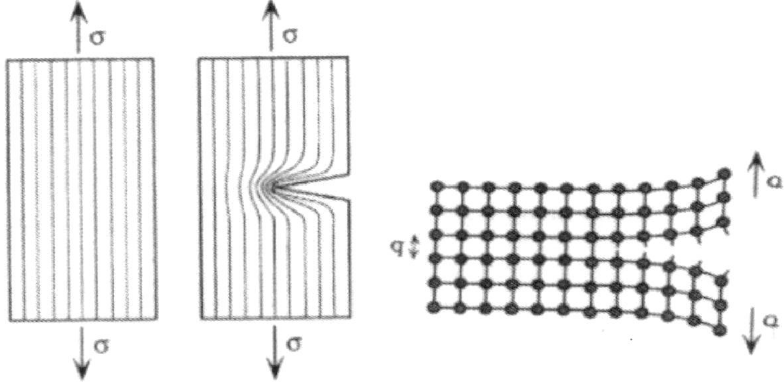

**Figure A1a.** *Propagation of a crack*     **Figure A1b.** *Force flow and stress concentration*

## A2. Elementary failure modes and stress situations

The examination of the fracture topography very often makes it possible to detect, after rupture, the failure mechanism and the type of crack propagation produced.

One generally notices:

– a smooth and silky zone corresponding to the crack propagation by fatigue; or

– a zone with crystalline or apparent grains, corresponding to brittle fracture. Any cracking can be brought back to the one of the three simple stress modes or their superposition. There are thus three elementary failure modes of cracking (Figure A2).

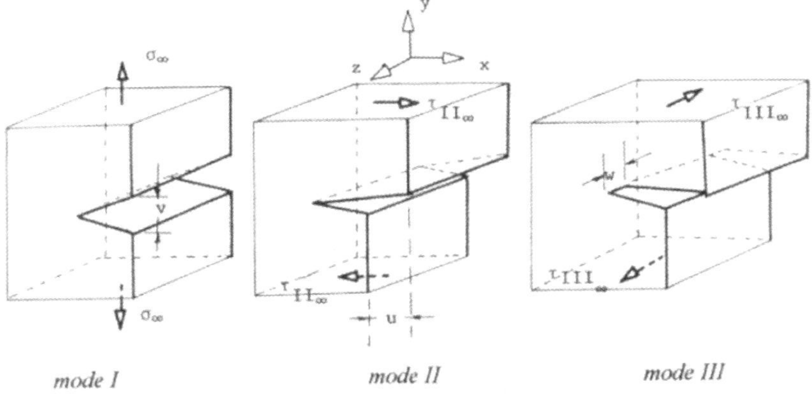



**Figure A2.** *Failure modes and related stress situations*

Mode I: mode of crack opening, where displacements with the lips of the crack are perpendicular to the direction of crack propagation.

Mode II: mode of in-plane shear, where displacements with the lips of the crack are parallel to the direction of crack propagation.

Mode III: mode of out-of-plan shear, where displacements with the lips of the crack are parallel to the crack front.

## A3. Foundations of fracture mechanics

In a homogenous material subjected to an axial stress, the effort is transmitted of one atom to the other while following tension fields that are parallel (Figure A1b).

In a material which has a notch, the tension fields must circumvent this notch, which leads to a concentration of these lines in the vicinity of the notch, therefore a stress concentration in this area, called a crack tip.

The fracture mechanics studies the interaction between geometrical discontinuity (crack) and the neighboring continuous medium, as well as the evolution of this discontinuity. From a mechanical point of view, we can distinguish schematically, in a cracked medium, three successive zones (Figure A3):

– Elaboration zone (zone 1): it is at the crack tip and in the wake left by the crack during its propagation. The study of this zone is very complex because of the significant stresses that can strongly damage the material. It is discontinuous within the meaning of solid mechanics. The classical theory of the fracture mechanics reduces this zone to a point for the plane problems and to a curve for the three-dimensional problems.

– Singular zone (zone 2): in which the displacements, strains and stresses fields are continuous and have a formulation independent of the remote geometry of the structure. It is shown that in this zone the components of stress field are infinite in the vicinity of crack tip ( $r \rightarrow 0$ ). More exactly, the singularity is in term of ( $1/\sqrt{r}$ ) in linear elastic medium, (Figure A3). For realistic materials having a yield stress, there is a radius $r_p$ around the crack tip which determines the shape of the plastic zone. According to the value of $r_p$, we will say that the fracture is brittle for small $r_p$ and that it is ductile for large $r_p$. This distinction on the basis of parameter $r_p$ is very significant because it indicates the validity of the used theory:

- linear fracture mechanics can be applied for brittle fracture, and

- nonlinear fracture mechanics in the case of a rupture associated with ductile fracture.

– External zone (zone 3): including the far fields being connected on the one hand, at the singular zone, and on the other hand to the boundary conditions. In this zone, the displacements, strains and stresses fields vary little and can be approximated by polynomials commonly used in the finite element method.

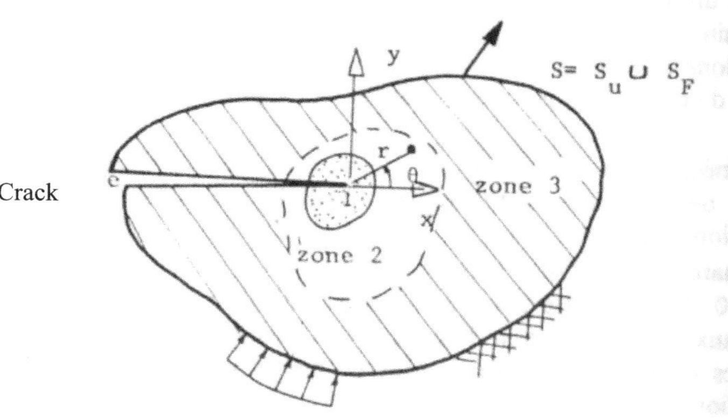


**Figure A3.** *Different zones near the crack tip*

Various methods of analysis make it possible to study the displacements, strains and stresses fields in the vicinity of a crack. In principle, there are two types of approaches:

– direct approaches that are founded on the use of the Airy stress function. These approaches solve plane problems and call upon the search for analytical functions. In particular the Williams expansion is representative of this type of approaches (see Chapter 10);

– energy approaches that are based on the energy analysis of the continuous medium containing a crack. The method considers the total energy balance by integrating the energy release rate due to a virtual increase in crack length.

## A4. Parameters characterizing the singular zone

In general, four parameters can be used to study the characteristics of the singular zone: the *stress intensity factor*, K, the *energy release rate*, G, the *J-integral* and the *crack opening displacement* COD. K and COD are the parameters of the direct local approach, whereas G and J are the parameters of the global energetic approach.

### A4.1. *The stress intensity factor (SIF), K*

In the singular zone, the stress field presents a singularity in $r^{-1/2}$ at the crack tip. The intensity of the singularity is characterized by the parameters called stress intensity factors (SIFs), noted $K_I$, $K_{II}$ and $K_{III}$ for each elementary mode respectively (see Figure A2). Using the theory of elasticity, IRWIN showed that in plane strain or plane stress, u, v and w displacements following the x, y and z axis and the stresses $\sigma_{ij}$ in this singular zone can be expressed according to the stress intensity factors as follows:

$$u = \frac{K_I}{2\mu}(\frac{r}{2\pi})^{\frac{1}{2}}\cos\frac{\theta}{2}(k - \cos\theta) + \frac{K_{II}}{2\mu}(\frac{r}{2\pi})^{\frac{1}{2}}\sin\frac{\theta}{2}(k + \cos\theta + 2)$$

$$v = \frac{K_I}{2\mu}(\frac{r}{2\pi})^{\frac{1}{2}}\sin\frac{\theta}{2}(k - \cos\theta) - \frac{K_{II}}{2\mu}(\frac{r}{2\pi})^{\frac{1}{2}}\cos\frac{\theta}{2}(k + \cos\theta - 2)$$

$$\sigma_x = \frac{K_I}{(2\pi r)^{\frac{1}{2}}}\cos\frac{\theta}{2}(1 - \sin\frac{\theta}{2}\sin\frac{3\theta}{2}) - \frac{K_{II}}{(2\pi r)^{\frac{1}{2}}}\sin\frac{\theta}{2}(2 + \cos\frac{\theta}{2}\cos\frac{3\theta}{2})$$

$$\tau_{xy} = \frac{K_I}{(2\pi r)^{\frac{1}{2}}}\cos\frac{\theta}{2}\sin\frac{\theta}{2}\cos\frac{3\theta}{2} + \frac{K_{II}}{(2\pi r)^{\frac{1}{2}}}\cos\frac{\theta}{2}(1 - \sin\frac{\theta}{2}\sin\frac{3\theta}{2}) \qquad (A\text{-}1)$$

$$\sigma_y = \frac{K_I}{(2\pi r)^{\frac{1}{2}}}\cos\frac{\theta}{2}(1 + \sin\frac{\theta}{2}\sin\frac{3\theta}{2}) + \frac{K_{II}}{(2\pi r)^{\frac{1}{2}}}\sin\frac{\theta}{2}\cos\frac{\theta}{2}\cos\frac{3\theta}{2}$$

with      $k = 3 - 4v$ in plane strain

        $k = \dfrac{3 - v}{1 + v}$ in plane stress

        $r, \theta$: the radius and the angle in polar coordinates, see Figure 3.
        $\mu$: the shear modulus

$$\mu = \frac{E}{2(1 + v)} \qquad (A\text{-}2)$$

$v$: Poisson's ratio and $E$: Young's modulus

In the case of out of plane loading, the only displacement component w is considered. The displacement and stress components are expressed as follows:

$$w = \frac{2K_{III}}{\mu}(\frac{r}{2\pi})^{\frac{1}{2}}\sin\frac{\theta}{2}$$

$$\tau_{xz} = -\frac{2K_{III}}{(2\pi r)^{\frac{1}{2}}}\sin\frac{\theta}{2} \qquad (A\text{-}3)$$

$$\tau_{yz} = -\frac{2K_{III}}{(2\pi r)^{\frac{1}{2}}}\cos\frac{\theta}{2}$$

Factors $K_I$, $K_{II}$ and $K_{III}$ are independent of r and $\theta$. They depend only on the distribution of the external loading for a given body and geometry of the crack. They are proportional to the discontinuity of the displacement of the lips of the crack. The following expressions constitute their definitions:

$$K_{I} = \lim_{r\to 0}(\frac{E}{8C}\sqrt{\frac{2\pi}{r}}[v])$$

$$K_{II} = \lim_{r\to 0}(\frac{E}{8C}\sqrt{\frac{2\pi}{r}}[u\ ]) \tag{A-4}$$

$$K_{III} = \lim_{r\to 0}(\frac{E}{8(1+v)}\sqrt{\frac{2\pi}{r}}[w])$$

with: $C=1$      in plane stress

       $C=1 - v^2$ in plane stress

       u, v and w are the displacements of the crack lips corresponding to each elementary mode.


### A4.2. *The energy release rate, G*

Griffith (Ref 1) was the first to tackle the problem of the cracked bodies from an energy perspective. By the analysis of the energy balance, the energy release rate, denoted G, was introduced. It is defined by the energy necessary to make the crack fronts extend the crack length by a unit length. It corresponds to the decrease of the total potential energy W of the cracked body when it passes from an initial configuration with a given crack length, to another configuration where the crack is increased by a unit of length da:

$$G = -\frac{dW}{da} \tag{A-5}$$

where $W = W_e + W_{ext}$

with:

    $W_{ext}$: potential energy of external forces

$$W_e = \int_V w_e \, dV \quad \text{elastic strain energy}$$

$$w_e = \int_0^\varepsilon \sigma_{ij} \, d\varepsilon_{ij} \quad \text{density of the elastic strain energy}$$

For a crack in 2-D medium with a thickness b:

$$G = \frac{1}{b} \frac{dW}{\Delta a} \tag{A-6}$$

$\Delta a$, being the variation of the crack length.

Using the stress field in the singular zone, we can relate $G$ to the stress intensity factors:

$$G = \frac{(K_I^2 + K_{II}^2)}{E'} + \frac{K_{III}^2}{2\mu} \tag{A-7}$$

with $E' = E$ in plane stress

$E' = E/(1 - v^2)$ in plane strain

$G$ is a negative value because of the reduction of the potential energy related to the crack growth.

### A4.3. *The J-integral*

Another way of characterizing the singularity of the stress field in the vicinity of the crack is through the study of certain contour integrals which we can deduce from the law of conservation of energy. In a 2-D cracked linear elastic medium, Rice in 1968 (Ref [2]) used for the first time within the framework of the fracture mechanics an integral of contour, defined as follows:

$$J = \int_\Gamma \left\{ w_e n_1 - \sigma_{ij} n_j \frac{\partial u_i}{\partial x_1} \right\} ds \tag{A-8}$$

where $w_e$ is the density of strain elastic energy and $\Gamma$ is a contour around the crack.

The crack is supposed as straight following the crack axis. $\vec{n}$ is the normal vector to the contour, $\sigma_{ij} \, n_j$ is the applied stress to the contour and $u_i$ is the corresponding displacement. (see Figure A4).

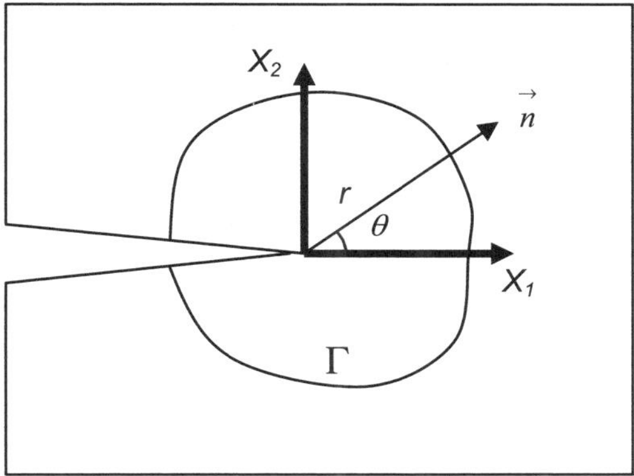


**Figure A4.** *Integral path for the J integral*

In the case of a homogenous, linear and nonlinear elastic solid, this integral is independent of the contour of integration. Rice interpreted the J-integral as the difference in potential energy W of two cracked bodies submitted to the same boundary conditions but from which the lengths of cracks differ a length $\Delta a$, which is expressed by:

$$J = -\lim_{\Delta a \to 0} \frac{W(a+\Delta a) - W(a)}{\Delta a} = -\frac{dW}{\Delta a} \tag{A-9}$$

If we compare with G defined previously, we can say that in the case of an elastic material or in the case of a very weak plastic deformation near the crack zone, and we can then write:

$$J = G = -\frac{dW}{\Delta a} \tag{A-10}$$

### A4.4. *The crack opening displacement (COD)*

The COD is the vector of displacement at the crack tip. It represents the sweeping force that makes the crack propagate. Measurements of the COD were limited to the crack of mode I which propagates along a fixed direction which corresponds to the axis of the initial crack. When the loading situation is of a mixed mode, these measurements are based on the vector CTD (Crack tip-Displacement) which is a combination of CTOD (Crack tip-Opening-Displacement) corresponding to the loading in mode I and of CTSD (Crack tip-Sliding-Displacement) which is the vector of displacement due to the slip of crack lips corresponding to the loading in mode II. In other words, when the crack is subjected to the mixed mode loading, the spacing of the crack lips can be divided into two parts: $\delta_I$ and $\delta_{II}$. $\delta_I$ is horizontal displacement, and $\delta_{II}$ is vertical displacement.

As a closure to this section, it should be noted that the majority of the conventional parameters of fracture mechanics (K, G, J and COD) can be employed to predict the direction of crack propagation under loadings in mixed mode, however, the use of these parameters was mainly limited to the failure in linear elastic medium. They lead to very good predictions for fragile elastic materials containing a real crack. Nevertheless, these criteria produce less good predictions whenever the material is ductile, or when the loading is repeated, particularly in the presence of overloads or of residual stresses.

## A5. Asymptotic stress field in elastic-plastic media

In the case of a crack in an elastic-plastic material under mixed mode loading, Shih and German (Ref [3]) showed that the stresses, strains and displacements fields near the crack tip are dominated by the HRR (Hutchinson, Rice and Rosengren) singularity, and can be characterized by two parameters, the J-integral and the mixity parameter $M^p$. They showed that for the material following Ramberg-Osgood hardening rule, the asymptotic stress field is given by:

$$\sigma_{ij} = \sigma_0 \left( \frac{EJ}{\alpha \sigma_0^2 I_n r} \right)^{\frac{1}{n+1}} \tilde{\sigma}_{ij} \left( \theta, M^p \right)$$

$$(A-11)$$

where $\alpha$ is a material constant, $n$ is the strain hardening coefficient, $\sigma_0$ is the yielding stress, and $I_n$ is an integral constant. The dimensionless functions $\tilde{\sigma}_{ij}$ depend only on $\theta$ and the degree of mixed mode stress field near the crack tip, expressed by the parameter $M^p$. $M^p$ (denoted mixity parameter in short) varies from zero to one. When $M^p = 0$, it is the case of pure mode II and when $M^p = 1$, it is the case of pure mode I. It is defined as follows:

$$M^p = \lim_{r \to 0} \frac{2}{\pi} \tan^{-1} \left| \frac{\sigma_{\theta\theta}(\theta = 0)}{\sigma_{r\theta}(\theta = 0)} \right|$$

$$(A-12)$$

It should be noted that the J-integral, the energy release rate, G and the crack opening displacement can also be defined in an elastic plastic media. These parameters allow us to elaborate fracture criteria under such conditions.

## A5. References


1. A.A. Griffith, "The phenomena of rupture and flow in solids" *Phil. Trans. Roy. Soc. Of London,* A221, 1920, pp.163-197

2. J.R. Rice, "A path independent integral and the approximate analysis of strain concentrations by notches and cracks" *J. Appl. Mech.*, 1968, pp.379-386

3. C.F. Shih and M.D. German, "Requirement for a one parameter characterization of crack tip field by the HRR singularity" *Int. J. Fracture* 17, 1981, pp. 27-43

# Appendix B

# Spreadsheet for Fatigue Life Estimates

| | | | ESTIMATED FATIGUE LIFE |
|---|---|---|---|
| FIELD | Confidential | | |
| ITEM | Turret structure | | |
| DETAIL | HS2 | Turret table lug at intersection with cover pl | 281 Years |
| LOAD | L4 | | |
| DATE | 05.07.05 | | |
| SIGN | Tom Lassen | | |

## CONVENTIONS

USE OF COLORS:

| | |
|---|---|
| BLUE | TEXT HEADINGS |
| RED | NOT TO BE CHANGED  (PROTECTED) |
| GREEN | TO BE GIVEN BY THE USER |

## CLASSIFICATION

Choose Joint type end environment from table

Choose Classification Factor from table

| Joint Environm. | $S_0$ Mpa | N Cycles | SB<S0 Log K | m | SB>S0 Log K | m |
|---|---|---|---|---|---|---|
| | | | | | | |
| PA | 53 | 1.00E+07 | 12.182 | 3 | 15.637 | 5 |
| PCP | 84 | 1.03E+06 | 11.784 | 3 | 15.637 | 5 |
| PFC | - | - | 11.705 | 3 | 11.705 | 3 |
| TA | 67 | 1.00E+07 | 12.476 | 3 | 16.127 | 5 |
| TCP | 95 | 1.75E+06 | 12.175 | 3 | 16.127 | 5 |
| TFC | - | - | 12 | 3 | 12 | 3 |
| USER | 115.8 | 1.00E+06 | 12.192 | 3 | 16.32 | 5 |

USER:    DNV C-class

| S-N Curve | Factor |
|---|---|
| | |
| B | 0.64 |
| C | 0.76 |
| D | 1 |
| E | 1.14 |
| F | 1.34 |
| F2 | 1.52 |
| G | 1.83 |
| W | 2.54 |

Selected values are:

| Joint | S0 | N | SB<S0 | | SB>S0 | | Class | Factor |
|---|---|---|---|---|---|---|---|---|
| Environm. | Mpa | Cycles | Log K | m | Log K | m | | |
| PCP | 84 | 1,026,000 | 11.784 | 3 | 15.637 | 5 | F | 1.34 |
| | | | | | | | | |

Give actual thickness   T (mm):   120

Choose type of thickness correction   3

O Option Button 7          No Correction

O Option Button 8          HSE Correction      (T/16)        (T/16)^0.33      22

● Option Button 9          DNV Correction      (T/22)^0.25  (C-class)      0.25

Take account of cut-off stresses?                                        FALSE

☐ Check Box 10                              Cut-off limit:      0


LOAD SPECTRUM

Fill in the three first columns of the table
Give multiplication factor for the stress ranges: 0.026

| ID | FORCE | n | Prob | TPEN | FAC | TOTFAC | DELSIG | n | N | D |
|---|---|---|---|---|---|---|---|---|---|---|
| 1 | 50 | 6.0E+05 | 1 | 1.528233 | 0.03484 | 0.053244 | 2.662182 | 596,321.6 | 3.24E+13 | 1.84E-08 |
| 2 | 100 | 8.4E+04 | 1 | 1.528233 | 0.03484 | 0.053244 | 5.324365 | 84,147.67 | 1.01E+12 | 8.31E-08 |
| 3 | 150 | 3.1E+04 | 1 | 1.528233 | 0.03484 | 0.053244 | 7.986547 | 31,059.23 | 1.33E+11 | 2.33E-07 |
| 4 | 200 | 1.6E+04 | 1 | 1.528233 | 0.03484 | 0.053244 | 10.64873 | 15,887.18 | 3.17E+10 | 5.02E-07 |
| 5 | 250 | 9.7E+03 | 1 | 1.528233 | 0.03484 | 0.053244 | 13.31091 | 9,711.361 | 1.04E+10 | 9.36E-07 |
| 6 | 300 | 6.5E+03 | 1 | 1.528233 | 0.03484 | 0.053244 | 15.97309 | 6,452.424 | 4.17E+09 | 1.55E-06 |
| 7 | 350 | 4.5E+03 | 1 | 1.528233 | 0.03484 | 0.053244 | 18.63528 | 4,548.905 | 1.93E+09 | 2.36E-06 |
| 8 | 400 | 3.4E+03 | 1 | 1.528233 | 0.03484 | 0.053244 | 21.29746 | 3,361.176 | 9.89E+08 | 3.40E-06 |
| 9 | 450 | 2.6E+03 | 1 | 1.528233 | 0.03484 | 0.053244 | 23.95964 | 2,575.297 | 5.49E+08 | 4.69E-06 |
| 10 | 500 | 2.0E+03 | 1 | 1.528233 | 0.03484 | 0.053244 | 26.62182 | 2,023.837 | 3.24E+08 | 6.24E-06 |
| 11 | 550 | 1.6E+03 | 1 | 1.528233 | 0.03484 | 0.053244 | 29.28401 | 1,618.792 | 2.01E+08 | 8.04E-06 |
| 12 | 600 | 1.3E+03 | 1 | 1.528233 | 0.03484 | 0.053244 | 31.94619 | 1,313.222 | 1.3E+08 | 1.01E-05 |
| 13 | 650 | 1.1E+03 | 1 | 1.528233 | 0.03484 | 0.053244 | 34.60837 | 1,079.479 | 87,316,090 | 1.24E-05 |
| 14 | 700 | 9.0E+02 | 1 | 1.528233 | 0.03484 | 0.053244 | 37.27055 | 898.9104 | 60,279,669 | 1.49E-05 |
| 15 | 750 | 7.6E+02 | 1 | 1.528233 | 0.03484 | 0.053244 | 39.93273 | 757.7373 | 42,692,810 | 1.77E-05 |
| 16 | 800 | 6.5E+02 | 1 | 1.528233 | 0.03484 | 0.053244 | 42.59492 | 645.5688 | 30,917,981 | 2.09E-05 |
| 17 | 850 | 5.5E+02 | 1 | 1.528233 | 0.03484 | 0.053244 | 45.25710 | 554.7444 | 22,833,182 | 2.43E-05 |
| 18 | 900 | 4.8E+02 | 1 | 1.528233 | 0.03484 | 0.053244 | 47.91928 | 479.8148 | 17,157,283 | 2.80E-05 |
| 19 | 950 | 4.2E+02 | 1 | 1.528233 | 0.03484 | 0.053244 | 50.58146 | 417.0149 | 13,093,117 | 3.18E-05 |
| 20 | 1,000 | 3.6E+02 | 1 | 1.528233 | 0.03484 | 0.053244 | 53.24365 | 363.7701 | 10,131,204 | 3.59E-05 |
| 21 | 1,050 | 3.2E+02 | 1 | 1.528233 | 0.03484 | 0.053244 | 55.90583 | 318.2911 | 7,938,063 | 4.01E-05 |
| 22 | 1,100 | 2.8E+02 | 1 | 1.528233 | 0.03484 | 0.053244 | 58.56801 | 279.2794 | 6,290,681 | 4.44E-05 |
| 23 | 1,150 | 2.5E+02 | 1 | 1.528233 | 0.03484 | 0.053244 | 61.23019 | 245.7348 | 5,036,999 | 4.88E-05 |
| 24 | 1,200 | 2.2E+02 | 1 | 1.528233 | 0.03484 | 0.053244 | 63.89238 | 216.8417 | 4,071,504 | 5.33E-05 |
| 25 | 1,250 | 1.9E+02 | 1 | 1.528233 | 0.03484 | 0.053244 | 66.55456 | 191.9081 | 3,319,793 | 5.78E-05 |
| 26 | 1,300 | 1.7E+02 | 1 | 1.528233 | 0.03484 | 0.053244 | 69.21674 | 170.3364 | 2,728,628 | 6.24E-05 |
| 27 | 1,350 | 1.5E+02 | 1 | 1.528233 | 0.03484 | 0.053244 | 71.87892 | 151.6102 | 2,259,395 | 6.71E-05 |
| 28 | 1,400 | 1.4E+02 | 1 | 1.528233 | 0.03484 | 0.053244 | 74.54110 | 135.2868 | 1,883,740 | 7.18E-05 |
| 29 | 1,450 | 1.2E+02 | 1 | 1.528233 | 0.03484 | 0.053244 | 77.20329 | 120.9925 | 1,580,597 | 7.65E-05 |
| 30 | 1,500 | 1.1E+02 | 1 | 1.528233 | 0.03484 | 0.053244 | 79.86547 | 108.4155 | 1,334,150 | 8.13E-05 |

# Appendix C

# CG – Crack Growth Based on Fracture Mechanics

**General information**
The background for Appendix C is found in Chapter 6
The aim of this appendix is to provide the reader with a tool for fatigue crack growth prediction
The user can play with the input parameters to reveal their influence on the fatigue crack growth and final life
A quasi-stochastic analysis is added for a *damage tolerance* analysis
Some calculations are carried out in macros that must be activated by buttons
The macros are written in Visual Basic

**Guidance for the user**
The colors are indicating the contents of the cells:
Blue indicates the issue the user is dealing with (Heading)
Green indicates where the users are to enter the input data
Red indicates values not to be changed but chosen by using the menus
Yellow indicates values that are used by the macros when these are activated by the macros buttons

**Units**
**Units are Mpa for stresses and meters for dimensions**
**The only exception is the POD curve for which the crack depth is given in millimeters**

**1 Identification of structure**

| **Name:** | Industrial case Chapter 6 | **Type:** | Steel pipe |
| **Loading mode:** | Riser loads | **Date:** | 06.06.2006 |

**2 Definition of the welded joint or component geometry**
You will find some standard geometries numbered from 1 to 6 in Figure 1
You can choose them with the menu in Table 1
In addition you must give some specific dimensions

## Standard geometries

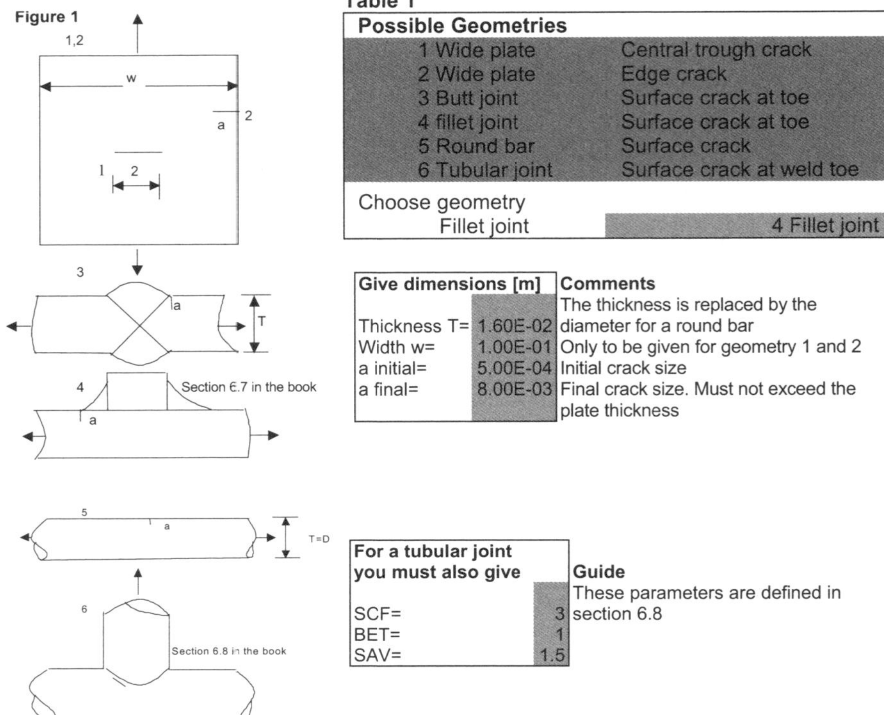

**Figure 1**

**Table 1**

| Possible Geometries | |
|---|---|
| 1 Wide plate | Central trough crack |
| 2 Wide plate | Edge crack |
| 3 Butt joint | Surface crack at toe |
| 4 fillet joint | Surface crack at toe |
| 5 Round bar | Surface crack |
| 6 Tubular joint | Surface crack at weld toe |
| Choose geometry | |
| Fillet joint | 4 Fillet joint |

| Give dimensions [m] | | Comments |
|---|---|---|
| | | The thickness is replaced by the |
| Thickness T= | 1.60E-02 | diameter for a round bar |
| Width w= | 1.00E-01 | Only to be given for geometry 1 and 2 |
| a initial= | 5.00E-04 | Initial crack size |
| a final= | 8.00E-03 | Final crack size. Must not exceed the plate thickness |

| For a tubular joint you must also give | | Guide |
|---|---|---|
| | | These parameters are defined in section 6.8 |
| SCF= | 3 | |
| BET= | 1 | |
| SAV= | 1.5 | |

The chosen geometry of the joint has a geometry function as shown in Figure 2

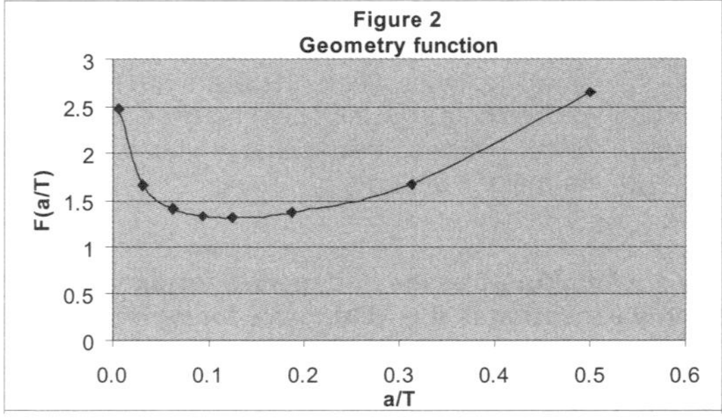

**Figure 2**
**Geometry function**

## 3 Material data

The main parameters are C and m in the Paris law. The menu 1 and 2 are the mean and mean plus two standard deviations for C given I for steel
The units are Mpa and m
To the right these are converted to Mpa and millimeter

If you have more exact values you can give them in the green cells

The parameters you have chosen give a growth curve as shown in Figure 3

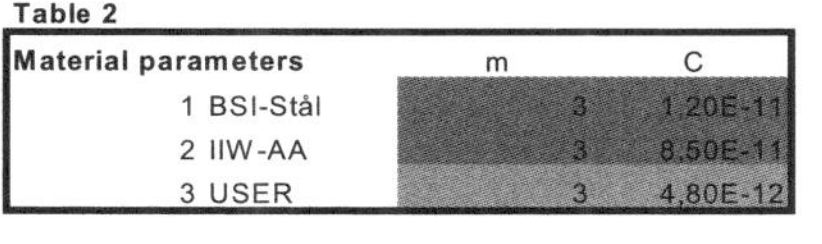

Table 2

| Material parameters | m | C |
|---|---|---|
| 1 BSI-Stål | 3 | 1,20E-11 |
| 2 IIW-AA | 3 | 8,50E-11 |
| 3 USER | 3 | 4,80E-12 |

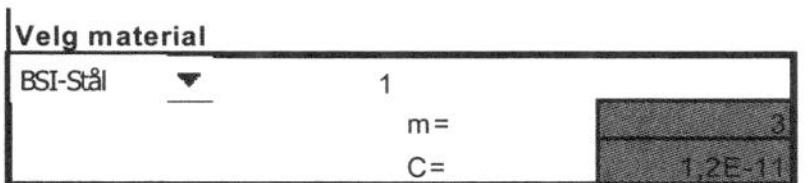

Velg material

| BSI-Stål ▼ | 1 | |
|---|---|---|
| m= | | 3 |
| C= | | 1,2E-11 |

**Converted values**
mm/cycles
C= 3.80E-13

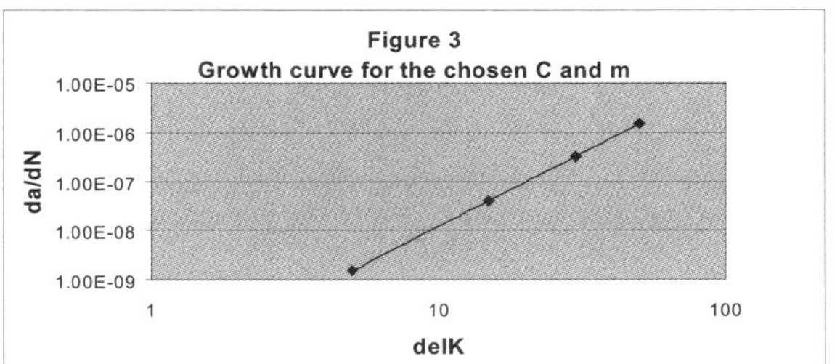



## 4 Load spectrum

The load spectrum may either be given as the equivalent stress range (equivalent stress range = TRUE) or as a histogram of stress ranges
Hence, you may either give one single value in column G or several lines (maximum 10) in columns B and C

☑ Equivalent stress?    TRUE

| ID | DELSIG | n | n*DELSIG^m | | |
|---|---|---|---|---|---|
| 1 | 20 | 100 | 800000 | Equivalent stress range: | 21 given as one value |
| 2 | 50 | 200 | 25000000 | | |
| 3 | 60 | 300 | 64800000 | Equivalent stress range: | 105.390288 calculated from given histogram |
| 4 | 70 | 400 | 137200000 | | |
| 5 | 80 | 500 | 256000000 | | |
| 6 | 90 | 600 | 437400000 | | |
| 7 | 100 | 700 | 700000000 | | |
| 8 | 110 | 800 | 1064800000 | | |
| 9 | 120 | 900 | 1555200000 | | |
| 10 | 130 | 1000 | 2197000000 | The following stress range is used in the calculation | |
| | | 5500 | 6438200000 | DELSIG= | 21 |

## 5 Calculation of crack growth

### A: Simplified estimate based on constant F function

| | | |
|---|---|---|
| F= | 1.41125 | (constant at a/T=0.001) |
| 1-m/2 | -0.50 | |
| nominator | -33.5410197 | |
| denominator | -8.69E-07 | |
| life | 3.86E+07 | |

## B: Calculation based on integration of the Paris law (Press Macro button)

Numerical results

| a(mm) | Cycles |
|---|---|
| 0.50 | 0.00E+00 |
| 0.875 | 9.34E+06 |
| 1.25 | 1.58E+07 |
| 1.625 | 2.05E+07 |
| 2 | 2.39E+07 |
| 2.375 | 2.65E+07 |
| 2.75 | 2.85E+07 |
| 3.125 | 3.00E+07 |
| 3.5 | 3.12E+07 |
| 3.875 | 3.20E+07 |
| 4.25 | 3.27E+07 |
| 4.625 | 3.33E+07 |
| 5 | 3.37E+07 |
| 5.375 | 3.40E+07 |
| 5.75 | 3.43E+07 |
| 6.125 | 3.45E+07 |
| 6.5 | 3.46E+07 |
| 6.875 | 3.47E+07 |
| 7.25 | 3.48E+07 |
| 7.625 | 3.49E+07 |
| 8 | 3.49E+07 |

The results are shown graphically on Figure 4

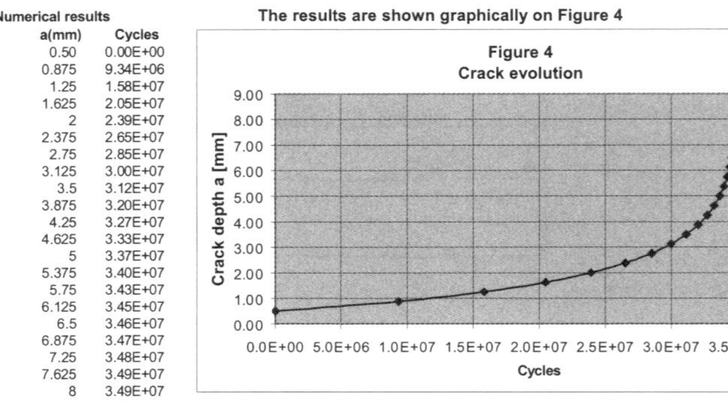

Figure 4
Crack evolution

## 6 Reliability of an inspection program: damage tolerance analysis

Estimate of the performance of a chosen inspection program (see section 7.6.4.1 in the book)

You must specify the POD curve for the inspection technique by giving P0, g and ab

Give parameters for POD

| P0= | 0.9 |
|---|---|
| g= | 1 |
| ab | 1 |

Your parameters give an POD curve as shown in Figure 5

Give inspection interval

| Interval | 5.00E+06 |
|---|---|

Figure 5
POD curve

Have a look at Figures 4 and 5 before choosing an interval

Based on the given values you can check the reliability by pressing the macro button

**Results:**
8.00   0.89917931   0.89917931
**The inspection program has**   4   **effective inspections**
**The first is scheduled at**   1.50E+07   **at an expected crack depth of 1.20 millimeters**
**The last is carried out at**   3.00E+07   **at an expected crack depth of 3.12 millimeters**
**Total reliability**   0.960144

**The first and last effective inspections are indicated in Figure 6**

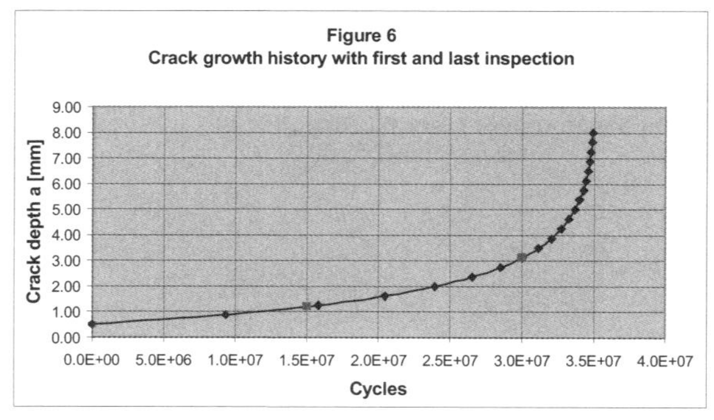

Figure 6
Crack growth history with first and last inspection

# Appendix D

# CI – Crack Initiation Based on Coffin Manson

**General information**

The background for Appendix D is found in Chapters 2 and 9
The aim of this appendix is to provide the reader with a tool that predicts the crack initiation phase
The user can play with the input parameters to reveal their influence on the initiation phase
Some calculations are carried out in macros that must be activated by buttons
The macros are written in Visual Basic

**Guidance for the user**

The colors are indicating the contents of the cells:

Blue indicates the issue the user is dealing with (Heading)
Green indicates where the users are to enter the input data
Red indicates values not to be changed but chosen by using the menus
Yellow indicates values that are used by the macros when these are activated by the macros buttons

**Units**
**Units are Mpa for stresses and millimeters for dimensions**
**Angles are given in degrees**

1 Identification of structure

| | | | |
|---|---|---|---|
| **Name:** | Welded structure | **Type:** | Fillet joint |
| **Loading mode:** | Axial | **Date:** | 06.06.2006 |

2 Definition of the welded joint or component geometry

You will find some standard geometries numbered from 1 to 6 in Figure 1
You can choose them with the menu in Table 1
In addition you must give some specific dimensions

**Figure 1**

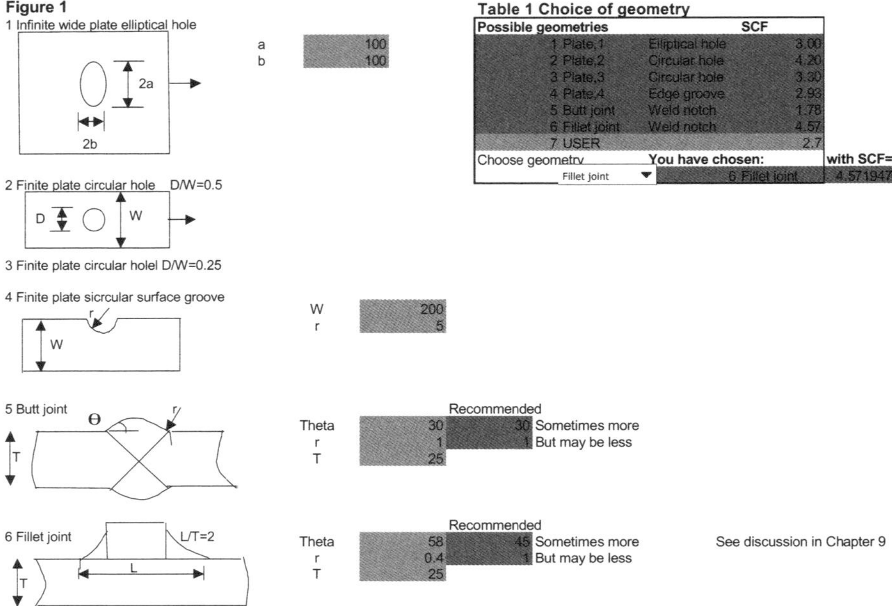



## Table 1 Choice of geometry

| Possible geometries | | SCF |
|---|---|---|
| 1 Plate,1 | Elliptical hole | 3.00 |
| 2 Plate,2 | Circular hole | 4.20 |
| 3 Plate,3 | Circular hole | 3.30 |
| 4 Plate,4 | Edge groove | 2.93 |
| 5 Butt joint | Weld notch | 1.78 |
| 6 Fillet joint | Weld notch | 4.57 |
| 7 USER | | 2.7 |

| Choose geometry | You have chosen: | with SCF= |
|---|---|---|
| Fillet joint ▼ | 6 Fillet joint | 4.57194751 |

## 3 Material data

You must specify the parameters in the Ramber Osgood law and in the Coffin Manson equation
The data given for 1 C-Mn steel are explained in Chapter 9. The parameters correspond to a fixed HB of 202
if you have other HB values you can use 2 Steel-HB and give the actual HB in the green field to the right
You can also specify all parameters freely in the USER line

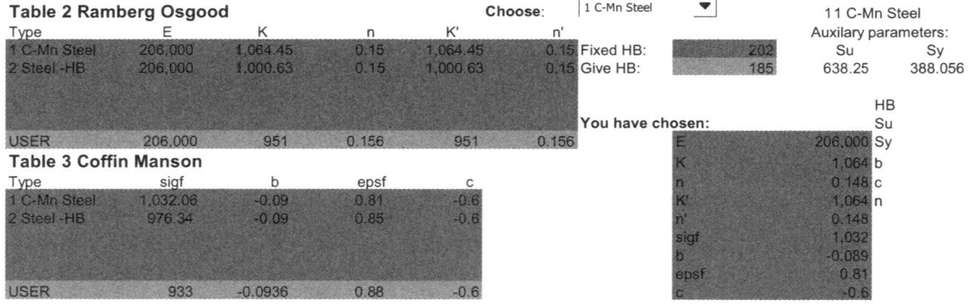

**Table 2 Ramberg Osgood**            Choose: 1 C-Mn Steel ▼            11 C-Mn Steel

| Type | E | K | n | K' | n' | | | Auxilary parameters: | |
|---|---|---|---|---|---|---|---|---|---|
| 1 C-Mn Steel | 206,000 | 1,064.45 | 0.15 | 1,064.45 | 0.15 | Fixed HB: | 202 | Su | Sy |
| 2 Steel -HB | 206,000 | 1,000.63 | 0.15 | 1,000.63 | 0.15 | Give HB: | 185 | 638.25 | 388.056 |
| USER | 206,000 | 951 | 0.156 | 951 | 0.156 | | | | |

| You have chosen: | | |
|---|---|---|
| | | HB |
| | | Su |
| E | 206,000 | Sy |
| K | 1,064 | b |
| n | 0.148 | c |
| K' | 1,064 | n |
| n' | 0.148 | |
| sigf | 1,032 | |
| b | -0.089 | |
| epsf | 0.81 | |
| c | -0.6 | |

**Table 3 Coffin Manson**

| Type | sigf | b | epsf | c |
|---|---|---|---|---|
| 1 C-Mn Steel | 1,032.06 | -0.09 | 0.81 | -0.6 |
| 2 Steel -HB | 976.34 | -0.09 | 0.85 | -0.6 |
| USER | 933 | -0.0936 | 0.88 | -0.6 |

Your parameters (E, K, n, K', n') give a material behavior as shown in Figure 2

Your choice of parameters (sigf', b, epsf', c) give an initiation curve as shown in Figure 3

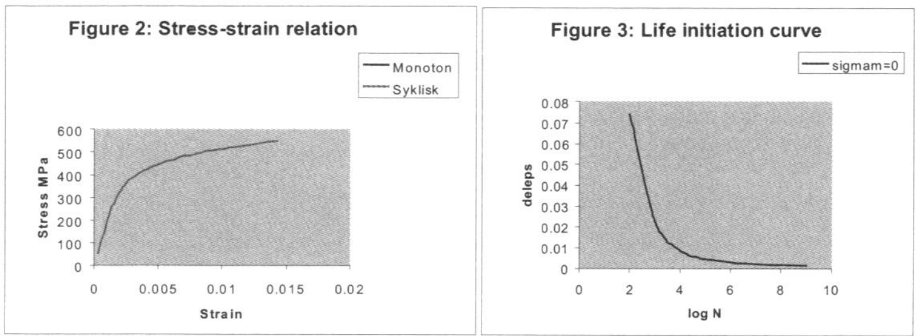

## 4 Definition of loading

You must specify the nominal stress on the component (excluding the notch effect)
The associated strains are calculated

Give:             Calculated:
sigmin      9     sigmamax           89      epsmaxt     0.000432092
sigrange    80    sigmamean          49      delepst     0.00038835
                  R         0.1011236        delepse     0.00038835
                                             delepsp     4.86079E-10

The nominal stress-strain relation is given in Figure 4

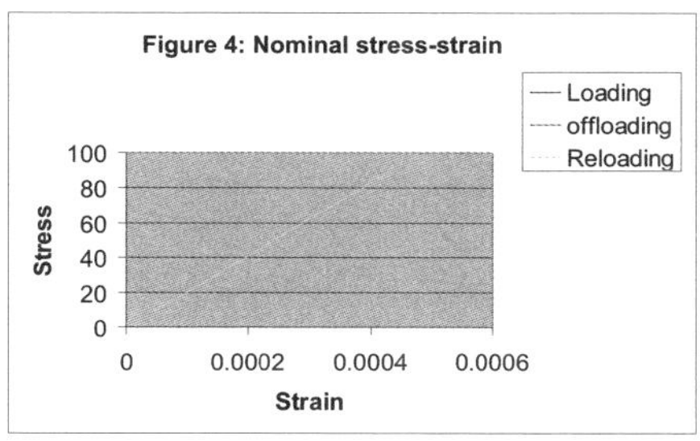



## 5 Determination of the hysteresis loop at the notch

Kt is calculated in Table 1. You must give the residual stresses at the notch

Kt (Kf):          4.57          sigmamax      475.312143  epsmaxt     0.00664962
sigres            400           delsig        364.372078  delepst     0.00178226
                                sigmamean     293.126104

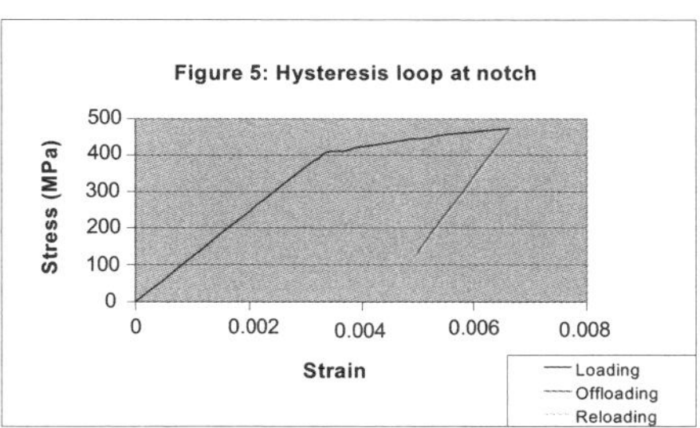



## 6 Time to crack initiation

You can now carry out the time estimate for crack initiation at the notch by pressing the macro button
The calculations are based on the hysteresis loop in Figure 5 and the lie curve in Figure 3

300292.054

Estimated time to crack initiation: 5.80E+06 cycles
For a C-Mn steel the estimate corresponds to the number of cycles to reach a rack depth of 0.1 millimeter and the weld toe.

# Index

## D

Damage tolerance 111, 221, 255-264, 288
Database 166-167, 278, 279, 288, 292, 293, 295, 296, 297, 298, 299, 306

## E

Elementary
    reliability models 207-212
    statistical methods 55-59
Energy spectrum 62, 64, 69-72, 130
Engineering critical assessment 141, 161
Environment 25-26, 83-84, 159, 303, 307, 355
Experimental work (testing) 37
    test preparation 41-46
    test specimen 38-40, 50-55, 92, 129, 160, 166, 273, 278, 293, 310
    test setup 38-40, 50-55
    test instrumentation

## F

Failure rate function 209, 210, 211, 212, 249
Fatigue
    crack growth 25, 55, 149, 156, 160, 161, 176-178 221, 288, 315, 319, 340, 362, 364, 365, 366-370, 375
    life prediction 8, 41, 75, 92, 107, 111, 119, 130, 200, 261, 263, 274, 314
    limit 6, 78, 83, 84, 86, 113, 121, 174, 175, 176, 273, 274, 275, 276, 277, 280, 281, 283, 284, 287, 299, 300, 301, 306, 307
    loading 12, 19, 61-73, 118, 126, 199, 261, 319, 326
    notch factor 103, 290, 292, 373

testing 5, 38, 41, 43, 62, 63, 94, 288
Fracture
    criterion 153, 154-156, 183
    mechanics 33, 37, 139-195, 212, 213, 215, 220-234, 287, 289, 294, 296, 306, 307, 319, 343, 349, 370
    toughness 152, 158, 256, 330

## G

Geometry function 22, 145, 146, 148, 149, 150, 151, 161-165, 170, 172, 184, 185, 192, 230
Grain size 44, 294
Grinding 30, 82, 83, 101, 112, 117, 122, 187, 191, 225

## H

Hammer peening 82, 122
Hot spot stress 46, 131

## I

Improvement techniques 82-83
Initiation phase 6, 9, 15, 25, 30, 34, 112, 287, 288, 292, 293, 296, 301, 307
Inspection
    planning 225-231, 303-305
    strategy 189-190
    updating 231-235
Intrusion 15, 16, 27, 28, 29, 49, 161, 250

## J

Joints
    cruciform 48, 49, 374
    plated 30-34
    tubular 34-35, 96-98, 176-178

## K, L

Lack of penetration 32
Linear
    damage accumulation 84-88
    elastic fracture mechanics 142-152, 230, 319
Load spectrum on
    an exceedance diagram format
    a histogram format 84-86
    Weibull distribution format 88-91
Loading frequency 45
Lognormal distribution 207, 208-209, 249
Longitudinal welds 33, 99

## M

Machining
Markov chains 6, 200, 202, 203, 204, 235-255, 267
Material properties 9, 25
Miner's sum 87, 88, 189, 217
Misalignment 25, 81-82
Mixed mode loading 333
Monte Carlo simulation 6, 200, 206, 207, 213, 219-220, 231, 245, 247, 255, 260, 264, 267, 307

## N

Non-destructive inspection (NDI) 112, 176, 191, 223, 224, 295
Non-welded material 112
Notch stress intensity factor (N-SIF) approach 311, 313, 314, 315
Notches 20, 25, 106, 142, 158

## O

Offshore structures 8, 130-135, 217-218
Overfill 20, 115, 116

Overload 5, 7, 87, 153, 159, 355, 356, 357, 359-362, 365, 366-371, 374-375

## P

Parameters in
    Paris law 186, 199, 200, 229, 373
    Manson-Coffin equation 23
    Ramberg-Osgood curve 24, 290
Paris law 53, 161, 175, 179, 183, 186, 192, 215, 235, 294, 306, 326, 328, 334, 344, 356, 367
Percentile curves 57, 58, 76-84
Plates 3, 41, 43, 80, 102, 105, 112, 114, 121, 122, 145, 149, 151, 152, 167, 179, 183, 263
Principal stress 15, 27, 45, 91, 98, 104, 106, 107, 109, 322
Probability
    density function 68, 203, 208, 277, 293
    of detection (POD) 191, 199, 221, 223, 224, 226, 239, 256, 259, 303
    of failure 8, 77, 199, 202, 203, 205, 206, 207, 208, 211, 215, 217, 219, 220, 221, 227, 231, 238, 239, 240, 241, 249, 253, 254, 257, 260, 263, 267, 304, 370, 374, 375
Propagation 12, 22, 80, 83, 87, 139, 242, 248, 292, 319

## Q

Quality 6, 25, 30, 31, 32, 34, 38, 75, 82, 102, 112, 118, 124, 130, 141, 166, 176, 209, 224, 229, 236, 243, 267, 275, 276, 279, 341

## R

Rain-flow counting 63, 65, 70
Random fatigue limit model 277-278, 279-283